Conversion of Système International to Imperial Unit~~s~~

VARIABLE	SI UNIT	~~EQUIVAL~~ENT
Length	kilometer	...mi
	meter	1.094 yd
	meter	39.37 in
	meter	3.281 ft
Area	hectare	2.471 A
	square kilometer	0.39 mi²
	square meter	1.196 yd²
	square centimeter	0.15 in²
Volume	cubic meter	1.308 yd³
	cubic meter	35.3 ft³
	cubic centimeter	0.06 in³
	cubic kilometer	0.24 mi³
	liter	1.06 U.S. qt
	liter	0.88 imperial qt
	liter	0.264 U.S. gal
	liter	0.22 imperial gal
Velocity	kilometers/hour	0.621 mph
	meters/second	3.281 ft/sec[1]
	meters/second	2.24 mph
Pressure	standard atmosphere	14.7 psi
	bar	14.5 psi
Mass	kilogram	2.205 lb
	gram	0.035 oz
Temperature	Celsius	$°C = 5/9\ (°F - 32)$

Common abbreviations used in the Imperial System

mi	mile	mph	miles/hour	gal	gallon
yd	yard	oz	ounce	qt	quart
in	inch	lb	pound	A	Acre
ft	feet	psi	pounds/in²		

CLIMATOLOGY
An Atmospheric Science

CLIMATOLOGY:
AN ATMOSPHERIC SCIENCE

JOHN J. HIDORE
The University of North Carolina at Greensboro

JOHN E. OLIVER
Indiana State University

MACMILLAN PUBLISHING COMPANY
New York

MAXWELL MACMILLAN CANADA
Toronto

MAXWELL MACMILLAN INTERNATIONAL
New York   *Oxford*   *Singapore*   *Sydney*

Cover photo: William Lesch
Editor: Paul F. Corey
Production Editor: Linda Hillis Bayma
Art Coordinator: Peter A. Robison
Photo Editor: Chris Migdol
Text Designer: Susan E. Frankenberry
Cover Designer: Thomas Mack
Production Buyer: Pamela D. Bennett
Illustrations: Academy ArtWorks, Inc.

This book was set in Times Roman by V&M Graphics and was printed and bound by Arcata Graphics/Halliday. The cover was printed by Phoenix Color Corp.

Macmillan Publishing Company
866 Third Avenue
New York, New York 10022

Macmillan Publishing Company is part of the
Maxwell Communication Group of Companies.

Maxwell Macmillan Canada, Inc.
1200 Eglinton Avenue East, Suite 200
Don Mills, Ontario M3C 3N1

Library of Congress Cataloging-in-Publication Data

Hidore, John J.
    Climatology : an atmospheric science / John J. Hidore, John E. Oliver.
        p.   cm.
    Includes bibliographical references (p.   ) and index.
    ISBN 0-02-354515-1
    1. Climatology. 2. Atmospheric physics.   I. Oliver, John E.
II. Title.
QC981.H5917   1993
551.5—dc20
                                                                    92-7399
                                                                    CIP
Printing: 1 2 3 4 5 6 7 8 9   Year: 3 4 5 6 7

Part-opening photo credits: Part One, Tim Davis/Photo Researchers, Inc.; Part Two, National Center for Atmospheric Research/National Science Foundation; Part Three, Reuters/Bettmann.

Map insert credit: Maryland Cartographics

Color plate credits: Plates 1 and 3, NASA; Plate 2, Joel Susskind/NASA/Goddard Space Flight Center; Plate 4, The Skin Cancer Foundation; Plate 5, Reuters/Bettmann; Plates 6 and 8, NOAA/NESDIS/NCDC/SDSD; Plate 7, © 1990 Warren Faidley/Weatherstock; Plate 9, Chip and Jill Isenhart/Tom Stack & Associates; Plate 10, Stephen J. Krasemann; Plate 11, Richard Steedman/The Stock Market; Plate 12, M. P. L. Fogden/Bruce Coleman, Inc.; Plate 13, D. Cavagnaro; Plate 14, Barry Griffiths/Photo Researchers, Inc.; Plate 15, John Shaw/Bruce Coleman, Inc.; Plate 16, M. P. Kahl/Bruce Coleman, Inc.; Plate 17, Hope Day/Bruce Coleman, Inc.

PREFACE

This book is written for students seeking an introduction to the processes of climate changes through time and from place to place. *Climatology* is an appropriate book for college students taking their first course in atmospheric science. It contains a substantial section on basic atmospheric processes as well as extended material on regional climatology and climatic change.

Because many users of this book will not have had rigorous mathematical training, this text keeps mathematics, statistics, and physical principles at a basic level. Students are referred to specialized books and manuals for more technical aspects of atmospheric science. Technical terminology has been kept to a minimum as well. In all sciences, however, there is an essential vocabulary. To assist the reader with this terminology, we have included a glossary of key terms at the end of the book.

A full-color insert provides a number of false color satellite images which illustrate climatic phenomena. They are called false color images because the original data collected by satellites are in digital form, and colors are arbitrarily assigned to the numbers, or groups of numbers. In addition to the satellite images, there are photographs pertinent to some of the topics discussed in the text included in the insert.

The purpose of this book is threefold. First, it provides a basic foundation in principles of meteorology. Part One covers the fundamentals of the energy balance, the hydrologic cycle, atmospheric motion, and storm systems. One chapter considers how human activity affects the energy balance, including the impact of increasing carbon dioxide and other gases on the greenhouse effect and the depletion of stratospheric ozone. Another chapter examines interannual variations in weather—persistent changes in weather that last for up to several years—such as changes in atmospheric temperatures due to varying solar radiation, sea-surface temperature changes, and volcanic eruptions.

The second purpose of this book is to present the fundamental elements in regionalization of climate. Part Two classifies and analyzes regional climates. The section begins with a chapter on climatic classification at different regional scales. This is followed by chapters on the tropical, midlatitude, and polar climates. A separate chapter discusses the climate of North America, with special emphasis on the United States. Since much of Part Two is geographic, a map insert is included, consisting of a map showing the distribution of world climates and a world political map.

The third purpose of *Climatology* is to consider the question of climatic change. Part Three contains three chapters on climatic change. The first chapter explains the nature of climates in the past and the methods by which they are established. The second chapter considers some of the causes of climatic change. The final chapter discusses the long-term hazards associated with global warming, acid rain, and ozone depletion.

The 1980s saw a number of significant climatological problems develop that captured the world's attention; these problems have continued into the 1990s. Acid rain, global warming, and ozone depletion are among the climate-related challenges, which are worldwide in scope and affect all living organisms. Humans are as sensitive to climate now as at any time in history and are affected by any major change in the atmosphere.

Climatological principles and climatic data are used to solve a host of real problems. To relate the climatological principles presented in this book to these problems, *Climatology* includes a number of applied studies. These applied studies consider such processes as the commercial use of wind energy and the effects of cold and altitude on human physiology. The studies do not cover all aspects of applied climatology but provide examples of the relationship of humans to climatic variation.

The ultimate goal of this book is to provide those interested in climatology with an up-to-date yet readily comprehensible account of the characteristics, causes, and effects of climate. To attain this goal, we had to be selective in the material we presented. It is our hope that in making this selection, we have provided information, concepts, and ideas that will both inform readers and further their interest in climatology.

We would like to acknowledge Adam W. Burnett of Colgate University, Scott A. Isard of the University of Illinois at Champaign-Urbana, Brent R. Skeeter of Salisbury State University, Dale J. Stevens of Brigham Young University, and Philip W. Suckling of The University of Northern Iowa for their many positive suggestions on the manuscript.

CONTENTS

CLIMATOLOGY
An Atmospheric Science

Part One

PHYSICAL AND DYNAMIC
CLIMATOLOGY

THE BASIS OF MODERN CLIMATOLOGY

Chapter Overview

Atmospheric Science: Aerology and Meteorology

Atmospheric Science: Climatology
- The Study of Climatology
- Applied Climatology

The Standard Atmosphere
- Atmospheric Chemistry
- Constant Gases
- Variable Gases

Vertical Structure of Earth's Atmosphere
- Troposphere
- Stratosphere
- The Upper Atmosphere
- Ionosphere

Earth's atmosphere is a fundamental and critical part of the planet's environment. Together with the lithosphere (the solid Earth), the hydrosphere (water in all its forms), and the biosphere (living organisms), it makes Earth the habitable, hospitable place that it is.

The **atmosphere** serves several major functions on the planet. Beyond providing the reservoir of gases required for life, it also plays a critical role in the distribution or redistribution of energy over Earth. The atmosphere provides an insulating layer around Earth that raises the mean temperature of the surface from $-23°C$ ($-9°F$) to near $15°C$ ($59°F$). It shields the surface from the large doses of ultraviolet radiation that would destroy most life forms. It also serves as a major transporter of heat horizontally across Earth's surface and vertically away from the surface. Tropical regions receive far more energy than polar regions; atmospheric circulation helps to equalize this imbalance by moving heat from the warmer to the colder areas. In addition to transferring energy, the atmosphere transfers water from ocean to land.

Atmospheric Science: Aerology and Meteorology

Because of the complexity of the gaseous envelope that surrounds the planet, atmospheric scientists often divide its study into specific areas of interest. Some divide it into three fields: aerology, meteorology, and climatology.

Aerology is essentially the study of the free atmosphere through its vertical extent. Initially, the major goal of aerology was identifying atmospheric structure and the amount and distribution of its parts. Today aerology deals mostly with the chemistry and physical reactions that occur within the various atmospheric layers. The word "aerology" is less widely used now than it once was, and its content is frequently considered part of meteorology. We will deal with this aspect of the atmosphere only briefly in later chapters.

Meteorology is the study of the atmosphere's motion and other phenomena to aid in forecasting weather and explaining the processes involved. It deals largely with the status of the atmosphere over a short period and uses principles of physics to interpret the atmosphere. Part One of this book considers some of the meteorological aspects of the atmosphere.

Atmospheric Science: Climatology

Climatology is the study of atmospheric conditions over periods of time measured in years or longer. It includes the study of the kinds of weather that occur at a place. Climatology concerns the most frequently occurring types (the average weather) as well as the infrequent and unusual types. Because dynamic change in the atmosphere causes variation and occasional extremes that have long-term as well as short-term impact, climatology is also partly meteorology.

Atmospheric science has grown at a geometric rate paralleling the growth in the human population (see Table 1.1). The need to understand the atmosphere has grown at least as fast as the development of technology.

TABLE 1.1
Significant events in the development of climatology.

DATE	EVENT
ca. 400 B.C.	The influence of climate on health is discussed by Hippocrates in *Airs, Waters, and Places*.
ca. 350 B.C.	Weather science is discussed in Aristotle's *Meteorologica*.
ca. 300 B.C.	The text *De Ventis* by Theophrastus describes winds and offers a critique of Aristotle's ideas.
ca. 1593	Galileo describes the thermoscope. (The first thermometer is most likely attributed to Santorre, 1612.)
1662	Francis Bacon writes a significant treatise on the wind.
1643	Evangelista Torricelli invents the barometer.
1661	Boyle's law on gases is propounded.
1664	Weather observations begin in Paris; although often described as the longest continuous sequence of weather data available, the records are not homogeneous or complete.
1668	Edmund Halley constructs a map of the trade winds.
1714	The Fahrenheit scale is introduced.
1735	George Hadley writes his treatise on trade winds and effects of Earth rotation.
1736	The Centigrade scale is introduced. (It was first formally proposed by du Crest in 1641.)
1779	Weather observation begins at New Haven, CT, the longest continuous sequence of records in the United States.
1783	The hair hygrometer is invented.
1802	Lemark and Howard propose the first cloud classification system.
1817	Alexander von Humboldt constructs the first map showing mean annual temperature over the globe.
1825	August devises the psychrometer.
1827	Beginning of the period during which H. W. Dove developed the laws of storms.
1831	William Redfield produces the first weather map of the United States.
1837	Pyrheliometer for measuring insolation is constructed.
1841	Movement and development of storms are described by Espy.
1844	Gaspard de Coriolis formulates the "Coriolis force."
1845	Berhaus constructs first world map of precipitation.
1848	Dove publishes the first maps of mean monthly temperatures.
1862	Renou drafts first map (showing western Europe) of mean pressure.
1879	Supan publishes a map showing world temperature regions.
1892	Beginning of the systematic use of balloons to monitor free air.
1900	The term *classification of climate* is first used by Köppen.
1902	Existence of the stratosphere is discovered.
1913	The ozone layer is discovered.
1918	Beginning of the development of the polar front theory by V. Bjerknes.
1925	Beginning of systematic data collection using aircraft.
1928	Radiosondes are first used.
1940	Nature of jet streams is first investigated.
1960	United States launches the first meteorological satellite, Tiros I.
1978	U.S. Congress passes the National Climate Act.
1978	The United States bans the use of CFCs as aerosol propellants.
1987	The Montreal Protocol limiting the production of CFCs is signed by more than 30 nations.
1989	The state of Vermont bans the use of CFCs in auto air conditioners.

Scientific analysis of the atmosphere began in the seventeenth century with the design of instruments to measure atmospheric conditions. The instruments provided data that helped researchers develop laws applying to the atmosphere. Galileo invented the precursor to the thermometer in 1593, and Evangelista Torricelli invented the barometer in 1643. In 1661, Boyle discovered the basic relationship between pressure and volume in a gas.

The eighteenth century brought improved and standardized instruments, and extensive data collection and description of regional climates began. Explanation of phenomena through the study of the physical processes began in the nineteenth century. The most widely recorded data are surface temperature and precipitation. The longest complete climatic records exist for temperature and precipitation. Records range in length from about 325 years for measurements in Central England to about 200 years for stations in Europe and the United States. Most of the stations that make up today's observational network have kept records for less than 100 years.

In 1817, Alexander von Humboldt constructed the first map that showed temperatures using isotherms. Soon after, in 1827, H. W. Dove explained local climates using polar and equatorial air currents.

Table 1.2 lists the variables measured most frequently today. Nearly 10,000 stations throughout the world record at least one of these variables. They are part of a primary land-based system of observational stations. Unfortunately, the stations are not evenly distributed; large parts of the world have but a sparse network of recording stations. Since the oceans occupy almost three-quarters of Earth's surface, climatic data over the oceans is of prime importance. Archives of surface data from Earth's oceans began in 1854. Through international agreement, the major maritime nations began a regular program of recording atmospheric and oceanic data from merchant and military ships. Nonetheless, long-term data from the oceans exist only for the popular sea lanes. For large oceanic areas only limited data are available.

In the early 1800s, the only data for conditions at high altitudes came from mountain observatories. In 1885, balloons became more widely used to monitor air currents and temperatures. During World War I (1914–18), the use of balloons, kites, and aircraft multiplied rapidly. A similar interest in high-altitude conditions occurred in World War II (1939–45). The need for upper-air observation prompted a group of countries to establish a worldwide network of stations to accomplish that. Balloon-carried instruments, called radiosondes, are released into the atmosphere at specified times to simultaneously sample atmospheric conditions. Today, a network of about 1000 stations using radiosondes routinely measures air temperature, dewpoint temperature, and wind direction and velocity.

TABLE 1.2
Commonly observed atmospheric variables.

Air temperature
Barometric pressure
Cloud type, height, and amount
Current or prevailing weather
Dew point temperature
Precipitation
Sunshine
Wind velocity and direction

Besides the parameters listed in Table 1.2, some weather stations observe other variables. These variables include intensity of solar radiation and duration of sunshine, soil temperature, and evaporation; air pollution data; and water quantity and quality. About 100 stations measure solar radiation and 150 monitor sunshine duration. Many agriculture research stations measure evaporation.

In recent years, satellite imagery has been used to derive climatic data. Remote sensing provides quantitative values for temperature, humidity, and wind. Satellites are also important in studying the distribution and variation in cloud and snow covers. Chapter 10 includes methods of using satellite images.

While weather data are gathered at many places, the data are of little value unless they are readily available to potential users. Fortunately, a number of agencies and organizations collect and process original data. The international agency responsible for worldwide climatic data is the World Meteorological Organization (WMO) located in Geneva, Switzerland. This United Nations agency publishes *Meteorological Services of the World*, a directory of the 150 member nations and the agency responsible for climatic data in each country. Several scientific groups, both national and international, also maintain collections of data (Table 1.3).

The official library for records generated by government weather services in the United States is the National Climate Data Center in Asheville, NC. This center publishes summaries of national and world climatic data, both in printed and computerized form.

Besides national and international agencies, many textbooks and almanacs contain general climatic data. With the widespread use of personal computers, several companies have assembled daily climatic data for most U.S. stations and placed the information on compact disks.

A new dimension in the study of the atmosphere was added when satellites were placed in orbit with the express purpose of sensing the atmosphere. The first was Tiros in 1960, followed in 1966 by the launching of the geostationary ATS-1 (Applications Technology Satellite), the first in a series of weather satellites. A geostationary (or Earth-synchronous) satellite orbits Earth at a height of about 35,900 km (22,300 mi). At this altitude, the velocity necessary to maintain orbit is equal to Earth's rotational velocity. Even though Earth and the satellite are both in motion, the effect is that the satellite

TABLE 1.3
Worldwide climate data centers.

TYPE OF DATA	LOCATION
Meteorology	National Climate Data Center Asheville, NC
Meteorology	Molodezhnaya Moscow
Glaciology	National Snow and Ice Data Center Boulder, CO
Oceanography	National Oceanic and Atmospheric Administration Washington, DC
Solar-Terrestrial Physics	National Geophysical Data Center Boulder, CO
Ozone	Atmospheric Environment Services Ontario, Canada

TABLE 1.4

Elements of the National Climate Program Act of 1978.

The programs shall include, but not be limited to, the following elements:

1. assessment of the effect of climate on the natural environment, agricultural production, energy supply and demand, land and water resources, transportation, human health, and national security.
2. basic and applied research to improve the understanding of climatic processes, natural and human-induced, and the social, economic, and political implications of climatic change
3. methods for improving climate forecasts
4. global data collection on a continuing basis
5. systems for the dissemination of climatological data and information
6. measures for increasing international cooperation in climatology
7. mechanisms for climate-related studies
8. experimental climate forecast centers
9. submission of five-year plans

appears to remain in place over a given point on earth. The working satellites are the Geostationary Operational Environmental Satellite (GOES)/Synchronous Meteorological Satellite (SMS) series.

International programs for weather and climate studies have also been developed. The Global Atmospheric Research Program (GARP), also called the Global Weather Experiment, for example, is a concerted research effort by more that 140 countries, WMO, and the International Council of Scientific Unions. The goal of this program is to test the practical limits of weather forecasting and determine the statistical properties of the atmosphere's general circulation. This knowledge will lead to a better understanding of the physical basis of climate.

An important event in the recent development of climatology is the National Climate Program Act. Signed into law in September 1978, this act's stated purpose is "to establish a national climate program that will assist the nation and the world to understand and respond to natural and man-induced climate processes and their implications." The program (see Table 1.4) acknowledges the importance of climate and our understanding of it.

The Study of Climatology

The study of regional climatology began at least as early as ancient Greece. In fact, the word **climate** comes from a Greek word meaning "slope." In this context, it refers to the slope, or inclination, of Earth's axis. It specifies an Earth region at a particular place on that slope—that is, the location of a place in relation to parallels of latitude. This mathematical derivation represents one of many contributions to mathematical geography by philosophers such as Eratosthenes and Aristarchus. The Greek search for knowledge about the world also resulted in written works on the atmosphere. The first climatography was *Airs, Waters, and Places* written by Hippocrates in 400 B.C. In about 350 B.C., Aristotle wrote *Meteorologica*, the first treatise on meteorology.

The academic world did not resume the Greeks' interest in the atmosphere for many hundreds of years. Although Arab scholars of the ninth and tenth centuries expanded upon the Greek writings, interest in the atmosphere didn't really resume until the middle of the fifteenth century with the Age of Discovery. Extended sea voyages and development of new trading areas led to descriptive reports of climates outside of Europe. Many of these descriptions were fanciful and provided the basis for long-standing misconceptions about parts of the world.

Since climatology analyzes atmospheric conditions at different locations, it is also geographical. British climatologist E. T. Stringer (1975) wrote:

> The variations in the Earth's surface have profound effects on the interchange of heat, moisture, and momentum between land, water, and atmosphere, and are vital in determining specific climatic conditions; here local empirical observation as well as meteorological theory is absolutely necessary. Climatology thus does not belong entirely within the fields of meteorology or geography. It is a science—really an applied science—whose methods are strictly meteorological but whose aims and results are geographical.

Part Two of this text examines regional climatology and climatic regions.

Maps also play a significant role in the depiction of climates and climatic data. Many maps use *isolines,* lines joining locations of equal value to show distribution of the elements. The name given to the isoline depends upon the climatic element shown on the map (Table 1.5). (Regional climates [Map A] and locations of countries and cities [Map B] are shown in the insert following page 272.)

Climatic studies can take a number of approaches (see Figure 1.1). Climatography consists of the basic presentation of data in written or map form. As the names imply, physical and dynamic climatology relate to the physics and dynamics, respectively, of the atmosphere. Physical climatology deals largely with energy exchanges and physical processes. Dynamic climatology is more concerned with atmospheric motion and exchanges that lead to, and result from, that motion. Some scientists consider the study of past and

TABLE 1.5
Isolines used in climatology.

ISOLINE	LINES OF
Isallobar	Equal pressure tendency showing similar changes over a given time
Isamplitude	Equal amplitude of variation
Isanomaly	Equal anomalies or departures from normal
Isobar	Equal barometric pressure
Isocryme	Equal lowest mean temperature for specified period (e.g., coldest month)
Isohel	Equal sunshine
Isohyets	Equal amounts of rainfall
Isokeraun	Equal thunderstorm incidence
Isomer	Equal average monthly rainfall expressed as percentages of the annual average
Isoneph	Equal degree of cloudiness
Isonif	Equal snowfall
Isophene	Equal seasonal phenomena (e.g., flowering of plants)
Isoryme	Equal frost incidence
Isoterp	Equal physiological comfort
Isotherm	Equal temperature

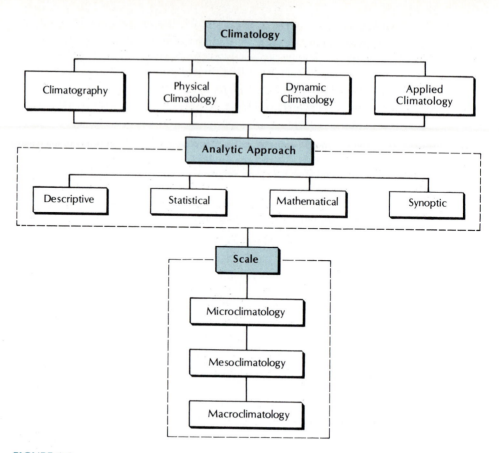

FIGURE 1.1

Subgroups, analytic methods, and scales of climate study.

Source: J. E. Oliver, *Climatology: Selected Applications* (Silver Spring, MD: V. H. Winston, 1981), 4. Used by permission.

future climates to be a subgroup of climatic studies. However, methods used in studying past and future climates are the same as those in other aspects of climatology.

The approaches suggested in Figure 1.1 are largely self-explanatory, with the possible exception of synoptic climatology. As shown in Chapter 10, the object of the synoptic approach is to relate local and regional climates to atmospheric circulation. The scale of studies shows that climatic investigation covers areas ranging from very large (macroclimates) to very small (microclimates).

Applied Climatology

Application of climatology to the behavior and activities of people has its roots in the distant past. As farming and fishing developed and expanded, for example, knowledge about seasonal weather and climate became integrated into the fabric of everyday life. Even classical writing shows the importance of climate to people: In Greece, Hippocrates (circa 420 B.C.) developed ideas to explain the influence of climate on human behavior. In Rome, Strabo (64 B.C. − 20 A.D.) attempted to explain how climate affected the rise and strength of the Roman Empire. Thereafter, a long series of historic works related climate to humans and the human world. Unfortunately, many of these works attempted to show how the environment, and often the climate, determined the nature of human activity.

This "determinism" is the philosophical doctrine that human action is not free, but is determined by external forces—such as climate—acting upon the human will. Determinists often make sweeping generalizations based upon limited data while ignoring evidence contrary to their viewpoint. The extreme bias of some determinists' writing produced a reactionary backlash in the 1930s that made the study of physical relationships between humans and climate a hazardous academic pursuit.

Fortunately, antideterminism did not enter into studies relating military activities and climate, and this aspect of applied climatology grew rapidly to address the operational requirements of World War II. The worldwide nature of that conflict required equipment and clothing designed to withstand a wide variety of climatic types and weather conditions. The Persian Gulf conflict of 1991 is a good example of the military adjusting to climate. The U.S. military, which had last fought a major war in a tropical jungle, suddenly found itself in a Middle Eastern desert. A massive effort was undertaken to supply new camouflage uniforms using desert colors and to repaint vehicles being sent to the area.

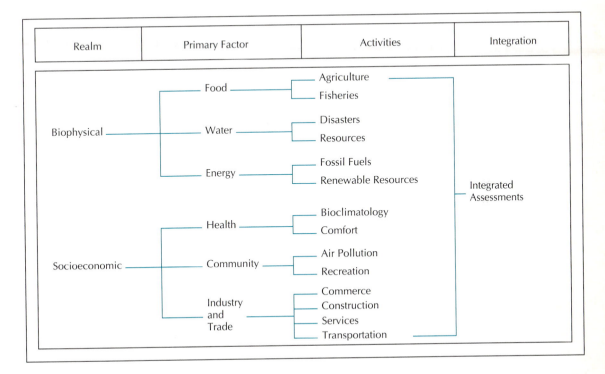

FIGURE 1.2
Subdivisions of applied climatology.

The increased availability of climatic data, the results of decades of research, and the application of statistical methods using computer technology mean that applied climatology now is "the use of both archived and real-time climatic data to solve a variety of social, economic and environmental problems" (Smith 1987). Applied climatology is the scientific application of climatic data to specific problems within such areas as forestry, agriculture, and industry; it can involve the application of climatic data and theory to other disciplines, such as geomorphology and soil science.

Figure 1.2 separates applied climatology into two major areas of study or realms. The major activities associated with the biophysical realm involve food, water, and energy. The three factors represent the role of climate in studies of the physical and plant world as applied to human life. The socioeconomic realm concerns the place of climate in the social and economic well-being of people. Main activities are those dealing with human health; community activities; and commerce, trade, and industry. Some problems do not fall neatly into any one of these categories. Similarly, we must consider components of each to solve an applied climatic problem or to examine the impact of a climatic event. Such studies are called integrated assessments. As we will see in examples later, such assessments often require the construction of scenarios and models.

The very broad realms and activities can be further categorized (see Figure 1.3). Food production can be divided into agriculture or fisheries, and each of those categories involves highly specific activities. Within agriculture, climate related to hazards (e.g.

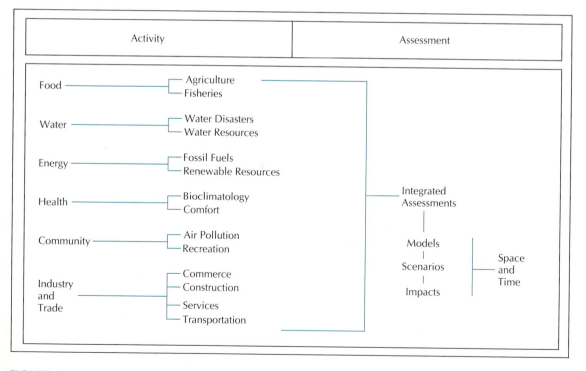

FIGURE 1.3
Activities and assessment in applied climatology.

frosts, droughts), irrigation needs, and livestock (stress under climatic extremes) are areas of concern for the applied climatologist.

Given the broad scope of applied climatology, it is not possible to provide detailed information for each of the identified activities. However, many examples of the application of climatic analysis to human problems appear throughout this book.

The Standard Atmosphere

Earlier in this chapter we said that climatology is partially meteorologic. Since weather is the basic ingredient in climate, any study of climate should begin with an introduction to the atmosphere and its workings.

The atmosphere is continually changing and is never exactly the same at any two different points in time or space. Despite that constant change, a certain set of conditions can describe the atmosphere most of the time. Knowing the most frequent conditions in a vertical column through the atmosphere provides a basis for examining the extent of change, and scientists have constructed a model atmosphere to accomplish that. In the United States this model is the **U.S. standard atmosphere** (see Table 1.6). This standard atmosphere represents ". . . an idealized, steady-state representation of Earth's atmosphere from the surface to 1000 km (600 mi), as it is assumed to exist in a period of moderate solar activity" (NOAA 1976, p. 1). The standard atmosphere does not represent an average condition, but a steady-state condition—one in which the atmosphere is in balance with the processes governing it. The standard atmosphere defines profiles of chemistry, temperature, pressure, density, and several other variables.

Atmospheric Chemistry

The atmosphere consists of a mixture of gases which, in its normal state, is colorless, odorless, and tasteless. Some of the particles in the atmosphere are single atoms, such as argon and helium. Others are molecules consisting of atoms of two or more elements, like water vapor and carbon dioxide.

Chemists in the eighteenth century first showed that the atmosphere includes many gases. The first gas studied in detail was carbon dioxide (CO_2). Its discovery in 1752 is somewhat surprising since, compared to nitrogen and oxygen, the atmosphere contains only a small amount of carbon dioxide. Gaseous nitrogen, discovered by Daniel Rutherford in 1772, was at first called "mephitic air." Shortly after this, Joseph Priestley isolated oxygen, which he called "dephlogistated air."

Over time other gases were discovered, the last of which was argon, isolated in 1894. Eventually the gases were given their modern names and their relative proportion by volume in the atmosphere was determined. As shown in Table 1.7, nitrogen and oxygen make up the bulk of the atmosphere. Other gases are present in very small but important amounts.

The gases making up the atmosphere fall into two groups: constant gases (those relatively constant by volume) and variable gases (those whose volumes vary in the atmosphere).

TABLE 1.6
The standard atmosphere.

ALTITUDE (km)	TEMPERATURE (°C)	PRESSURE (mb)	P/P_0*	DENSITY (kg/m³)	D/D_0*
30.00	−46.60	11.97	0.01	0.02	0.02
25.00	−51.60	25.49	0.03	0.04	0.03
20.00	−56.50	55.29	0.05	0.09	0.07
19.00	−56.50	64.67	0.06	0.10	0.08
18.00	−56.50	75.65	0.07	0.12	0.09
17.00	−56.50	88.49	0.09	0.14	0.12
16.00	−56.50	103.52	0.10	0.17	0.14
15.00	−56.50	121.11	0.12	0.20	0.16
14.00	−56.50	141.70	0.14	0.23	0.19
13.00	−56.50	165.79	0.16	0.27	0.22
12.00	−56.50	193.99	0.19	0.31	0.25
11.00	−56.40	226.99	0.22	0.37	0.30
10.00	−49.90	264.99	0.26	0.41	0.34
9.50	−46.70	285.84	0.28	0.44	0.36
9.00	−43.40	308.00	0.30	0.47	0.38
8.50	−40.20	331.54	0.33	0.50	0.40
8.00	−36.90	356.51	0.35	0.53	0.43
7.50	−33.70	382.99	0.38	0.56	0.45
7.00	−30.50	411.05	0.41	0.59	0.48
6.50	−27.20	440.75	0.43	0.62	0.50
6.00	−23.90	472.17	0.47	0.66	0.54
5.50	−20.70	505.39	0.50	0.70	0.57
5.00	−17.50	540.48	0.53	0.74	0.60
4.50	−14.20	577.52	0.57	0.78	0.63
4.00	−11.00	616.60	0.61	0.82	0.67
3.50	−7.70	657.80	0.65	0.86	0.70
3.00	−4.50	701.21	0.69	0.91	0.74
2.50	−1.20	746.91	0.74	0.96	0.78
2.00	2.00	795.01	0.78	1.01	0.82
1.50	5.30	845.59	0.83	1.06	0.86
1.00	8.50	898.76	0.89	1.11	0.91
0.50	11.80	954.61	0.94	1.17	0.95
0.00	15.00	1013.25	1.00	1.23	1.00

Source: J. M. Morgan and M. D. Morgan, *Meteorology,* 3d ed. (New York: Macmillan, 1991), 541.
*P/P_0 = ratio of air pressure to sea-level value; D/D_0 = ratio of air density to sea-level value.

Constant Gases

The constant gases remain in the same proportion in the atmosphere upward to an altitude of about 80 km (48 mi). The three most important constant gases are nitrogen (78% by volume), oxygen (21%), and argon (0.93%). Nitrogen is by far the most abundant of the gases, but it is relatively inactive in the atmosphere. Argon is also inactive, but it is present only in small amounts. Oxygen is present in large quantities and is very active in the chemical processes of the physical and biological environment.

TABLE 1.7
Chemical composition of the dry atmosphere below 80 km.

GAS	PARTS PER MILLION	PERCENTAGE OF TOTAL
Nitrogen	780,840.0	78.1
Oxygen	209,460.0	20.9
Argon	9,340.0	0.9
Carbon dioxide	350.0	
Neon	18.0	
Helium	5.2	
Methane	1.4	
Krypton	1.0	
Nitrous oxide	0.5	
Hydrogen	0.5	
Xenon	0.09	
Ozone	0.07	

Note: Data from U.S. Department of Commerce, NOAA, *United States Standard Atmosphere* (Washington, D.C.: Government Printing Office, 1976).

Variable Gases

Variable gases, as the term implies, change as a proportion of total atmospheric gases from time to time and place to place. The most important variable gases are water vapor and CO_2. Water-vapor content varies from near zero to a maximum of about 4% by volume, but this small relative volume is extremely important. Carbon dioxide exists in amounts that average near 0.03%. Water vapor is, of course, the source of all precipitation that falls on Earth. Water vapor, CO_2, and dust all absorb solar radiation. Ozone, another variable gas, is in the lower atmosphere in small amounts, the average being about one part per million. Another variable element of the atmosphere that frequently acts like a gas is the particulate matter suspended in the air (aerosol). This includes soil particles, smoke residue, ocean salt, bacteria, seeds, spores, volcanic ash, and meteoric particles (see Table 1.8). The primary source of the solid particles is Earth's surface. Particulate matter decreases rapidly and changes in composition with altitude. High-altitude particles accumulate from meteoric dust, volcanic erup-

TABLE 1.8
Major ingredients in a steady-state tropospheric aerosol.

SOURCE	TOTAL (TONS)	PERCENTAGE OF TOTAL
Vegetation	1.7×10^7	25.8
Dust rise by wind	1.6×10^7	24.1
Sea spray	7.6×10^6	11.9
Forest fires	6.2×10^6	9.9
Sulfur cycle	5.5×10^6	8.6
NO_x NO_3	5.5×10^6	7.7
Nitrogen cycle (ammonia)	3.9×10^6	6.0
Combustion and industrial	1.7×10^6	2.6
Anthropogenic sulfates	1.7×10^6	2.6

tions, and nuclear explosions in the atmosphere. The overall amount of particulate matter in the atmosphere varies from as little as 100 parts per cm^3 to several million parts per cm^3.

Both particulate matter and water vapor are important in the atmosphere. They are largely responsible for the day-to-day variation in solar energy reaching Earth's surface. Solid particles, moreover, serve as nuclei for condensation of water vapor and thus are necessary for precipitation.

Vertical Structure of Earth's Atmosphere

Many changes take place with height in the atmosphere. Some changes occur continuously with height, and others occur at well-defined altitudes. We can divide the atmosphere vertically into several different zones based upon these changes (see Figure 1.4).

Troposphere

The lowest zone, the **troposphere**, is a turbulent zone in which temperature decreases fairly uniformly with increases in height. Worldwide, the average rate of change in temperature with altitude (the lapse rate) is 6.5°C per 1000 m (5.5°F/1000 ft).

The troposphere has two subzones, the lower and upper troposphere. The lower troposphere extends upward to about 3 km above the surface; it is the zone with the most friction between Earth and the atmosphere. It is also the zone most affected by the daily changes in surface conditions. Temperature inversions frequently occur in this zone. A temperature **inversion** exists when the temperature increases rather than decreases with increases in height.

The upper troposphere extends to a mean height of 11 km (7.2 mi). It is less affected by the daily changes near the surface and by friction at the surface. The primary changes in this zone result from the secondary circulation (atmospheric storms) and seasonal changes in energy.

The water-vapor content of the atmosphere at any point in time and space depends upon the air temperature and to a lesser extent on atmospheric pressure, proximity to a moisture source, and the history of the air mass. Water-vapor content is normally highest close to the surface for two reasons: First, most of the water in the atmosphere gets there as a result of evaporation from the ocean and from evapotranspiration. Second, air temperature is normally highest near the surface.

The highest water-vapor content recorded in the atmosphere is on the shores of the Red Sea, where the temperature averages 34°C (93°F) in summer. At air temperatures from 35°C down to −5°C (95°–23°F), the amount of water vapor the air can hold decreases by half for every 10°C (18°F) drop in temperature. The recorded range of water vapor in the atmosphere near the surface varies from a low of 0.1 parts per million (ppm) in the Antarctic and Siberia to a high of 35,000 ppm along the Persian Gulf. Due to the combination of cooler temperatures and distance from the moisture source, mean water-vapor content decreases rapidly with altitude to a height of 12–15 km (7.2–9.0 mi). Here it reaches a level of 2–3 ppm (Figure 1.5).

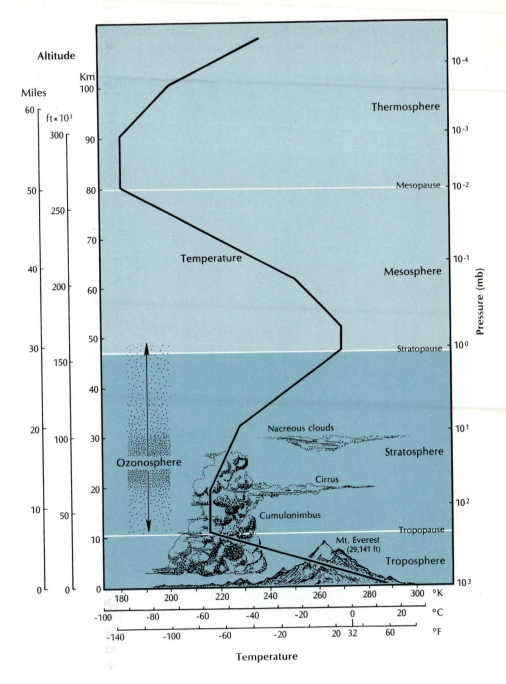

FIGURE 1.4
Vertical structure of the atmosphere.

The atmosphere is an unconfined gaseous fluid resting on Earth's surface. The mass and pressure of the air decrease with height. Atmospheric pressure at any point on the surface or in the atmosphere is the force per unit area exerted by the mass of the atmosphere above that point. Air close to the surface is subject to the mass of gases above.

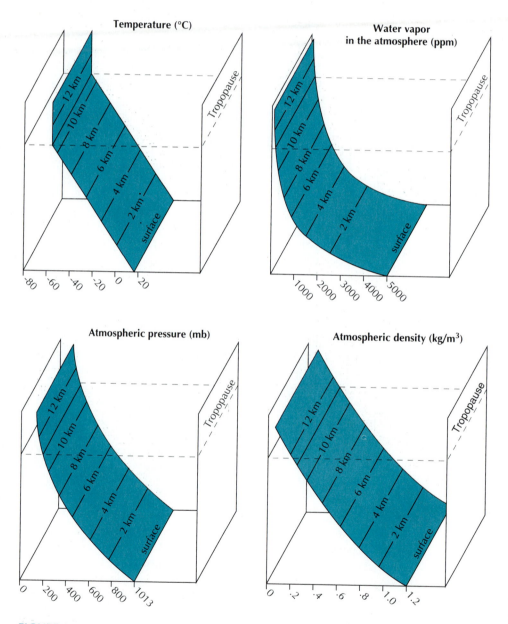

FIGURE 1.5

Vertical distribution of temperature, water vapor, pressure, and density in the lower 14 km of the atmosphere.

Hence, the greatest portion of the mass of the atmosphere lies near the surface. Atmospheric pressure decreases geometrically with height; that is, it decreases at a decreasing rate. Surface pressure is on the order of 1000 mb. In the lower stratosphere, at 15 km (9 mi) above sea level the pressure is only 100 mb. Of the total mass, about 50% lies below 5500 m (18,000 ft), 84% below 13 km (7.8 mi), and 99% below 30 km (18 mi). The density of the atmosphere decreases with height at a geometric rate, but the gases stay in roughly the same proportion up to heights of at least 80 km (48 mi). The gases decrease in density away from Earth and extend for a distance of at least 1000 km (600 mi).

A distinct boundary layer known as the **tropopause** marks the upper reaches of the troposphere. A series of overlapping layers at different heights actually makes up the tropopause. This boundary zone is significant in several ways. First, it marks the upper limit of most turbulent mixing started from the surface. It also represents a cold point in the vertical temperature structure of the atmosphere. Finally, it marks the upper limit of most of the water in the atmosphere. Very little moisture penetrates the tropopause except through severe thunderstorms, which will go as high as 20 km (12 mi). The amount of water vapor in the atmosphere above the tropopause is very small. At the equator, the tropopause is at a height of 16–17 km (9–9.6 mi). The temperature averages $-70°$ to $-85°C$ ($-94°$ to $-121°F$), and the pressure averages only 100 mb, a tenth of sea-level pressure. Above the North and South Poles, the tropopause is at a height of 9–12 km (5.4–7.2 mi), with a temperature averaging $-50°$ to $-60°C$ ($-58°$ to $-76°F$), and at a pressure of some 250 mb.

Stratosphere

Above the tropopause is the **stratosphere**, named for the layered nature of the air at these levels. The stratosphere is relatively stable, relatively dry, and has relatively little vertical motion. Some high-velocity winds occur just above the tropopause, but otherwise winds are noticeably absent.

The temperature at the tropopause in the midlatitudes is about $-55°C$ ($-72°F$). The temperature remains nearly the same up to about 20 km (12 mi). In the upper region, extending to some 50 km (31 mi), temperature increases with height by as much as 4°C per km (2.2°F/1000 ft). However, the temperature at the top of the stratosphere is near that at the bottom.

The Upper Atmosphere

Above the stratosphere is another boundary layer, the **stratopause**. As is the case with the tropopause, there may not be a single boundary layer but instead a series of overlapping layers making up a transition zone.

Above the stratopause is a layer identified by a temperature decrease with altitude, the **mesosphere**. Beginning at an elevation of about 48 km (29 mi), the decline in temperature continues outward to the mesopause near 80 km (48 mi). Up to the mesopause, the mixture of gases is about the same as at the surface. For this reason the lower 80 km of the atmosphere is termed the **homosphere**. Above 80 km, the composition of the air begins to change: Gases tend to stratify on the basis of molecular weight. This layer is termed the **heterosphere**. From 80 to 220 km (130 mi), molecules of nitrogen largely make up the atmosphere. Layers of helium and oxygen exist outward from there.

Ionosphere

The mesosphere merges with a zone called the **ionosphere**. As the name implies, the ionosphere contains ionized gases and free electrons resulting from absorption of solar

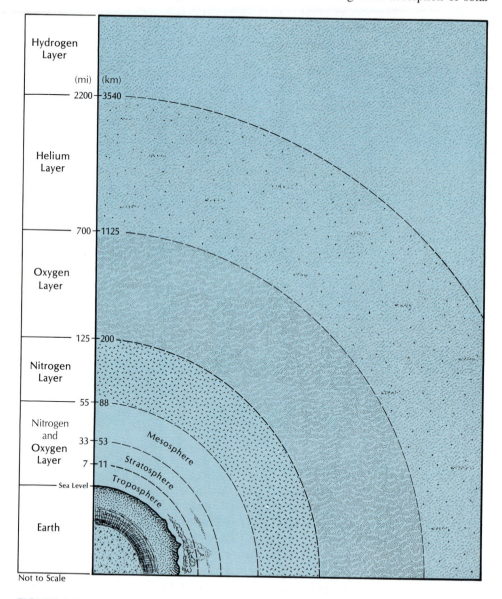

FIGURE 1.6

Chemical zones in the atmosphere.

radiation. Short-wave radiation from the sun is absorbed, and the energy causes the electrons to split from the atoms of nitrogen and oxygen. The proportion of ionized particles is large because the density of particles is low and solar energy is high. The higher in the ionosphere, the fewer the particles, the more solar energy available, and the more ionization occurs.

Incoming solar radiation begins to interact with the atmosphere at altitudes beyond 500 km (310 mi). This outer region is also where gaseous particles escape Earth's gravity. Particles are far enough apart that some of them have sufficiently high velocity in a direction away from Earth to escape to space. The atmosphere is very thin at this altitude, but it is here that incoming space vehicles and meteorites begin to heat due to friction.

Above this, to perhaps 1125 km (675 mi), atomic oxygen is prevalent. Beyond this layer of atomic oxygen, helium is most common out to 3540 km (2124 mi). Still further out, hydrogen atoms predominate. The boundaries between the zones are not clearly defined. The heights represent the altitudes above which a different chemistry predominates (Figure 1.6).

Summary

Climate is the total of weather at a place for a particular period. Study of climate is often divided into climatography, synoptic climatology, dynamic climatology, applied climatology, and climate forecasting.

The study of climatology began in ancient times but received considerable impetus from information gained in the great voyages of discovery. During this period instruments were developed that could help humans observe the atmosphere and record data. The development of weather satellites improved climatic interpretation and understanding. Similarly, the creation of the National Climate Program has promoted study.

Climatologists gather, store, and analyze climatic data. Most climatic data come from weather monitoring stations and are stored by both government and private agencies. Computers and computer techniques now aid in data analysis, including summarizing and time-series analysis. The modeling of past and future climates is growing rapidly in importance.

Climate has always played a major role in the history of humans. The climate was more favorable in some areas than in others for the development of the species. Even in places where the climate was favorable for humans, it was not static. Climate changes with time, sometimes for the better for us and sometimes for the worse. The climate of the planet is far different today than what it was a million or a billion years ago.

The atmosphere changes with time and from place to place. There are also changes which take place with altitude. To provide a baseline against which to evaluate these changes, a model has been developed called the standard atmosphere. This standard atmosphere establishes the attributes of the atmosphere as it is most frequently found.

In the 1990s, climate affects more people directly than at any other time, simply because more people inhabit the planet than at any time in Earth's history. The billions of people living on the planet today depend on Earth to provide their food supply as it always has. When the climate changes, it can cause major changes in food production, often for the worse.

THE ENERGY BALANCE

Chapter Overview

Energy is the capacity for doing work. It can exist in a variety of forms; those forms can change from one to another (Table 2.1). The transfer of energy from place to place is of major importance in climatology. Transfer can occur in three ways: conduction, convection, and radiation.

Conduction consists of energy transfer directly from molecule to molecule when the molecules are densely packed and contact one another. Energy always moves from an area of more energy to an area of less energy, similar to the way water flows from high places to lower places.

Convection is the transfer of energy when an energy-laden substance moves from one location to another. Both convection and conduction need a physical substance—a solid, liquid, or gas—in which to operate.

Solar radiation is the only means of energy transfer through space without the aid of a material medium. Between the sun and Earth, where a minimum of matter exists, radiation is the only important means of energy transfer. The sun is the major source of energy for our planet—it is so important that most climatological phenomena are related to it.

Comprehending the significance of solar energy first requires understanding three properties of solar radiation. The first is the nature of solar energy, the second is energy's effect on the Earth-atmosphere system, and the third is changes that solar energy undergoes in the system.

TABLE 2.1
Energy forms and transformations.

FORMS	
Radiation	The emission and propagation of energy in the form of waves
Kinetic energy	The energy due to motion: one-half the product of the mass of a body and the square of its velocity
Potential energy	Energy that a body possesses by virtue of its position and that is potentially converted to another, usually kinetic energy
Chemical energy	Energy used or released in chemical reactions
Atomic energy	Energy released from an atomic nucleus at the expense of its mass
Electrical energy	Energy resulting from the force between two objects having the physical property of charge
Heat energy	A form of energy representing aggregate internal energy of motions of atoms and molecules in a body

Examples of Transformations

Atomic energy (Sun) $\longrightarrow$ Radiation (Sunlight) $\longrightarrow$ Heat (Earth surface) $\longrightarrow$ Radiation (Terrestrial)

Radiation (Sunlight) $\longrightarrow$ Chemical energy (Photosynthesis) $\longrightarrow$ Food chain

Potential energy (Water vapor) $\longrightarrow$ Kinetic energy (Raindrop) $\longrightarrow$ Heat (Friction)

The Nature of Radiation

Every object above the temperature of absolute zero radiates energy to its environment. It does so in the form of electromagnetic waves that travel at the speed of light. Energy transferred in the form of waves varies according to wavelength, amplitude, frequency and speed (Figure 2.1). Using wavelengths as a criterion, radiant energy exists along a spectrum from very short to very long (Figures 2.2a and 2.2b).

The characteristics of the radiation emitted by an object depend mainly upon its temperature. The amount of radiation varies as the fourth power of the absolute temperature in degrees Kelvin (°K). The Kelvin scale is based upon the concept of absolute zero, which is the theoretical temperature at which all molecular motion would cease. Absolute zero equals −273°C or −460°F. The hotter an object is, the greater the flow of energy is from it. The **Stefan Boltzmann law** expresses this relationship as follows:

$$F = KT^4$$

where

F = flux of radiation in langleys per minute
T = temperature in degrees Kelvin
K = a constant of 0.813×10^{-10} calories/cm^2/min

For example, the average temperature at the surface of the sun is 6000°K (11,000°F). The average temperature of Earth is 288°K (59°F). The temperature at the surface of the sun is more than 20 times as high as that of Earth. Since 20 raised to the fourth power is 160,000, the sun emits 160,000 times as much radiation per unit area as Earth. The sun emits radiation in a continuous range of electromagnetic waves ranging from long radio

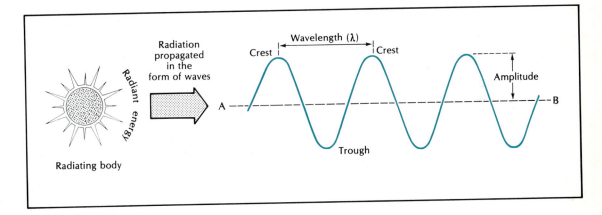

FIGURE 2.1
Radiant energy is transferred in the form of waves. The waves are defined in terms of wavelength, amplitude, frequency, and speed. Wavelength is the distance between wave crests. Amplitude is half the height difference between crest and trough. Frequency is the number of waves past a point in space per unit time. Speed is the distance a wave travels per unit time.

waves of 10^5 m (62.5 mi) down to very short waves such as gamma rays that are less than 10^{-4} μm (4^{-10} in.) in length.

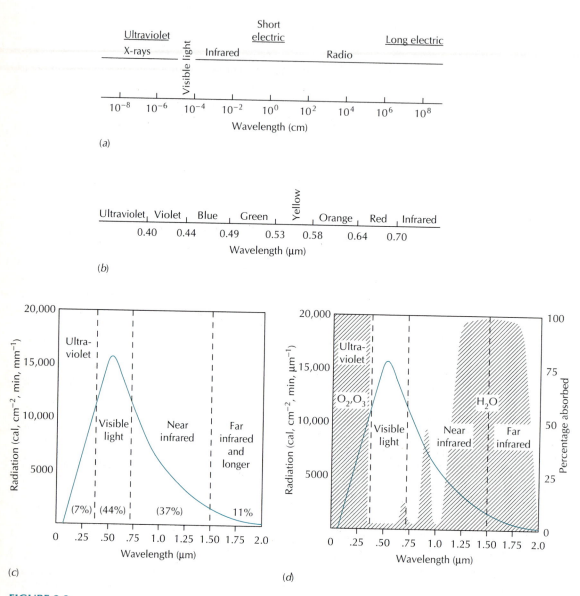

FIGURE 2.2

The major bands of solar radiation: (a) The electromagnetic spectrum of solar radiation. (b) The range between 10^{-4} and 10^{-5} cm is a highly significant band. In this band the energy is visible to the human eye. Wavelengths in the range of visible light are given in micrometers (μm), where 1 μm = 10^{-4} cm. (c) The portion of the electromagnetic spectrum containing most of the solar radiation. (d) The absorption bands of Earth's atmosphere.

Another law of radiant energy (Wien's law) states the wavelength of maximum radiation is inversely proportional to the absolute temperature. In other words, the higher the temperature, the shorter the wavelength at which maximum radiation occurs:

$$\lambda_{max} = K/T$$

where

K = constant (2897)

T = temperature in °K

For the sun:

$$\lambda_{max} = K/T$$
$$= 2897/6000$$
$$= 0.48 \ \mu m$$

For Earth:

$$\lambda_{max} = K/T$$
$$= 2897/288$$
$$= 10 \ \mu m$$

At the temperature of the sun's surface, the maximum radiation ranges from 0.4 to 0.7 μm. This is precisely the range of radiation that the human eye perceives; accordingly, we call it the visible range (Figure 2.2*b*). As a good example of evolutionary processes, human eyes have evolved to take maximum advantage of the part of solar radiation in greatest abundance. Individuals do not all see the same range of radiation, since some see shorter or longer wavelengths than others.

The Solar Source

The sun is a gaseous mass with a diameter 109 times that of Earth. The sun rotates on its axis about once a month. Since it is gaseous, the entire sun does not rotate at the same speed. The equator rotates every 25 Earth days, while latitude 70° rotates in 33 days. Also, since the sun is a gaseous mix, it has no sharp boundaries. Because different sections of the sun have different characteristics, we can arbitrarily divide it into four major parts: the core, photosphere (or visible surface), chromosphere, and corona (Figure 2.3).

The source of the sun's energy is nuclear fusion. A nuclear reaction in the core of the sun changes hydrogen into helium. The nuclear reaction started when the sun formed from cosmic debris 4.5 billion years ago. As the dust and gases began to collect, the mass forming the sun grew in size. The internal pressure due to gravity became so great that temperatures grew hot enough for the fusion process to begin. The reaction converts four protons (hydrogen nuclei) into a helium nucleus. The process converts some matter into energy. The combined mass of four hydrogen atoms is 4.032 and that of a helium atom is 4.003. The difference in mass, 0.029, changes to energy according to Albert Einstein's famous equation:

$$E = mc^2$$

where

E = energy released

m = atomic mass

c = the speed of light

FIGURE 2.3

Internal structure of the sun and surface features.

Source: Reprinted with the permission of Merrill, an imprint of Macmillan Publishing Company from *Earth Science*, 6th ed., by E. J. Tarbuck and F. K. Lutgens. Copyright © 1991 by Macmillan Publishing Company.

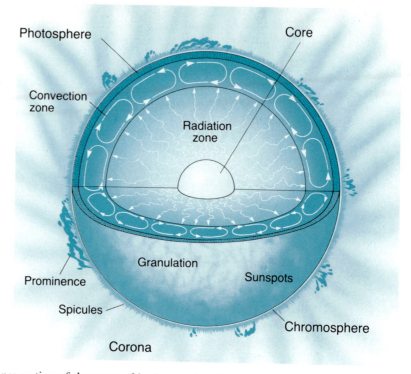

While the proportion of the mass of hydrogen that changes to energy is a small 0.7%, the amount is very large. The sun is consuming 600 million metric tons of hydrogen per second, meaning 4 million tons of hydrogen change into energy every second. The remaining mass changes to helium and adds to the core, which is growing larger at the expense of hydrogen. The probable life of a star such as our sun is 10 billion years. Since the sun is now nearly 5 billion years old, it has 5 billion years to continue as an energy source for Earth.

Photosphere

The photosphere is the bright outer layer of the sun that emits most of the radiation, particularly visible light. It consists of a zone of burning gases 300 km (200 mi) thick. The photosphere is an extremely uneven surface with many small bright areas called granules. Their brightness results from bursts of extremely hot gases that form in the photosphere and well up to the surface. Darker areas surrounding each granule represent cooler gases. The hot gases spread out, cool on reaching the surface, and settle back below the surface. The net effect of all of these granules is convection of hot gases outward and cooler gases inward. It is much like the boiling of a large kettle. The granules average 1000 km (620 mi) in diameter. The elements in the photosphere are hydrogen (90%) and helium (10%). At the radiative surface, the effective temperature is on the order of 6000°K (11,000°F). The surface temperature determines both the amount of energy emitted and the wavelengths at which it occurs.

Chromosphere

Just above the photosphere is the chromosphere, a relatively thin layer of burning gases. This region of gases is under little pressure and extends away from the sun as much as 1 million km (600,000 mi). It is here that the particles reach velocities high enough to leave the sun. The electrons and protons escape the sun in a stream called the solar wind. When this stream reaches Earth's environment, the magnetic field surrounding this planet diverts most of the stream toward the two poles. One feature of solar activity that occasionally affects Earth is the solar flare, a sudden and explosive burst of energy near sunspot clusters lasting a matter of hours. Flares emit huge quantities of radiant energy and atomic particles. The atomic particles add to the solar wind. When these particles reach Earth, they disturb the ionosphere. This disruption in turn interrupts radio and satellite communications and triggers the aurora.

Sunspots

Some of the most prominent visible features of the surface of the sun are dark areas known as sunspots. Sunspots appear as dark areas because they are about 1500°C (2700°F) cooler than the surrounding chromosphere. Science credits Galileo with determining that they were a feature of the sun, not clouds between Earth and the sun. They start as small areas about 1600 km (1000 mi) in diameter. The individual sunspot has a lifetime ranging from a few days to a few months. Each spot has a black center, or umbra, and a lighter region, or penumbra, surrounding it.

Because records of sunspots exist for many years and sunspots are relatively easy to see, they are one indicator of solar activity. The number of sunspots appears to follow an approximate 11-year cycle. First, the sunspots increase to a maximum, with 100 or so visible at one time. Over a period of years, the number diminishes until only a few or none occur. Change from the minimum to the maximum number (and vice versa) takes about 5.5 years.

Just how the sunspot cycle influences weather and climate is a matter of controversy. Some researchers claim to have found that weather patterns in a particular part of Earth follow the sunspot cycle closely. In other cases, no relationships exist. Even the physical process by which sunspots might affect weather is not clear. Radiation from sunspots corresponds in wavelength to that of X rays; this radiation has different effects on different parts of the atmosphere. A further complication is that sunspot activity coincides with solar wind intensity, and it is difficult to separate the effects of each.

The Atmosphere and Solar Radiation

The energy emitted by the sun passes through space until it strikes some object. The intensity of radiation reaching a planet follows a basic physical law known as the inverse square law. This law states that the area illuminated, and hence the intensity, varies with the squared distance from the light or energy source. Thus, if the intensity of radiation at a given distance X is one unit, at a distance of 2X the intensity of the light will be one-fourth that of X. This law controls the intensity of solar energy intercepted by planets in the solar system. Earth receives only about 1/2,000,000,000 of the sun's energy output.

We know the amount of energy radiated by the sun (100,000 cal/cm^2/min) and the mean Earth-sun distance of 149.5 million km (93 million mi). That information also tells us the amount of radiation intercepted by a surface at right angles to the solar beam at the outer limits of the atmosphere. This quantity of radiation, approximately 1.94 cal per cm^2 per min, is the **solar constant**. An often-used measure of solar energy is the langley, which is 1 g-cal per cm^2.

The solar constant is the basic amount of energy available at the outer limits of Earth's atmosphere. Several processes deplete the solar radiation as it passes into the atmosphere. These processes include reflection, scattering, absorption, and transmission.

Reflection

Earth and its atmosphere reflect part of the solar radiation back to space, with natural surface reflecting varying amounts. Reflectivity (or **albedo**) is expressed as a percentage of the incident radiation reaching the surface. Clouds are by far the most important reflectors in Earth's environment. Clouds' reflectivity ranges from 40% to 90%, depending upon their type and thickness.

Water covers the largest area of Earth's surface. The reflectivity of water depends on the angle of the solar beam and the roughness of the water surface. Reflectivity decreases as the sun gets higher in the sky and as the water's surface gets rough. People out in boats before 10 A.M. and after 2 P.M. can get sunburned even when wearing broad-brimmed hats, since hats do not protect from radiation reflected from the water. Water reflects as little as 2% of the radiation when the solar angle is 90° and the water is choppy.

Earth's land surface reflects only 40%–50% of solar radiation. Reflectivity of land surfaces varies with the type of surface cover. Fresh snow reflects over 75% and dry sand over 35% of the incident radiation. Table 2.2 provides examples of the reflectivity of various natural surfaces. The planet Earth has an albedo of 30% that represents the mean reflectivity from the ocean, land, and atmosphere (see Plate 1 in the photo insert).

Scattering

Scattering is the process by which small particles and molecules of gases diffuse part of the radiation in different directions. The process changes the direction of the radiation fairly randomly. The English scientist Lord John Rayleigh developed the explanation for scattering, thus the effect is called **Rayleigh scattering**.

The amount and direction of scatter depend upon the ratio of the radius of the scattering particle to the wavelength of the energy. Furthermore, the amount of scatter is inversely proportional to the fourth power of the wavelength. This means that in a given set of conditions, the shorter wavelengths scatter more readily than do longer wavelengths. For example, radiation of a wavelength twice as long as that of another wavelength scatters only $\frac{1}{16}$ as much.

The most obvious effect of scattering in the atmosphere is sky color: Our sky appears blue because of scattering of radiation in the shorter wavelengths of visible light. As radiation in the visible range enters the outer regions of Earth's atmosphere, small gas

TABLE 2.2
Typical albedos of various surfaces to solar radiation.

TYPE OF SURFACE	ALBEDO (PERCENT)
Fresh snow	75–95
Clouds	
Cumuliform	70–90
Stratus	60–84
Cirrostratus	44–50
Planet Venus	78
Old snow and sea ice	30–40
Dry sand	35–45
Planet Earth	30
Desert	25–30
Concrete	17–27
Savanna	
Dry	25–30
Wet	15–20
Grass-covered meadow	10–20
Tundra	15–20
Dry, plowed field	5–25
Asphalt road or parking lot	5–17
Green field crops	3–15
Deciduous forest	10–20
Coniferous forest	5–15
Moon	7
Water	
Angle of inclination of the sun:	
0°	99+
10°	35
30°	6
50°	2.5
90°	2

molecules scatter the shortest wavelengths—the violet—first. Normally, the atmosphere scatters and absorbs the violet so the sky doesn't have a violet color. Blue scatters next. This randomly diffused radiation in the blue range scattered through the lower atmosphere gives it its color. Most of the radiation in the visible range has wavelengths greater than the diameter of particles in the dry atmosphere. This energy passes through without being changed.

Scattering also explains the orange and red colors seen at dawn and sunset (Figure 2.4). During these times the radiation passes through the atmosphere at a very low angle. Thus it passes a long distance through air close to the ground. The lower atmosphere contains not only the dry gases but water vapor, solid particles, organic material, and salt. These larger particles scatter the longer-wavelength radiation. In fact, the atmosphere

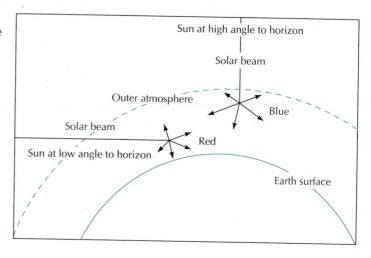

FIGURE 2.4

Sky color depends on the lengths of the path that solar radiation takes through the atmosphere. The shortest wavelengths are scattered first. When the radiation is perpendicular to Earth, the sky appears blue. When the sun is low in the sky, the blue is all filtered out and yellow, orange, and red predominate.

absorbs most of the radiation in the shorter wavelengths (violet to green). Only the longest wavelengths, orange and red, pass through.

In early morning and late evening, the bottoms of clouds often have a red tint, the result of red light reflected from the clouds. The colors associated with shorter wavelengths scatter before the light reflects from the clouds.

Spectacular sunsets occur when the atmosphere contains a lot of dust. Dust scatters the radiation the way gas particles do. Dust from storms on the surface or from volcanic eruptions causes unusually colorful sunsets. Colorful sunsets in the fall of 1991 followed the eruption of Mount Pinatubo in the Philippines in June. The massive explosion of the island of Krakatoa in 1883 also preceded brilliant sunsets for several years until the dust settled. (See Plates 5 and 6.)

Since air's density, moisture content, and particulates decrease rapidly with height in the atmosphere, scattering decreases rapidly as well. As a result, on a clear day at higher elevations in mountain regions the sky is a much darker blue or violet. The sky appears black to astronauts above Earth's atmosphere as well as from the surface of the moon. Since the moon has no atmosphere to scatter the radiation, the sky has no color.

Rayleigh scattering is not the only scattering mechanism. Gustave Mie developed another theory for scattering in 1908. Mie's theory states that molecules with a larger ratio of diameter to wavelength than those that give rise to Rayleigh scattering produce scattering of light at all wavelengths. More light scatters in a forward, or continuing, direction than in a backward direction. That would explain why the sky is often the darkest blue shortly after a rain: The rain washes out the larger particles of debris and removes much of the moisture, yielding less scattering and a darker sky. On a clear day, the sky is a darker blue directly overhead than near the horizon. This is because a direct path through the atmosphere overhead allows less overall scattering. Large particles scatter the light coming from near the horizon. The bright white color of the sky on a hazy day is due to the scattering of all of the visible light as it passes through the haze (Figure 2.5).

In polar regions, solar radiation scattered downward toward Earth (sky radiation) is a significant part of the total radiation. During winter, when the sun does not come above the horizon, sky radiation is the main source of radiant energy.

FIGURE 2.5
Scattering of the full visible spectrum on a hazy day. The haze scatters all radiation to make it appear that white light comes from all directions.

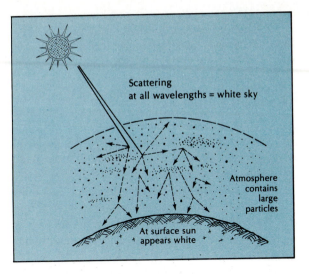

Scattering
at all wavelengths = white sky

Atmosphere
contains
large
particles

At surface sun
appears white

Absorption

Absorption retains incident radiation and converts it to some other form of energy. Most often it changes to sensible heat, which raises the temperature of the absorbing object. For example, sunlight striking the side of a house is absorbed and heats the wall.

Gas molecules, cloud particles, haze, smoke, and dust absorb part of the incoming solar radiation. Such absorption is selective, for gases absorb only in certain wavelengths. Because each gas has a characteristic absorption spectrum, different gases absorb different portions of the electromagnetic spectrum. The two most common gases in the atmosphere, nitrogen and oxygen, absorb ultraviolet radiation. Triatomic oxygen (ozone) absorbs shorter wavelengths than nitrogen and oxygen (Figure 2.2*d*).

Nitrogen does not absorb much incoming solar radiation. The maximum solar radiation is at 0.5 μm, and nitrogen does not absorb well in this frequency. Oxygen (O_2) and ozone (O_3) absorb well at wavelengths below 0.3 μm, with most of the absorption occurring in the ionosphere. Water vapor absorbs fairly well in the infrared range but not in the range of maximum solar radiation. These three gases make up over 99% of atmospheric gases. None are good absorbers in the visible range, where most solar radiation occurs.

Transmission

Reflection, scattering, and absorption deplete the solar beam as it passes through the atmosphere. Transmissivity is the proportion of the solar radiation ultimately passing through the atmosphere. Transmissivity depends upon the state of the atmosphere and the distance the solar beam must travel through it. The relative distance the solar beam travels through the atmosphere is the path length or *optical air mass*. The path length has a value of 1 when the sun is directly overhead, or 90° above the horizon. The path length increases as the angle of the sun above the horizon decreases, and it can be calculated for any sun angle. For example, if the sun is 30° above the horizon, the path length has a value of 2, since the solar beam has twice the distance to travel through the atmosphere.

The amount of solar radiation reaching a unit area of the surface, the **insolation,** is made up of energy transmitted directly through the atmosphere and scattered energy. These two radiation sources make up the global solar radiation. On days of thick cloud cover, no direct radiation reaches the surface. Solar energy reaches the surface only as scattered or diffuse radiation.

Figure 2.6 is a simple model of the planetary energy balance that is similar to that of the standard atmosphere. It represents a composite of the planet over a period of years. The model shows the percentages of the total annual energy inflow reflected, scattered, absorbed, and transmitted. Reflected energy plays no part in planetary heating. For the planet, a composite (or average) reflectivity of the atmosphere, ocean, and land masses is about 30%.

Clouds are the most important element in reflecting solar radiation back to space. The brilliant white of cloud tops you see when flying above them is reflected radiation of all wavelengths. Both Earth and Venus are very bright reflectors due to cloud cover; in fact, the cloud cover of Venus makes it the brightest of the planets as seen from Earth—the second brightest object in the sky after the moon. While the moon appears bright on a clear night, it has an albedo of only 7%. The moon lacks a cloud cover, and the rocks on its surface are of relatively dark color. Earth appears much brighter from space than the moon does, reflecting more than four times as much sunlight.

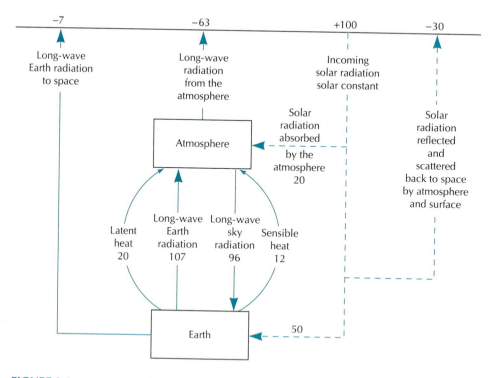

FIGURE 2.6

A model of the disposition of solar energy, Earth energy, and atmospheric energy.

Of the incoming solar radiation, 22% is scattered, with most of it reaching the ground. However, of that 22% scattered, over a fourth goes back to space. Like reflected energy, the radiation returned to space plays no part in heating the planet.

In the model, the atmosphere absorbs 20% of the incoming radiation. The stratosphere absorbs much of the shorter wavelengths, such as ultraviolet. Water vapor and CO_2 in the troposphere absorb the longer wavelengths.

Earth's Surface and Solar Energy

To maintain a steady-state temperature over a period of years, Earth's surface disposes of the energy it receives. It does this through one of three processes: radiation, evapotranspiration, or conduction-convection.

Radiation.   Earth, like all objects with a temperature above absolute zero, radiates energy. The planet has a mean temperature of 15°C (59°F). This is a much lower temperature than the sun, and so the wavelengths of radiation are much longer. Earth radiation reaches a maximum at 10 μm. Radiation from Earth's surface is in wavelengths 20 times as long as the incoming solar radiation.

Latent heat transfer.   Changing liquid water to water vapor (evaporation) requires considerable energy. This absorbed energy is stored in the water vapor until it changes back to liquid. Condensation in the atmosphere then adds this energy to the air. (For an explanation of the evaporation-condensation process, see Chapter 5.)

Sensible heat transfer.   Conduction moves some energy from Earth to the air above. Because air is a very poor conductor of heat, the process heats only a few centimeters of air in this fashion. This warmed layer of air then moves upward by convection, and the conduction process continues at the surface. Conduction-convection is the main heating process in the lower atmosphere.

The Greenhouse Effect

Life is possible on Earth solely because of a process known as the greenhouse effect. If not for a complex interchange between Earth's surface and the atmosphere, the mean temperature of the atmosphere near the surface would be −23°C (−9°F). No water would exist in the liquid or gaseous forms. A selective absorption process raises the mean temperature of the atmosphere near the surface to 15°C (59°F). Solar radiation is of fairly short wavelengths, and the atmosphere is relatively transparent to this radiation. The solar energy received at Earth's surface is processed and eventually radiated or otherwise released to the atmosphere.

Nitrogen and oxygen are poor absorbers of both short-wave solar radiation and long-wave Earth radiation. Several of the variable gases including water vapor and CO_2 are relatively transparent to solar radiation in the visible range, but they are good absorbers of Earth radiation. Clouds are even better absorbers of Earth radiation. Figure 2.7 shows the radiation curve for Earth and the bands in which the atmosphere absorbs this radiation. Earth radiates energy at wavelengths in the infrared band of about 3–30 μm. Water vapor absorbs this radiation in the bands from 5–8 and beyond 13 μm. Carbon dioxide also absorbs radiation at 4 μm and from 13 to 17 μm. The result is that most Earth radiation that escapes to space does so in a narrow band from 8 to 13 μm. This is the atmospheric window for Earth radiation. Figure 2.8 also shows some absorption by upper-atmospheric ozone at about 10 μm. Thick clouds are effective absorbers of Earth radiation over the entire range, even in the bands from 8 to 13 μm. So when layers of cloud over 100 m (330 ft) thick cover the sky, the atmosphere is able to absorb most of Earth's radiation.

The gases and liquid water that absorb Earth radiation also are good radiators of energy. The atmosphere radiates part of the energy absorbed to space and part back to Earth's surface. Nearly two-thirds of atmospheric radiant energy is directed back to the surface, providing an additional energy source to direct solar radiation. Atmospheric radiant energy is the largest source of radiant energy absorbed at Earth's surface—nearly double the amount of energy received directly from the sun. This additional atmospheric radiation raises the mean temperature at the surface to 15°C (59°F).

This energy exchange between Earth and the atmosphere is popularly called the greenhouse effect. It is more properly called the atmosphere effect. It is not how a greenhouse works, but the name is firmly entrenched. In a greenhouse, the solar energy coming through the glass warms the air inside. The glass roof traps this warmed air and it cannot escape from the surface as the outside air does.

FIGURE 2.7
Radiation spectrum for Earth. Note that the amount of energy emitted is much less than that of the sun, and that it is emitted at much longer wavelengths.

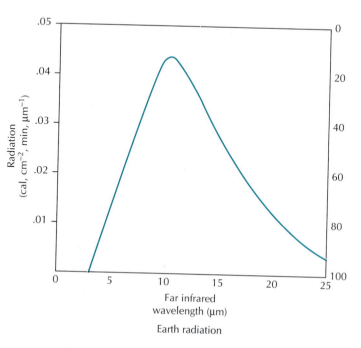

Earth radiation

FIGURE 2.8
Earth radiation and atmospheric
absorption.

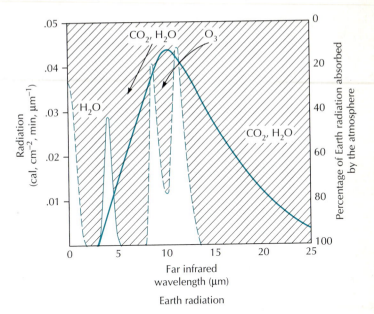

The Steady-State System

The energy flow is a continuous process. Energy from the sun enters the Earth–atmosphere system and ultimately returns to space. To remain at a steady-state temperature, incoming and outgoing energy must be in balance. If more energy entered than left, Earth would grow progressively hotter. If the balance favored the returning energy, it would get cooler.

Earth's surface receives energy from two main sources. One is direct and scattered solar radiation. Fifty percent of the solar radiation reaching the planet passes through the atmosphere to the surface. The atmosphere transmits 33% of incoming radiation, and an additional 17% reaches the surface as scattered energy. The second source of energy is atmospheric radiation. This is part of the greenhouse effect. The total amount of energy Earth receives from these two sources is 146 units, with each unit equal to 1/100 of the total solar radiation reaching the outer atmosphere.

Radiation, evaporation, and conduction balance the incoming 146 units of energy. Radiation to the atmosphere removes most of the energy received at the surface. A small amount (7 units) is radiated directly back to space. Evapotranspiration removes another 20 units of energy. Conduction-convection removes the remaining 12 units. If we consider the net loss of heat beyond the exchange of the greenhouse effect, Earth loses nearly equal amounts of heat by radiation and evaporation of water. Conduction removes a smaller amount. Convection moves heat rapidly away from the surface. Convection and evaporation-condensation move heat upward from the surface. In greenhouses, air warmed by conduction cannot escape and so it continues to warm.

The atmosphere is also in balance. The atmosphere receives energy by absorbing direct solar radiation, by absorbing Earth radiation, by conduction of heat at the surface, and by the heat of condensation. Of the energy received by the atmosphere, 77% comes from Earth, and radiation represents the largest single source of that energy. Only 13% of the total energy comes from direct absorption of solar radiation. As a result of the greenhouse effect, the prime heat source for the atmosphere is Earth's surface. Radiation to space and back to the surface balances the energy received by the atmosphere.

Energy received by Earth is counterbalanced by an equal amount of radiation to space. The largest amount of energy released is through radiation from the atmosphere, primarily from cloud tops. Reflection and scattering of solar radiation and direct radiation from the surface remove the remaining energy. Table 2.3 shows the inflow and outflow of energy for the surface, atmosphere, and planet.

Because Earth is in a steady-state balance of incoming and outgoing energy, its temperature undergoes small changes, but the mean temperature stays nearly the same. It varies only slightly around 15°C (59°F) at the surface, with daily, seasonal, and year-to-year changes. Certain mechanisms prevent major change and keep the system in a steady state. If Earth receives more energy than usual, for instance, the temperature must rise. If the temperature goes up, the planet radiates away energy at a higher rate (Wien's law). By the

TABLE 2.3
Energy balances of the Earth surface, atmosphere and planet. Units are in hundredths of the incoming solar radiation over a year.

ENERGY BALANCE OF EARTH'S SURFACE

Inflow		Outflow	
Solar radiation	50	Earth radiation	114
Sky radiation	96	Latent heat	20
Total	146	Conduction	12
		Total	146

ENERGY BALANCE OF THE ATMOSPHERE

Inflow		Outflow	
Solar radiation	20	Radiation to space	63
Condensation	20	Radiation to surface	96
Earth radiation	107	Total	159
Conduction	12		
Total	159		

ENERGY BALANCE OF EARTH

Inflow		Outflow	
Solar radiation	100	Reflected and scattered	30
Total	100	Sky radiation to space	63
		Earth radiation to space	7
		Total	100

same token, if Earth starts to cool, it loses less heat by evaporation, conduction, and radiation. This offsets the reduction in incoming energy. Since the energy flow is in a steady state, the planetary climate is steady over a period of years. One of two events must occur for the steady-state temperature of the lower atmosphere to change, and hence for the climate to change: a change in either the flow of energy to and from the planet or a change in the internal greenhouse effect. Chapter 4 will show how such changes do occur.

■ Applied Study

Solar Power

Worldwide price fluctuations in the cost of energy have resulted in the intensive investigation of alternative energy sources to replace or subsidize the use of fossil fuels. Of the many possible alternatives, direct use of solar energy shows great promise. In recent years, major advancements have been made at different levels of energy conversion. Photoelectric cells routinely power pocket calculators, satellites, other space vehicles and night lights for sidewalks and secured areas. In Australia, solar-powered automobiles compete in a race.

Using direct sunshine as an energy source can follow a number of approaches. In fact, with the exception of nuclear power and geothermal energy, all energy sources such as fossil fuels, wind, and falling water indirectly use solar radiation as the basic energy source. Active solar systems collect, store, and transform the solar radiation into some controllable form, often as electricity or steam.

Examples of active systems range from large government-sponsored schemes to small units in solar-heated homes. Active solar-energy systems for homes range from simple efficient placement of windows to relatively complex solar collectors and distribution systems. Strategic placement of windows on the south side of structures maximizes the amount of solar energy that can enter, as does use of glass on south-facing roofs. The floor, walls, and objects in the room absorb the solar radiation, which is converted to sensible heat and reradiated to the room. This form of heat does not pass out through the glass as readily as sunlight comes in. The warm air can then be moved through the house by an electric circulation system.

An alternative is to use a brick or stone wall on the south-facing side of the house and enclose the wall on the outside with glass or plexiglass in such a way as to keep the heat in. The wall heats from the sun, and air admitted from the base of the wall warms and passes up and into the house through vents in the wall. This system, though more expensive, is much more effective in circulating the heat energy.

Still more elaborate, but more effective, are roof-mounted solar collectors. These are shallow boxes painted black on the inside and covered with glass. Forced air through the box heats and is then circulated through the house. The box often incorporates a grid of metal water pipe. The solar energy heats the water in the pipes and then is sent either to radiators in the house or to hot water taps. In the Middle East, solar-powered water heaters are common. Since that region has a high percentage of clear days even in winter, the system provides hot water year-round.

An example of a large system is a solar furnace at Odeillo, France in the Pyrenees Mountains. This solar furnace is based upon a series of 56 mirrors that ultimately focus the radiation into a central opening in the oven only 30 cm (12 in.) in diameter. Temperatures in the oven reach 3500°C (6300°F). The oven is used to produce high-quality products such as special ceramics. The high quality results from the lack of contamination from the fuel that might alter the product.

Solar systems that generate electricity are more complex. In these systems, the solar collectors are backed by cadmium sulfide solar cells that convert sunlight to electrical current. A very large experimental facility for generating electricity, called Solar One, is located in the Mohave Desert near Barstow, CA. Like the solar oven in France, it utilizes an array of 2000 mirrors to track the sun during the day and focus the rays on a small target in a central tower. Water in the tower is heated to about 500°C (900°F). The superhot steam is directed through turbines which generate electricity. ■

Summary

The flow of energy into, through, and out of the environment is dynamic and complex. The source for the energy that drives the atmosphere is solar radiation. The amount of radiant energy ejected from the sun changes, but it is steady enough to call the solar constant. Energy reaching the atmosphere is disposed of in several ways. Some is reflected and scattered back to space, some is absorbed, some is scattered and transmitted through to the surface. This latter energy becomes the major heat source for the atmosphere. A built-in storage mechanism known as the greenhouse effect raises the temperature such that life can exist. The energy balance of the planet is a steady-state system that maintains a mean temperature at the surface of about 15°C (59°F).

ATMOSPHERIC TEMPERATURES

Chapter Overview

Temperature is the most widely used atmospheric measurement. Temperature represents the quantity of energy, or heat, present in a substance. At a more basic level, it is a measure of the kinetic energy produced by molecules moving in the body. The temperature of a substance measures the amount of heat per unit volume. The total heat in a substance depends not only on the temperature but on the mass. Raising the temperature of 25 g (0.9 oz) of water from 20° to 25°C (68°–77°F) requires five times more energy than raising 5 g (.15 oz) of water the same amount. The temperature in the container with 5 g of water is the same as that in the container with 25 g (0.9 oz), but the total heat is substantially different.

The temperature of substances, or within substances, determines the direction of the flow of energy. Heat always flows from the area of most heat (or highest temperature) to that of least heat. The study of atmospheric temperatures is the study of the ebb and flow of energy at a place and the variation in energy from place to place.

The Seasons

Temperature varies with time over Earth's surface due to changes in radiation received at the surface. Motions of Earth result in two periodic patterns of radiation influx and temperature: One produces the seasons, and the other the daily changes in radiation and temperature. Few places (if any) on Earth's surface are truly without seasons. Seasons are most often thought of as summer, fall, winter, and spring. This is because most people live in the midlatitudes, where temperature changes from summer to winter are very large. However, over large areas of Earth the change from hot to cold is not as important as the change from a rainy season to a dry one. The most important differences in energy received at various locations on Earth result from basic motions of the planet in space—rotation on its axis and revolution around the sun. Figure 3.1 is a schematic diagram of these Earth-sun relationships over a year.

In its revolution around the sun, Earth follows an elliptical orbit, so that the distance from Earth to the sun varies. Earth is usually closest to the sun (perihelion) on January 4 and most distant (aphelion) on July 4. The date varies as a result of leap year. The amount of solar radiation intercepted by Earth at perihelion is about 7% higher than at aphelion. This difference, however, is not the major force behind the change in seasons.

The most important element in producing the seasons is the amount of radiation a place receives through the year. The amount of radiation varies as the angle of the sun above the horizon (intensity) and the number of daylight hours (duration) change through the year. The intensity of solar radiation is largely a function of the *angle of incidence,* the angle at which solar energy strikes Earth. The angle of incidence directly affects both the energy received per unit area of surface and the amount of energy absorbed. Since Earth is nearly spherical, a curved surface is exposed to the radiation. Intensity of radiation is highest in latitudes where solar radiation is perpendicular to the surface (the solar equator) (see Figure 3.2). The intensity of radiation decreases north and south of the solar equator as the angle of incidence decreases.

The seasonal changes in energy result from the inclination of Earth's axis. Earth's axis is tilted 23°30′ from being perpendicular to the plane of the ecliptic (see Figure 3.3). As

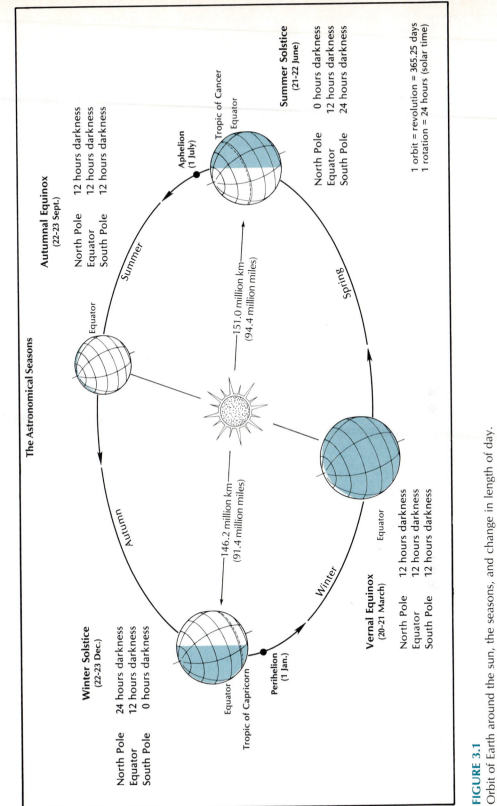

FIGURE 3.1
Orbit of Earth around the sun, the seasons, and change in length of day.

Earth revolves about the sun, the solar equator moves north and south through a range of 47°. The geographic equator (0° latitude) is the mean location of the solar equator. The solar equator moves north and south through the year between 23°30′ north (Tropic of Cancer) and 23°30′ south (Tropic of Capricorn). The two tropics represent the latitudes furthest from the equator where solar radiation is perpendicular to the surface sometime

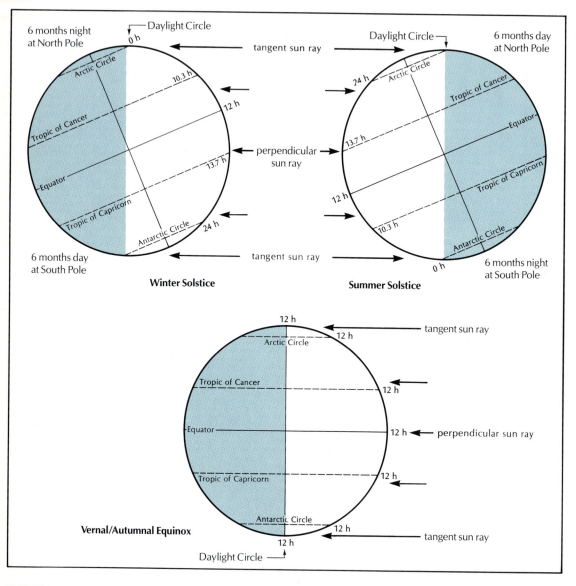

FIGURE 3.2

Earth's tilted axis results in variations of the angle of overhead sun and length of daylight. Diagram shows conditions at the equinoxes and solstices.

FIGURE 3.3

Angle of the rays of the sun and intensity of radiation. The same amount of energy is contained in both beams of radiation. The lower the angle of the beam, the larger the area over which the beam is spread.

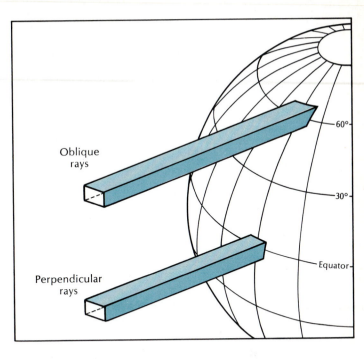

during the year. As Earth travels around the sun during a year, the intensity of solar radiation varies at all latitudes.

The geographic equator has the least variation in the angle of incidence. Here the sun is never more than 23°30′ from the zenith (the point directly overhead). All latitudes between the two tropics vary in their angle of incidence of between 23°30′ and 47°. The variation reaches 47° at the tropics. All places between 23°30′ and 66°30′ of latitude experience a change of 47° in the angle of solar radiation through a year's time.

The primary factor responsible for seasons is Earth's revolution about the sun and the inclination of Earth's axis to its orbital plane. These phenomena result in an imbalance of energy over Earth's surface through the year.

The change in the length of the daylight period strengthens the seasonal variation in temperature. Only during the spring and fall equinoxes are the length of daylight and darkness equal everywhere over Earth. Daylight is longer in the hemisphere that has the highest intensity of solar radiation. Thus, when the sun's vertical rays are north of the equator, the daylight period in the Northern Hemisphere is longer than 12 hours. On the summer solstice, the length of daylight increases throughout the Northern Hemisphere, from 12 hours at the equator to 24 hours at the Arctic Circle (see Tables 3.1–3.3). The imbalance in daylight and darkness accentuates the seasons more than the position of the solar equator alone would.

Two major aspects of the energy balance distinguish tropical environments from those of the midlatitudes and polar areas. First, the influx of solar energy in tropical areas is high throughout the year, even though it does vary. The intensity of solar radiation is high all year because the rays of the sun are always at a high angle. From this

TABLE 3.1
Ephemeris of the sun (declination of the sun on selected days of the year).

DAY	JAN.	FEB.	MAR.	APRIL	MAY	JUNE
1	−23°04′	−17°19′	−7°53′	+4°14′	+14°50′	+21°57′
5	22 42	16 10	6 21	5 46	16 02	22 38
9	22 13	14 55	4 48	7 17	17 09	22 52
13	21 37	13 37	3 14	8 46	18 11	23 10
17	20 54	12 15	1 39	10 12	19 09	23 22
21	20 05	10 50	−0 05	11 35	20 02	23 27
25	19 09	9 23	+1 30	12 56	20 49	23 25
29	18 08	—	3 04	14 13	21 30	23 17

DAY	JUL.	AUG.	SEPT.	OCT.	NOV.	DEC.
1	+23°10′	+18°14′	+8°35′	−2°53′	−14°11′	−21°40′
5	22 52	17 12	7 07	4 26	15 27	22 16
9	22 28	16 06	5 37	5 58	16 38	22 45
13	21 57	14 55	4 06	7 29	17 45	23 06
17	21 21	13 41	2 34	8 58	18 48	23 20
21	20 38	12 23	+1 01	10 25	19 45	23 26
25	19 50	11 02	−0 32	11 50	20 36	23 25
29	18 57	9 39	2 06	13 12	21 21	23 17

Source: U.S. Naval Observatory, *The American Ephemeris and Nautical Almanac for the Year 1950.*
Table by R.J. List (Washington, DC, 1950).

point of view, these are indeed environments without a winter. Second, tropical areas have very little change in the length of daylight, ranging from 11 to 13 hours through the year.

Polar areas differ from those of the tropics and midlatitudes in the annual distribution of solar energy. Winter months in polar regions may see no direct solar radiation. But some radiation, including that in the visible range, continues to find its way to the surface. In summer, solar intensity is low but its duration is long, so the total energy is quite large. Much of the energy goes to evaporate water or melt snow and ice.

The intensity of radiation in the polar regions is never very high for any length of time when compared to midlatitude or tropical systems. At the Arctic and Antarctic Circles, the sun never climbs more than 47° above the horizon and at the poles it is never more than 23.5° above the horizon. Refracted sunlight gives both polar areas more hours of radiation during the year than other parts of Earth. When the sun is below the horizon, reflection and refraction of the rays of sunlight produce twilight. These processes bring some sunlight to the surface until the sun is 18° below the horizon (astronomical twilight). Thus, high latitudes may have several months with no direct sunshine but almost continuous twilight. The stars, moon, and auroras provide light for these areas, so darkness is not as intense as it might be and total darkness seldom exists.

TABLE 3.2

Length of daylight for intervals of 1° of latitude.*

	0°	1°	2°	3°	4°	5°	6°	7°	8°	9°
0°	12:07 12:07	12:11 12:04	12:15 12:00	12:18 11:57	12:22 11:53	12:25 11:50	12:29 11:46	12:32 11:43	12:36 11:39	12:39 11:36
10°	12:43 11:33	12:47 11:29	12:50 11:25	12:54 11:21	12:58 11:18	13:02 11:14	13:05 11:10	13:09 11:07	13:13 11:03	13:17 10:59
20°	13:21 10:55	13:25 10:51	13:29 10:47	13:33 10:43	13:37 10:39	13:42 10:35	13:47 10:30	13:51 10:26	13:56 10:22	14:00 10:17
30°	14:05 10:12	14:10 10:08	14:15 10:03	14:20 9:58	14:26 9:53	14:31 9:48	14:37 9:42	14:43 9:37	14:49 9:31	14:55 9:25
40°	15:02 9:19	15:08 9:13	15:15 9:07	15:22 9:00	15:30 8:53	15:38 8:46	15:46 8:38	15:54 8:30	16:03 8:22	16:13 8:14
50°	16:23 8:04	16:33 7:54	16:45 7:44	16:57 7:33	17:09 7:22	17:23 7:10	17:38 6:57	17:54 6:42	18:11 6:27	18:31 6:10
60°	18:53 5:52	19:17 5:32	19:45 5:09	20:19 4:42	21:02 4:18	22:03 3:34	— 2:46	— 1:30	— —	— —

*The upper figure in each pair of figures represents the longest day, and the lower figure represents the shortest day. To find the length of daylight for 37 degrees, read down to 30° in the left column and across the row to 7°. The longest day at 37° is 14 hr, 43 min. Note that the two periods do not add up to 24 hr because daylight is measured when the rim of the sun, rather than the center of the sun, is visible.

TABLE 3.3
Total daily direct solar radiation reaching the ground (in cal per cm², transmission coefficient = 0.6).

Latitude	APPROXIMATE DATE							
	Mar. 21	May 6	June 22	Aug. 8	Sept. 23	Nov. 8	Dec. 22	Feb. 4
90° N		127	299	125				
80°	6	158	309	156	5			
70°	47	234	349	232	46			
60°	120	312	406	308	118	10		10
50°	202	376	450	372	199	58	19	58
40°	282	426	477	421	278	130	75	131
30°	350	453	481	449	345	213	152	215
20°	404	459	465	454	398	293	237	296
10°	436	444	428	439	430	366	323	370
0°	447	407	372	404	440	422	397	427
10° S	436	353	303	349	430	461	457	465
20°	404	282	222	279	398	475	497	480
30°	350	206	143	204	345	470	514	475
40°	282	125	70	124	278	441	509	445
50°	202	56	18	55	199	391	481	395
60°	120	10		10	118	323	434	327
70°	47				46	242	373	245
80°	6				5	164	330	166
90°						131	319	133

Source: Smithsonian Institution, Smithsonian Meteorological Tables, 1966. Values apply to a horizontal surface.

Seasonal Lag in Temperatures

Maximum and minimum temperatures lag seasonal changes. Earth's revolution causes maximum and minimum solar energy (outside the tropics) to occur at the time of the summer solstices in each hemisphere. Thus, June and December represent the time of maximum and minimum solar energy receipts in the Northern Hemisphere. The reverse holds true in the Southern Hemisphere. Also, the months of maximum and minimum solar energy are not the warmest or coldest. Table 3.4 shows that a lag of a month or more occurs between the time of maximum and minimum solar radiation and the warmest and coldest months. Naples, Italy, for example, experiences a two-month lag.

Daily Temperature Changes

The rotation of Earth on its axis produces alternating periods of day and night, resulting in daily variations in temperature.

TABLE 3.4
Sample data illustrating seasonal temperature lag.

| | AVERAGE TEMPERATURE AT SOLSTICE | | AVERAGE MONTHLY TEMPERATURE | |
	June	Dec.	Warmest Month	Coldest Month
Charleston, SC (33°N)	26°C (79°F)	11°C (51°F)	July 28°C (82°F)	Jan. 10°C (50°F)
Urbana, IL (40°N)	22°C (72°F)	0°C (32°F)	July 25°C (77°F)	Jan. −1°C (30°F)
Naples, Italy (41°N)	22°C (72°F)	11°C (51°F)	Aug. 25°C (77°F)	Jan. 9°C (48°F)
Moscow, Russia (43°N)	19°C (66°F)	−6°C (22°F)	July 21°C (70°F)	Jan. −8°C (17°F)
Edmonton, Canada (53°N)	14°C (57°F)	−8°C (18°F)	July 17°C (63°F)	Jan. −14°C (7°F)

Daytime Heating

After the sun rises above the horizon in the morning, the surface begins to heat. If the air is relatively calm, conduction rapidly moves heat from the surface to the boundary layer of air. This heat transfer takes place in a very limited laminar layer of air, often only a few millimeters deep. Gradually, the heat is distributed upward by diffusion of heated air molecules. On a hot, clear, still day in summer, a thermal gradient of 20°C (36°F) may exist in the lower 2 m (6.5 ft) of air. In tropical deserts, it may become impossible to see through a surveying instrument by 10 A.M. because the upward movement of heat so disturbs the air.

If wind is blowing, the upward movement of heat is much more rapid and the gradient in the lower 2 m (6.5 ft) is much less. Even with turbulence carrying heat away from the surface, daily changes in temperature normally do not extend above 1 km (0.6 mi).

As expected, temperatures are highest during the day—when large amounts of energy are flowing to Earth—and lowest at night. There is not, however, a one-to-one relationship between the time of highest solar input and highest temperature, because air temperature results largely from absorbing Earth radiation.

Highest daytime temperatures usually occur several hours after the time of maximum solar input, which is noon. The commonly held notion that the highest temperature occurs when solar input equals the flow of outgoing energy is not true. Measurements show that equilibrium between incoming and outgoing radiation often occurs about an hour and a half before sunset, not at the time of maximum temperature. During the early afternoon, Earth receives a steady flood of incoming long-wave radiation from the lower atmosphere (the greenhouse effect). When this flow of long-wave radiation from the atmosphere reaches a maximum, the day's highest temperatures occur.

The extent of the daily lag in maximum temperature varies. Under normal conditions the lag will be greatest when the air is still and dry and the sky free of clouds. In these instances the maximum temperature of the day may not occur until an hour or so before sunset. When humidity is high, or the atmospheric aerosol is unusually thick, the lag will not be as great. Minimum lags occur when cloud cover reduces the incoming and outgoing radiation.

Actual daily high and low temperatures can occur at any time of day if the wind shifts direction and feeds cold or warm air into the area. This is typical of temperature changes brought about by circulation around midlatitude lows and by the development of land and sea breezes.

Moist soil or a cover of vegetation also reduces the maximum temperatures through evaporation or transpiration. Evaporation is a major cooling mechanism taking 590 cal of heat from the environment for each gram of water that evaporates. Evaporation explains why air never gets as warm in summer over water as it does over land. Cities east of the Mississippi River generally have mean daily high temperatures and extreme highs 10°C (50°F) lower than cities in the western desert. The higher atmospheric humidity and a vegetation-covered surface reduce the temperature of the lower atmosphere.

Nighttime Cooling

Once the sun passes its zenith, radiation intensity starts to decrease. Sometime near sunset on a clear day the ground surface and the boundary layer of air begin to receive less energy than they emit and begin to cool. The ground surface cools more rapidly, since it is a better radiator than air. Soon the coldest air is close to the ground and temperature increases with height, the opposite of the standard atmospheric state. Earth radiation decreases through the hours of darkness, and the surface cools. How much cooling occurs depends on the length of the night, the humidity, and wind velocities: Cooling is greatest on a night when the humidity is low, the sky clear, and the wind calm. A wet surface retards cooling just as it retards warming. Dew, frost, and fog are major feedback mechanisms that prevent the atmosphere from cooling still further. When dew forms, 590 cal of heat are added to the atmosphere for each gram of water. This heat added by condensation offsets the heat lost by radiation.

Frost is an even more effective means of controlling cooling. For each gram of water that sublimates as frost, 680 cal of heat are added to the boundary layer of air. Fog likewise is a major block to cooling. Not only does the condensation add heat to the air, but the fog acts as a blanket, reducing radiative heat loss. In desert regions, dew and frost are very important in reducing nighttime cooling. Minimum daily temperatures occur in the early morning when incoming solar energy balances outgoing Earth radiation. Shortly after dawn, incoming solar rays provide enough energy to balance outgoing terrestrial rays.

Daily Temperature Range

The daily range in temperature is a function of both daytime heating and nighttime cooling. When conditions are good for rapid inflow of radiant energy in the daytime and outgoing radiation at night, the range will be large. The conditions favoring high daily ranges in temperature are clear skies, low relative humidity, and calm winds.

The magnitude of the daily change in temperature varies a great deal. Near the equator, the daily range exceeds the annual range. Near the two poles, the daily range is almost zero, since each pole generally has only one daylight and one nighttime period each year.

The daily temperature range is also small over the ocean, for several reasons. First is the high specific heat of water: It heats slowly and cools slowly. Second is the mixing of the surface water with the water below, which modifies heating and cooling. Third, solar radiation penetrates deeper into the ocean than into the land, so it distributes heat more evenly with depth.

The daily range in temperature at the surface is the result of solar heating during the day and Earth radiation at night. The daily range of temperature decreases rapidly upward from the ground surface to about 1 km (0.6 mi). Above about 1 km (0.6 mi), the change in air temperature from day to night is minimal.

In North America, daily ranges are a function of cloud cover and humidity. In the dryer regions, daily ranges may exceed 22°C (40°F), while in more humid regions, the normal range is nearer 17°C (30°F). The greatest known recorded daily range in temperature occurred in North Africa, where the temperature dropped from a high of 56°C (132°F) in the afternoon to 0°C (32°F) the following morning. The difference is an amazing 56°C (100°F) (see Figure 3.4).

■ Applied Study

Temperature Normals

Mean values of temperature data are frequently given. The mean, or average, represents a value that gives a reasonable approximation of what is the normal, where *normal* is the value at the center of the data distribution. It is derived by adding the individual measurements and dividing them by the number of data (sum of data/number of data).

Any number of means can be calculated depending upon the unit of time that the data cover. For example, to calculate the average daily temperature for Christmas Day, you sum the means for that day [(maximum + minimum) / 2] for every year for which data are available. Then divide this sum by the number of years of record. To determine monthly mean temperatures, calculate the mean of all the daily means for the month. The daily data form the basis for calculating all average temperatures for periods longer than one day.

The mean is a measure of the central tendency of the data set and does not tell how the data scatter about the mean. The mean of the set 2, 4, 6, 8, and 10 is 6. The same average occurs for 4, 5, 6, 7, and 8. Other statistical measures are used to account for scatter of the data about the mean. One such measure is the *range*. Obtain the range by subtracting the smallest number from the largest in the set. The range together with the mean provide more information about the data.

The *mean annual temperature* is the single value that best describes the temperature of a place. It is calculated as the mean of the 12 monthly means. Unfortunately, the mean annual temperature also hides a tremendous variation in temperatures throughout the year. The *mean annual range* of temperature adds considerable information about a place. This is the difference between the mean temperature of the coolest month and the warmest month. For example, the mean annual temperature at Quito, Ecuador, is 15°C (59°F). At Nashville, it is 15.2°C (59.5°F). However, the range at Quito is only 0.5°C (15.2–14.7°C), while at Nashville the range is 22.2°C (3.8–26°C). Thus, despite very similar average annual temperatures, the two locations experience different temperature regimes. The magnitude of the seasons in the two places influences their annual range: Near the equator, the length of day and intensity of radiation are similar throughout the year, so the annual range is small. Seasonality and the mean annual range both increase as we move toward the poles.

Climatic normals are also used. Climatic normals are the mean values over a fixed period of years. In the United States, we use a period of 30 years, and the period shifts every decade. The climatic normals now used as a reference are those for the period 1951–80; the next 30-year period used to represent the present climate will be from 1961–90. A 30-year period is long enough to hide year-to-year variations in weather. The period changes every 10 years because weather for the most recent years is closest to current conditions. Since climate changes, so do climate normals. ■

Temperature Extremes

Temperature extremes are temperatures that are furthest from normal and occur least frequently. To reach extreme temperatures, conditions for incoming and outgoing radiation must be optimal. High temperatures require high-intensity solar radiation; long hours of sunlight; clear, dry air; little surface vegetation; and relatively low wind velocity. These conditions are optimal near the Tropics of Cancer and Capricorn around the time of a solstice. Under these circumstances, the atmosphere is dominated by high pressure and clear skies. The highest recorded surface temperature, 58°C (136°F), was measured in El Aziz, Libya in September 1922. In North America, the highest temperature recorded was 57°C (134°F) in Death Valley, CA, in July 1913. At any location, record highs occur when the air is clear and dry.

Record lows occur under conditions that encourage loss of energy from the surface and lower atmosphere by direct radiation to space. These conditions occur with high-pressure systems near the poles. The coldest temperature ever recorded is −89°C (−128°F) in Antarctica at Vostock in 1983. The cold temperature is in part due to the high altitude of the ice mass. In the Northern Hemisphere, the lowest temperature recorded is −68°C (−76°F) at Verkoyansk, in what was the Soviet Union. In North America, the coldest temperature of record is −62°C (−80°F) at Prospect Creek, AK. Not surprisingly, records of both cold and heat occur away from the oceans, since the water greatly modifies temperature extremes.

Factors Influencing the Vertical Distribution of Temperature

The nature of the underlying surface influences the vertical distribution of temperature. For example, temperature decreases most rapidly with altitude over continental areas in summer. Consider the north-south temperature profile over the Northern Hemisphere up to a height of 23 km (14 mi). The profile is along the 80th meridian (which passes through eastern Canada, the United States, Cuba, and Panama) from the North Pole to the equator in January and July. The following are features of the vertical temperature distribution:

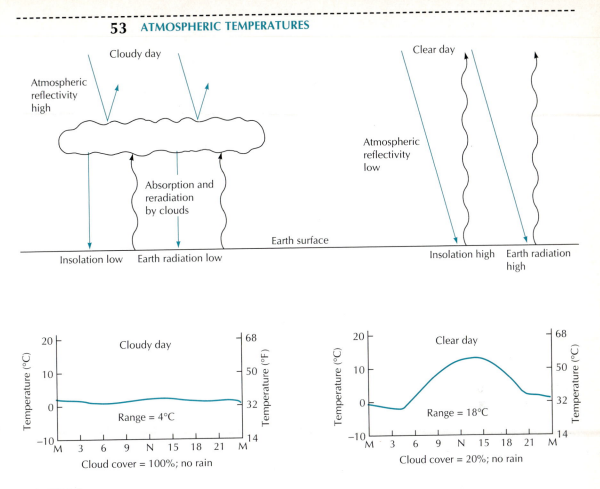

FIGURE 3.4
Effects of cloud cover and rain on two different days in February at Evansville, IN.

1. The tropopause is lower over high latitudes than low latitudes. A well-marked break occurs in the midlatitudes.
2. The north-south temperature gradients are much steeper in winter.
3. The strongest horizontal gradients are in the midlatitudes in both summer and winter. This is the region of maximum storm activity.
4. The coldest part of the troposphere occurs over the equator in the region of the tropopause.

The diurnal range of temperature at higher elevations is greater than at sea level. Figure 3.5 provides a graphic model of this effect. The reason for the greater range is that the atmosphere is less dense at higher altitudes. Maximum daily temperatures are about the same or slightly less than at low altitudes. The main difference occurs at night, when heat escapes much more readily at high elevations because of the lower density of gases.

Decreasing pressure with altitude changes the meaning of some values on temperature scales. The gas laws state the relationships between temperature and pressure. At higher elevations, pressure changes become important. Under reduced pressure, molecules of

FIGURE 3.5

Daily range of temperature in a mountain and a lowland station at places of similar latitude.

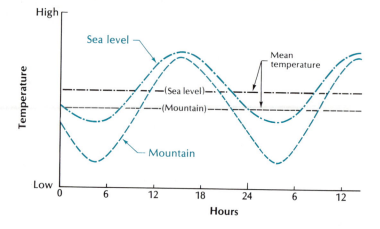

water vapor escape more easily from a water surface. Thus, at sea level, water boils at a temperature of 100°C (212°F). At Quito, Ecuador, at an elevation of 2824 m (9350 ft), water will boil at 90.9°C (196°F). At the top of Mount Everest, the boiling point of water is only 71°C (160°F).

Factors Influencing the Horizontal Distribution of Temperature

A number of factors control temperatures over Earth. They include latitude, surface properties, position with respect to warm or cool ocean currents, and elevation.

Latitude

The highest temperatures on Earth do not occur at the equator but near the Tropics of Cancer and Capricorn. A partial explanation of this occurrence is the migration of the vertical rays of the sun between 23.5° N and 23.5° S. The vertical rays of the sun move quickly over the equator but slowly as they progress north and south. Thus, between 6° N and 6° S, the sun's rays are nearly vertical for 30 days around the equinoxes. Between 17.5° and 23.5° N and S, near-vertical rays occur for 86 days around the solstice. The longer period of high sun and the concurrent longer days allow time for surface heat to accumulate. Thus, the zone of maximum heating and highest temperatures is near the tropics. The predominance of clear skies near the tropics enhances heating even more compared to the very cloudy equatorial belt.

If Earth were a homogenous body without the land–ocean distribution, its temperature would change evenly from the equator to the poles. It is possible to compare actual temperatures with hypothetical data for a uniform surface and identify areas of temperature anomalies. Anomalies represent deviations from temperatures that would result only from solar energy. Isolines called *isanomals* are drawn through points of equal temperatures to produce isanomalous temperature maps of the world.

Figures 3.6 and 3.7 are maps of temperature anomalies for January and July, respectively. Studying the maps yields general observations:

1. During winter in each hemisphere, the land masses have large negative anomalies—as much as $-20°C$ $(-4°F)$ below the hypothetical mean. By contrast, ocean areas show no anomalies or slightly positive ones.
2. During summer in each hemisphere, the largest anomalies also occur over the continents, but they are positive. In some areas, a small negative anomaly occurs over the oceans.
3. The patterns of the anomalous conditions over the oceans are distinctive in shape and are associated with ocean currents.
4. Small anomalies occur in the equatorial realms, and the highest anomalies occur in upper-middle latitudes.

These observations indicate that two groups of factors alter the zonal radiation climate and hence the patterns of temperature over the globe. The first group of factors involve primarily geographic location on Earth's surface. Location essentially determines the amount of solar radiation received. The other group of factors are temperature characteristics that result from the atmosphere and oceans moving energy. This horizontal movement of energy can substantially change the temperature of a place from what the mean radiation pattern provides.

Surface Properties

What happens to solar energy striking a surface depends largely upon the type of surface, especially its albedo. Surfaces with high albedo absorb less incident radiation, so less total energy is available. Polar ice caps remain intact, for example, because they reflect as much as 80% of the solar radiation falling on them.

Even if two surfaces have a similar albedo, the incident energy does not always result in similar temperatures, since the heat capacities of the surfaces may differ. The heat capacity of a substance is the amount of heat required to raise its temperature. **Specific heat** is the amount of heat (number of calories) required to raise the temperature of 1 g (.035 oz) of a substance through $1°C$ ($1.8°F$). As Table 3.5 shows, the specific heat of substances can vary appreciably.

TABLE 3.5
Specific heat of selected substances.

SUBSTANCE	SPECIFIC HEAT (CAL/G/°C)
Water	1.0
Ice (at freezing)	0.5
Air	0.24
Aluminum	0.21
Granite	0.19
Sand	0.19
Iron	0.11

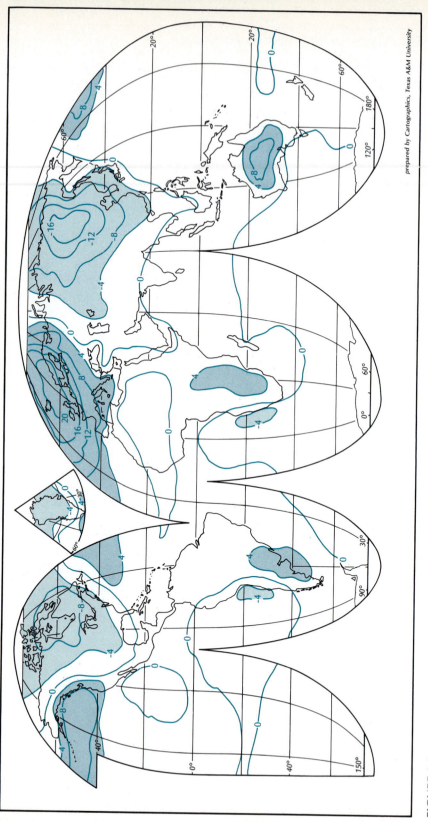

FIGURE 3.6
Isonomalies of temperatures (°C) in January.

prepared by Cartographics, Texas A&M University

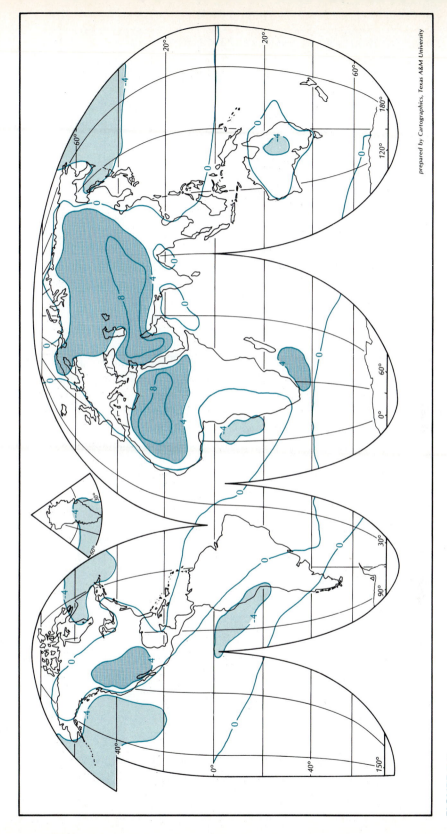

FIGURE 3.7
Isonomalies of temperatures (°C) in July.

prepared by Cartographics, Texas A&M University

Note that the specific heat of water is about five times greater than that of rock material and the land surface in general. In other words, raising the temperature of water 1°C (1.8°F) requires five times more heat than does achieving the same temperature increase for land. Applying the same amount of energy to a land surface and a water surface produces much hotter land than water. The difference increases due to the different heat conductivity of the Earth materials and water (Table 3.6).

Loose, dry soil is a very poor conductor of heat. Only a superficial layer will experience a rise in temperature from solar radiation. Water has only fair conductivity, but its general mobility and transparency permit heat to circulate well below the surface. Natural, undisturbed soil with vegetation cover may have daily temperature changes to a depth of 1 m. A quiet pool of water has daily temperature variations that can be measured to a depth of perhaps 6 m (20 ft).

Land masses heat much more rapidly than oceans in summer. Since this heat concentrates near the surface, however, it rapidly radiates to the atmosphere as winter approaches. This explains why land masses tend to experience extreme temperatures, while water bodies are more equable and show less change (Figure 3.8). Absorbed energy raises the temperature of the absorbing surface and evaporates moisture from that surface *(LE)*. This energy is then either radiated back to the atmosphere or passed to the air in the form of sensible heat *(H)*. The amount of sensible heat in the surface materials—and hence its temperature—depends in part upon the fraction of incident energy that is used to evaporate water. Evaporation is continuous over the oceans. On land, evaporation depends upon the amount of water available at the surface. The significance of this process is determined by comparing the ratio of sensible heat to latent heat for different parts of Earth. The Bowen ratio, which expresses this relationship, is *H/LE*. The larger the ratio, the more sensible heat (rather than latent heat) is passing to the atmosphere.

The much larger Bowen ratio for land areas further underscores land-sea temperature differences. In fact, so distinctive is this effect that climatologists use the term *continentality* to describe the continental influence of weather and climate.

The relative division of *LE* and *H* in the energy budget equation also varies at individual stations. Figure 3.9 provides data for cities with different climates: West Palm Beach, FL, and Astoria, OR, are in moist, coastal locations, and Yuma, AZ, is in a desert. Note the amount of net radiation that passes to *LE* in the moist coastal locations. By con-

TABLE 3.6
Thermal conductivity of selected substances.

SUBSTANCE	HEAT CONDUCTIVITY*
Air	0.000054
Snow	0.0011
Water	0.0015
Dry soil	0.0037
Earth's crust	0.004
Ice	0.005
Aluminum	0.49

*The number of calories passing through an area 1 cm² in a second when temperature gradient is 1°C/cm. It is expressed in cal/cm²/sec/°C/cm.

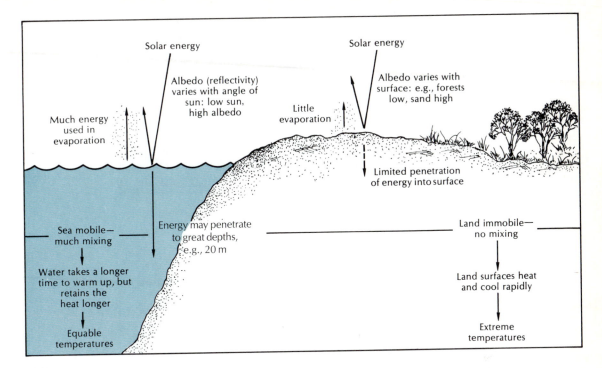

FIGURE 3.8
Solar energy striking land and sea surfaces divides in different ways. The water surfaces are conservative, warming and cooling slowly; the land masses experience large temperature changes.

trast, in the desert, much of the net radiation becomes sensible heat (*H*) resulting in high temperatures.

Coastal cities have smaller ranges in temperature than inland cities. Cities in which the wind blows onto the land have lower range than coastal cities, where the dominant wind direction is offshore.

Inland cities, particularly in the Northern Hemisphere, have the highest annual ranges in temperature. In North America, continentality is greatest in the Great Plains of the Northern United States and southern Canada. The greatest continentality of climate is found in eastern Europe. At Yakutsk, the annual range is 62°C (144°F). This represents a change in mean temperature of more than 13°C (24°F) per month from January to July.

Aspect and Topography

The steepness and direction a slope faces combine to determine its aspect. Differences between north-facing and south-facing slopes in the Northern Hemisphere illustrate the importance of aspect. A north-facing slope may have snow on it, while a south-facing slope is bare. The north slope gets less-intense radiation and, as the sun gets lower in the sky, will be in shadow long before the south-facing slope.

FIGURE 3.9
Average annual variation in the components of the surface energy budget at three different locations.
Source: W. D. Sellers, *Physical Climatology.* (Chicago: The University of Chicago Press, 1965), 106.

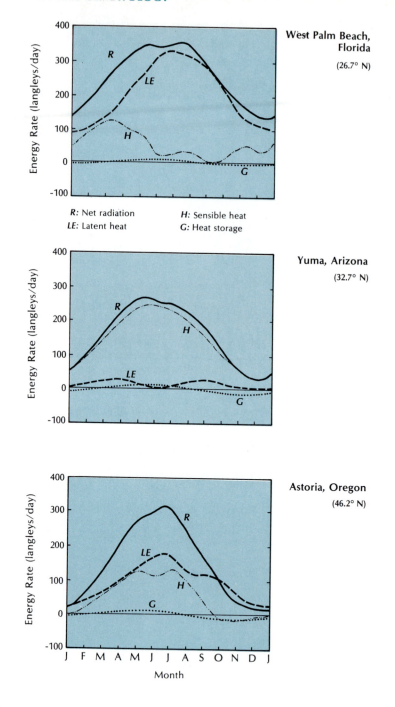

West Palm Beach, Florida
(26.7° N)

R: Net radiation   *H:* Sensible heat
LE: Latent heat   *G:* Heat storage

Yuma, Arizona
(32.7° N)

Astoria, Oregon
(46.2° N)

Slope aspect influences many natural phenomena. For example, the height of permanent snow and ice on mountains varies from one slope to another, as does the tree line. Similarly, the depth of snow and frost differ on north- and south-facing slopes.

TABLE 3.7
Theoretical temperatures in stationary atmosphere and actual temperatures by latitude.

	EQUATOR	10°	20°	LATITUDE 30°	40°	50°	60°	70°	80°
				Temperature in °C (°F)					
				Northern Hemisphere					
Planetary temp.	33(91)	32(89)	28(83)	22(72)	14(57)	3(37)	−11(12)	−24(−11)	−32(−26)
Actual temp.	26(79)	26(80)	25(78)	20(69)	14(57)	5(42)	−1(30)	−10(13)	−18(−1)
Difference	−7(−12)	−6(−9)	−3(−5)	−2(−3)	0	+2(+5)	+10(+18)	+18(+24)	+14(+25)
				Southern Hemisphere					
Planetary temp.	33(91)	32(89)	28(83)	22(72)	14(57)	3(37)	−11(12)	−24(−11)	−32(−26)
Actual temp.	26(79)	25(78)	22(73)	17(62)	11(53)	5(42)	−3(26)	−13(8)	−27(−17)
Difference	−7(−12)	−7(−11)	−6(−10)	−5(−10)	−3(−4)	+2(+5)	+8(+14)	+11(+19)	+5(+9)

Topography also plays an important role in climates of some lowlands. On a continental scale, mountain ranges that run north–south have a very different effect from those that run east–west. The lack of any extensive east–west mountain barrier in the United States permits polar and tropical air to penetrate long distances into the continent. One result of this unobstructed flow of air is the high incidence of tornadoes in the United States.

Dynamic Factors

A large imbalance of energy exists between tropical regions and the poles. This imbalance is partly alleviated by the transfer of latent heat, sensible heat, and heat stored in the water of the oceans. Table 3.7 provides the theoretical planetary temperature for sea level and the actual mean annual temperature for every 10° of latitude. The biggest differences are at the equator and at latitudes above 60°. Tropical latitudes are cooler than the theoretical value, while high latitudes are warmer. The differences between the actual and theoretical values result from the transport of energy over Earth by air and ocean currents. Every storm system, circulation pattern, and evaporation/precipitation event aids in moving heat from tropical regions toward the poles.

Temperatures over Earth's Surface

As we have seen, a number of factors influence the distribution of average temperatures over Earth's surface. Figures 3.10 and 3.11 show the distribution of mean July and January temperatures, respectively. Figure 3.12 shows the annual average range of temperature. Plate 2 shows the actual July and January temperatures in 1979.

In each case the general decline in temperatures from the equator to the poles is clear, illustrating the basic influence of latitude. However, temperatures sometimes do vary sig-

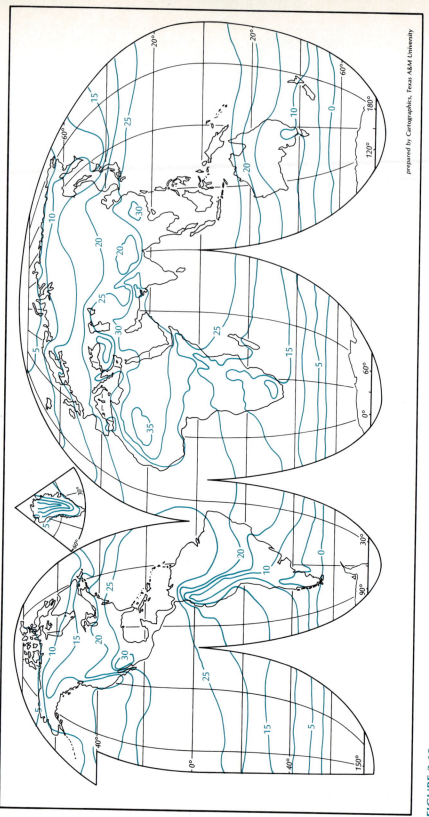

prepared by Cartographics, Texas A&M University

FIGURE 3.10

Distribution of average July temperatures over Earth (0°C).

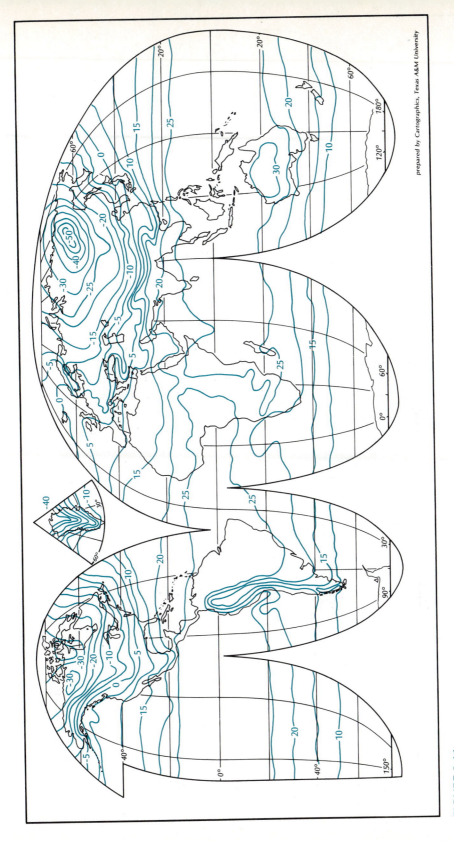

FIGURE 3.11
Distribution of average January temperatures over Earth (0°C).

prepared by Cartographics, Texas A&M University

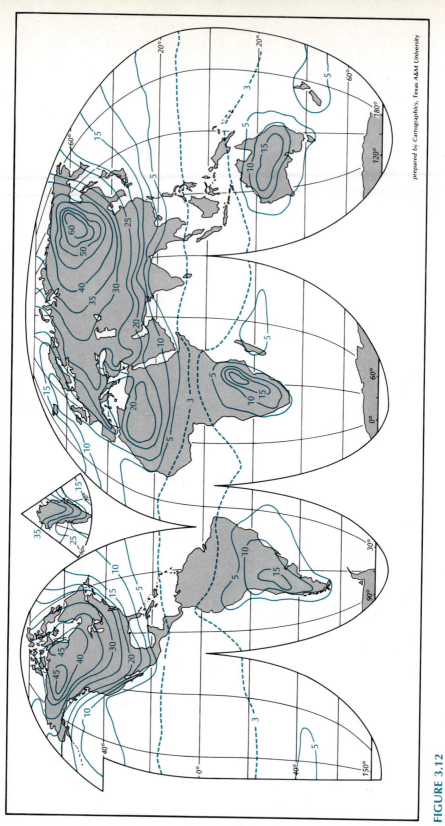

prepared by Cartographics, Texas A&M University

FIGURE 3.12
Distribution of mean annual range in temperature over Earth (0°C).

nificantly from a simple zonal pattern, the result of a combination of other temperature controls. For example:

1. Figures 3.10–3.12 show the extremes of temperature that occur over continental land masses. The largest land mass, Asia, experiences average temperatures that range from −40°C (−40°F) in January to more than 20°C (68°F) in summer. Figure 3.12 clearly shows the effect of the continent size on the average annual range of temperature.

2. The oceans exhibit the results of heat transport by ocean currents. The displacement of isotherms toward the poles shows the warming effects of the North Atlantic and North Pacific drifts. Cold ocean currents, such as the California and Canary currents, lower temperatures.

3. The figures show temperatures affected by altitude. Some world temperature maps have isotherms reduced to sea level, eliminating the effect of elevation. Notice, for example, the isotherms over South America. The Andes appear as a tongue of colder temperatures extending toward the equator.

4. The closest approximation to zonal temperatures occurs over the southern ocean and Antarctica. This is the most extensive area of homogenous surfaces encircling the globe without interruption.

5. Compare Figures 3.6 and 3.7 and note that the hemispheric temperature gradient is steepest in winter. For example, in January the Northern Hemisphere gradient is more than 60°C (110°F). In July, it is about 10°C (50°F). This pattern is highly significant for atmospheric circulation.

Summary

Temperatures vary through time and space over the planet. Temperatures change vertically through the atmosphere. The troposphere normally gets cooler with height. Temperature inversions occur when the temperature profile increases with height.

A number of significant changes in temperature take place with time. Among them are the seasons and daily changes. Both seasonally and daily, the highest temperature lags the maximum solar radiation. These lags are a result of the atmosphere's greenhouse effect.

Temperature changes rapidly from place to place. Among the factors contributing to these changes are latitude, nature of the surface, direction of slope with respect to the sun, and location in the general circulation. The net effect of these various factors is a global pattern of temperatures that generally reflects the latitude but contains some wide anomalies.

4

HUMAN IMPACT ON THE ENERGY BALANCE

Chapter Overview

A number of ongoing processes stand to affect Earth's energy balance in the coming decades. Of these processes, two are human in origin and involve changes in atmospheric chemistry that may ultimately cause Earth's temperature to rise. These processes are the increasing carbon dioxide (CO_2) content of the atmosphere and the introduction of chlorofluorocarbon compounds (CFCs) into the atmosphere.

Carbon Dioxide and the Greenhouse Effect

The quantity of CO_2 in the atmosphere varies with time, both daily and seasonally. During the day, photosynthesis withdraws CO_2 from the atmosphere; at night, respiration releases CO_2. Daily fluctuations during the growing season in the midlatitudes reach up to 70 parts per million (ppm). Variation in the rate of photosynthesis also causes a seasonal change in CO_2 in the midlatitudes. In the Northern Hemisphere, CO_2 content usually peaks in April and falls to a minimum in late September or October. The seasonal change in atmospheric CO_2 reflects a very important influence on the atmosphere: the metabolism of all living matter. The seasonal change in CO_2 concentration in the atmosphere results from the "pulse" of photosynthesis that occurs during the summer in the midlatitudes of both hemispheres. The difference ranges from about 5 ppm at Mauna Loa, HI, to more than 15 ppm in central Long Island, NY. The difference is smaller near the equator, where the biota are green most of the year, and is greatest in the midlatitudes, where the vegetation is largely deciduous. The difference is also smaller at higher elevations at all latitudes. These diurnal and seasonal changes in CO_2 have been a part of the natural atmosphere throughout time. Some irregular changes also take place over longer periods. These are due to changes in volcanic activity, the rate of chemical weathering, and Earth's biomass.

Historical Changes in Carbon Dioxide

The CO_2 content of the atmosphere dropped steadily through geologic time until about 50 million years ago; the levels of the present atmosphere have existed for the past 3–4 million years. Analysis of air bubbles in 160,000-year-old Antarctic ice cores shows that during the last 2 million years, atmospheric CO_2 varied over a relatively small range, between 200 and 280 ppm. The higher CO_2 levels occurred during the interglacials and lower levels during periods of glaciation. The most likely reason for this fluctuation is that cold water absorbs more CO_2 than does warm water. Given that, CO_2 content dropped during the periods of glaciation.

Data show that the CO_2 content of the atmosphere has been increasing since at least 1850 (Figure 4.1). The precise total increase is unknown due to the nature of early measurements. Early in the twentieth century, scattered measurements showed an average annual concentration of 316 ppm. The best data are available only after 1958, when Charles D. Keeling of the Scripps Institute of Oceanography set up a continuous monitoring station at Mauna Loa, HI. The CO_2 concentration increased from about 280 ppm in 1850 to about 350 ppm in 1988. The amount of increase at Mauna Loa averaged 0.8 ppm per year, ranging from 0.5 to 1.5 ppm per year. It is important to note that the rate of

FIGURE 4.1
Increase in atmospheric carbon dioxide.

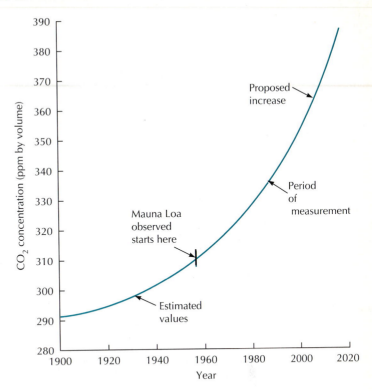

change is increasing: From 1958 to 1968, the increase averaged 0.7 ppm per year, and from 1968 to 1978, it was 1.3 ppm per year. A fourth of the total increase occurred in the decade 1967–77. In 1977–78, the increase was 1.5 ppm for the whole atmosphere and 2.0 ppm over Antarctica.

The CO_2 concentration of 350 ppm in 1988 was greater than at any time since 100,000 years ago. Present data suggest that CO_2 emissions will grow by about 1.6% annually from 1980 to 2025, and then decline to a growth rate of around 1% per year. Based upon present trends, it is probable that CO_2 will rise to 32% above the 1850 level by the year 2000. It is both possible and likely that the level of 400 ppm will be reached, a level not attained in the past million years. This level is significant even considering geologic time. Once the level rises, it will remain there for centuries, since we have no rapid means of removing it from the atmosphere.

Carbon Dioxide and the Hypothesis of Global Warming

The projected increase in CO_2 has considerable potential for changing Earth's energy balance. Carbon dioxide is partially responsible for the "greenhouse effect" of the atmosphere, because it is transparent to solar radiation but relatively opaque to infrared radiation. Earth's radiation is concentrated in the infrared band from 7 to 20 μm. Carbon dioxide absorbs radiant energy in the wavelengths in which Earth radiates heat away from the surface, and most of the absorption occurs in the 15- to-20-μm range. If CO_2

absorbs Earth radiation and the amount of CO_2 in the atmosphere is increasing, then the atmosphere is absorbing more solar energy. This absorption and reradiation back to the surface will shift the energy balance toward increased storage of energy, hence raising the temperature of Earth's surface and atmosphere.

Estimates of how much the temperature increase will be vary significantly because they come from different mathematical models of atmospheric processes. If CO_2 concentrations double from the preindustrial level of 280 ppm, mean world temperatures will likely increase about 1.5°–4.5°C (3°–8°F). The National Research Council (1983) placed the range for a doubling from 300 ppm to 600 ppm at 2°–3.5°C (4°–6°F). If mean global temperatures increase, warming will not be consistent over the globe: The subpolar latitudes will likely experience a much greater warming than equatorial regions—two or three times as much. In the portion of the Northern Hemisphere where the seasonal snow cover shifts from year to year, mean temperatures may increase 4°C (7°F) if CO_2 content doubles. We have yet to see evidence of any changes in the cover of Arctic Sea ice or snow cover on the adjacent land masses.

The Controversial Evidence of Global Warming

In testimony before the U.S. Senate in June 1988, James Hansen asserted that it could be said with about 99% confidence that Earth is getting warmer. Hansen and others have since begun to support the hypothesis of global warming. That support was based mainly on three pieces of evidence: (1) 1988 was the warmest year on record; (2) the four warmest years on record occurred in the 1980s; and (3) the 1980s are the warmest decade on record.

The arguments for global warming are countered by other analyses. If we consider historical data for the entire period since records have been kept, there may be no cause for alarm. Actual temperature changes, for example, do not reflect the impact of the increasing CO_2 content. Earth should have undergone a 0.5°C (0.9°F) rise in temperature based on the amount of fuel burned since 1850, but this has not happened. Careful studies of the historical record show that temperature change since the beginning of the industrial revolution is near zero. The year 1880 marks the first year for which we have enough data to compare temperatures. Critical study of the data show that little measurable warming has actually taken place. For example, urbanization has increased the temperature at about a third of the weather stations with old records. In addition, most of Earth is covered with oceans, for which we have little data.

The Complexity of the Energy Balance

The problem of global warming due to CO_2 is complex. Many factors affect atmospheric CO_2; in fact, natural feedbacks work to control CO_2 content. For example, CO_2 is soluble in seawater. The change in temperature that might occur with increased CO_2 depends to some extent on how well natural feedback mechanisms react to lower the CO_2 content. The faster the feedback mechanisms operate, the less the temperature increase is likely to be. Earth's biomass and the ocean have absorbed some 40% of the CO_2 placed in the atmosphere.

Part of the problem is that other factors cause the temperature to oscillate independently of the CO_2 in the atmosphere. One atmospheric component that affects global temperature is pollutants in the form of particulates, such as sulfur dioxide (SO_2). Sulfur dioxide particles serve as very small nuclei for cloud particles. Clouds made of very fine particles reflect more solar radiation than clouds made of large particles. Since SO_2 and CO_2 are both products of burning fossil fuels, they work in opposite directions on the energy balance: One leads to warming and the other to cooling. Increased reflectivity of the atmosphere may be offsetting as much as half of the potential greenhouse warming.

Other variables being constant, CO_2 content in the atmosphere will continue to increase during the forseeable future. On the other hand, continued increases in CO_2 will have less effect on global temperatures than previous increases, since each additional increment of CO_2 has less effect on the energy balance than the previous one.

Global Warming and Rising Sea Level

Global warming would increase sea level by adding water from melting mountain glaciers and the ice sheets of Greenland and Antarctica. Warmer temperatures also cause sea level to rise due to the thermal expansion of seawater. The historic reconstruction of sea level is difficult, but it is now generally agreed that over the past 100 years sea level has been rising at an average rate of 1–2 mm (.04–.08 in.) per year. Under present conditions, a temperature increase between 0.6° and 1.0°C (1.2°–1.8°F) may cause the mean sea level to rise 40–80 mm (1.6–3.2 in.) by the year 2025. The increase in temperature would also cause additional melting in the polar ice caps and mountain glaciers. The National Research Council (1983) estimated a total sea-level rise of 200–700 mm (8–28 in.) by the end of the next century. The buildup of CO_2 and the depletion of atmospheric ozone may increase the rate.

Sea level could rise rapidly if mean global temperatures rise 2°C (4°F) or more. With a rapid melting of sea ice, the sea-level rise could increase 1 or 2 m (3–6.5 ft) in the next several centuries. The west Antarctic ice sheet, which has its base on the sea floor, would respond rapidly to changes in water and atmospheric temperatures. The summer temperature of the ice sheet now averages −5°C (23 °F). If that temperature should rise above freezing, then very rapid melting would occur, producing a rise in sea level of 5–6 m (16–20 ft).

The major effect of rising sea level is flooding on some relatively flat coastlines. The lateral (or horizontal) effect of rising sea level can be significant. On emergent coastlines, a rise of 1 cm (0.4 in.) could increase the rate of beach erosion inland by 1 m (3.3 ft) (see Figure 4.2). It would also push the boundary between salt water and fresh water 1 km upstream along coastal rivers. The salt water would also displace fresh groundwater for a substantial distance inland. A rise in sea level damages the fresh-water supply of coastal cities. In many areas of coastal plains, a delicate balance exists between salt and fresh groundwater. Rising sea level tips the balance in favor of salinity, raising the costs of transporting fresh water or of desalinization. The small—but significant—change in sea level is already causing major problems in some areas. England and The Netherlands are taking remedial measures to reduce the risk of flooding from the sea.

A rapid rise in sea level of several meters would flood large sections of Earth's coastal plains, including many of the largest cities. Complete melting of the ice caps would

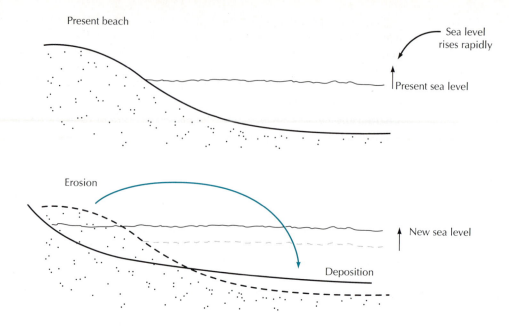

Present beach

Sea level rises rapidly

Present sea level

Erosion

New sea level

Deposition

FIGURE 4.2
Coastal erosion resulting from a rapid rise in sea level.

change sea level some 70 m (230 ft). However, the ice caps won't respond for thousands of years to a warming of several degrees in Earth's mean temperature. This period may be beyond the range of the possible effects of the CO_2 changes. We are not even certain whether warming of the seas would cause the ice caps to grow or to shrink. If the water warms slightly—but not enough to raise the temperature of the atmosphere over the ice above freezing—snowfall might increase and glaciers expand, thus lowering the sea level. We do not know that sea level will continue to rise. In some areas, the rise of coast-lines due to decompression from the last ice age is outpacing the rise in sea level. The net effect is one of a lower sea level.

Other implication of global warming are considered in Chapter 20, which places global warming in the broader context of future climates.

Stratospheric Ozone and Ultraviolet Radiation

Three atoms of oxygen combine to form a single molecule of **ozone**. Ozone normally is not abundant in the troposphere under natural conditions. It forms in smog by the action of sunlight on oxides of nitrogen and organic compounds. This ozone does not remain ozone for very long; it reacts with other gases in the atmosphere and changes to oxygen.

Ozone exists in the stratosphere even though the total amount is small. It is in a layer, or layers, between altitudes of 20–32 km (12–20 mi). This ozone layer absorbs ultraviolet radiation, and that absorption of radiant energy breaks the ozone (O_3) into normal oxygen molecules (O_2) and single oxygen atoms.

The quantity of ozone in the stratosphere changes through time due to a variety of factors. For example, the amount of ultraviolet radiation reaching the planet (and thus ozone broken down) varies with sunspot activity. Ultraviolet radiation is highest at the time of sunspot maxima and lowest at the time of sunspot minima. Ozone content also changes as stratospheric temperatures change. If stratospheric temperatures increase, ozone content increases, and if temperatures drop, ozone content drops.

Decline in Ozone

Measurement of global ozone began in August 1931 when scientists at Arosa, Switzerland, first measured the atmospheric absorption of ultraviolet radiation. Data show that the ozone content of the upper atmosphere has been declining in recent years. The first big change detected was after the April 1982 eruption of El Chichon, which discharged large quantities of debris into the stratosphere. It also coincided with El Niño, a major disturbance in the general circulation of the atmosphere. While these events may have been partially responsible for the decline in ozone, the ozone levels did not rebound after these events.

Chlorofluorocarbons

In 1974, scientists first publicized evidence suggesting that compounds known as chlorofluorocarbons (CFCs) had a depleting effect on stratospheric ozone layers. These compounds were synthesized in 1928 and promised to have many uses. They are odorless, nonflammable, nontoxic, and chemically inert. The primary compounds, $CFCl_3$ and CCl_2F_2, have been used as refrigerants in refrigerators and air conditioners since the 1930s (see Table 4.1). Since World War II, they have been used in deodorant and hair spray as propellants, in producing plastic foams, and in cleaning electronic parts. The United States, Japan, and Europe produce and consume most chlorofluorocarbons. In the United States, per capita use reached 1.1 kg (2.4 lb) per person per year in 1985. Automobile air conditioners (found in 90% of all new cars sold in the United States) use CFC-12, one of the most damaging of the CFCs. (Home refrigerators use a different CFC.) Automobiles are the largest single source of harmful CFCs, representing 26.6% of the CFCs released into the environment. In recognition of this problem, Vermont enacted legislation to outlaw cars with air conditioners using CFCs beginning with the 1993 model year.

Chlorofluorocarbons are not natural compounds. They do not react with most products dispersed in spray cans, are transparent to sunlight in the visible range, are insoluble in water, and are relatively inert to chemical reaction in the lower atmosphere. For these reasons the chemicals are a problem in the stratosphere. The average lifetime of a CFC-11 molecule is between 40 and 80 years. A CFC-12 molecule may last 80–150 years.

Chlorofluorocarbons rise into the upper atmosphere, where they break down by photodissociation. The dissociation takes place when the compounds are exposed to radiation of wavelengths of less than 0.23 μm. Ultraviolet radiation of this wavelength, or shorter, does not reach the troposphere because it is absorbed at altitudes of 20–40 km (12–24

TABLE 4.1
Common chlorofluorocarbons and halons.

COMPOUND (CHEMICAL FORMULA)	OZONE DEPLETION POTENTIAL*	ATMOSPHERIC LIFETIME (YEARS)	AMT. USED IN U.S.‡ (MILLONS OF KG)	AMT. USED WORLDWIDE‡ (MILLIONS OF KG)	MAJOR USES
CFC-11 ($CFCl_3$)	1.0	64	79.7	368.3	Rigid and flexible foams, refrigeration
CFC-12 (CF_2Cl_2)	1.0	108	136.9	455.0	Air conditioning, refrigeration, rigid foam
CFC-113 ($C_2F_3Cl_3$)	0.8	88	68.5	177.0	Solvent
Halon-1211 (CF_3BrCl)	3.0	25	2.8	7.1	Portable fire extinguishers
Halon-1301 (CF_3Br)	10.0	110	3.5	7.0	Total flooding fire extinguisher systems
HCFC-22† ($HCClF_2$)	0.05	22	99.2	Figures not available	Air conditioning

Source: EPA.

*Ozone-depleting potentials represent the destructiveness of each compound. They are measured relative to CFC-11, which is given a value of 1.0.
†Will not be limited by Montreal Protocol.
‡Consumption data based on 1985 figures.

mi). This photodissociation releases chlorine that interacts with oxygen atoms to reduce the ozone concentration. The process ends with the chlorine atom once again free in the atmosphere. Each atom of chlorine may persist for years, acting as a catalyst that may remove 100,000 molecules of ozone. The maximum rate of ozone destruction takes place at an altitude of 40 km (25 mi). The final means of disposing of the chlorine is a slow drift downward into the troposphere, where it combines with water molecules and precipitates out as hydrochloric acid.

Antarctic Ozone Hole

The most disturbing change in atmospheric ozone is that found over the Antarctic continent called the *ozone hole* (Figure 4.3). The ozone hole is a loss of stratospheric ozone over Antarctica that has occurred in September and October since the late 1970s. The hole appears in September when sunlight first reaches the region and ends in October when the general circulation brings final summer warming over Antarctica. During the

FIGURE 4.3

Altitudinal distribution of ozone.
Source: U.S.D.C. NOAA 1976 U.S.
standard atmosphere.

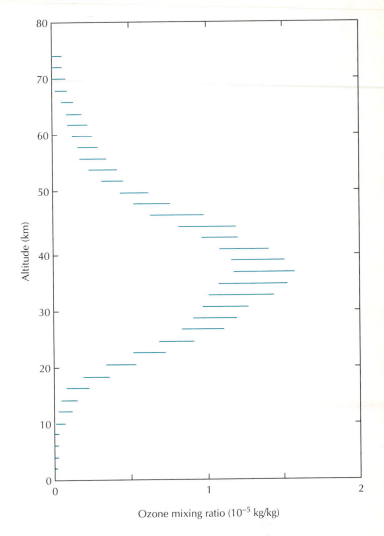

Ozone mixing ratio (10^{-5} kg/kg)

Antarctic spring, ozone decreases north from the pole to nearly 45° south latitude. In August and September 1987, the amount of ozone over the Antarctic reached the lowest level recorded to this date. On September 17, 1987, Nimbus-7 recorded a large area in which the ozone concentration was only about half that of the surrounding region. That fall the ozone hole, the area of maximum depletion, covered nearly half of Antarctica (see Plate 3).

In the Antarctic, a key chemical process involves molecules of chlorine monoxide (ClO) combining to form Cl_2O_2. When exposed to sunlight, the Cl_2O_2 breaks down into two chlorine atoms and an oxygen molecule. This frees the chlorine atoms to combine with ozone and repeat the process. In 1986 and 1987, ground-based measurements showed large concentrations of Cl_2O_2 between 17 and 23 km (10–14 mi), the altitude range where most of the ozone disappears. Further confirming the chemical reactions, the data showed that the ClO concentration decreased at night and increased rapidly after sunrise.

The analysis showed concentration of ClO of 100–500 times those of the midlatitudes. It also showed very little hydrochloric acid, a nondestructive reservoir of chlorine, and a low concentration of nitrogen compounds (which also form an inactive chlorine reservoir). The nitrogen compounds include nitric oxide, nitrogen dioxide, and nitric acid. Levels of about 10% those of the normal atmosphere were found. The ozone-depleted region extended from 12 km up to 23 km (7.4–14 mi). At the time, chlorine monoxide was concentrated above 18 km (11 mi).

In the winter over Antarctica, a very strong vortex of extremely cold, dry air keeps out warmer air from lower latitudes. This cold air gets even colder during the months without sunlight. Temperatures drop as low as $-84°C$ ($-119°F$). The extreme cold causes moisture to condense into ice crystals and nitric acid crystals to form. The aerosol of ice and nitric acid crystals forms very high, thin clouds in the polar stratosphere. These crystals play a very important role in the chemistry of CFCs and ozone, as each crystal provides a place for accelerated chemical reactions. Chemical processes are more rapid where they have a surface upon which to occur, and ice particles are the most efficient—some 10 times as efficient as the surface of water droplets. This partially explains the speed with which the process takes place in the Antarctic spring and why the process is less effective in low latitudes. The interaction begins when sunlight appears in the spring. The warming increases the rate of chemical reactions, and chlorine destroys ozone at a rapid rate. Spring occurs in September and October, during which time ozone drops until either no more ozone is left or the clouds evaporate. Up to 60% of the total ozone may be lost over the center of the Antarctic hole; up to 95% at some altitudes. Eventually, air from surrounding regions flows into the area and ozone levels recover somewhat. Polar stratospheric clouds disappear with the spring warming. The same process takes place elsewhere in the atmosphere, but at higher altitudes and at slower rates.

At least two elements are associated with the decrease in ozone over Antarctica. The obvious one is the increase in CFCs, and the other is meteorologic. The meteorologic element involves changes in seasonal circulation over the region and a high-altitude change in climate. On September 5, 1987, over an area of 3 million km^2 (1.1 million mi^2), the ozone decreased by 10% from previous weeks. This sharp decrease must have been due to the inflow of air low in ozone. Weather conditions also play a role in the size of the hole. Chemical processes operate over the region from 68° south to the South Pole. The ozone-depleted area extends out to latitude 45°. The circulation must carry the ozone-depleted air outward from the source area. A temperature change is taking place over Antarctica as well. The stratosphere over the Southern Hemisphere cooled by 2°C (3.6°F) between 1980 and 1985, while the stratosphere over the Antarctic cooled between 2° and 4°C (3.6°–7.2°F).

In the fall of 1989, the ozone content of the stratosphere over Antarctica once again plummeted (Figure 4.4). Before 1989, the worst year was 1987. The depletion of 1989 was as severe as 1987. In the fall of 1989, ozone almost completely disappeared from some zones in the stratosphere. In the 16- to 18-km (9.6–10.8 mi) zone, ozone dropped 90% from August to the first week of October. In both 1987 and 1989, the depleted geo-

FIGURE 4.4
Ozone depletion in the fall of 1989.

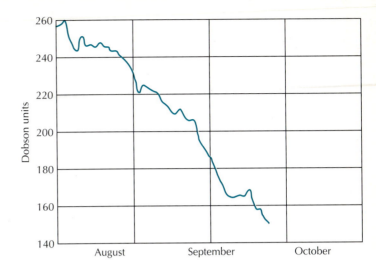

graphic area was twice the size of Antarctica. In 1987 and in 1989, the average ozone content dropped 40% during the spring. Ozone fell sharply again in the fall of 1990, the first time back-to-back bad years of decline occurred. In October 1990, heights between 15 and 18 km (9.3–10.8 mi) had practically no ozone. By October 4, a drop of almost 40% had occurred.

Despite the three years with severe drops in ozone, the stratosphere in the Southern Hemisphere's midlatitudes was stable during the winter and first few weeks of spring. Sunlight first reaches the South Pole about mid-September. The stable Antarctic atmosphere favored the development of a strong flow of upper-level winds around the Antarctic over the southern ocean. This high-altitude airstream separated the cold pool of air from the warmer air of the midlatitudes, allowing the cold pool to become exceptionally cold. Temperatures within the cold pool dropped to −85°C (−121°F).

Ozone Depletion in the Arctic

Ozone depletion is not as serious outside of the Antarctic because the stratospheric aerosols are less abundant and consist of liquid sulfuric acid droplets rather than ice. Because of this difference, the Arctic area has no ozone hole like that of the Antarctic. Temperatures are warmer and weather varies more in the Arctic, providing less-favorable conditions for the necessary chemical and circulation processes. Ozone levels in the high latitudes of the Northern Hemisphere have dropped 5% since 1971, however. Furthermore in 1988 researchers in Thule, Greenland measured increased concentrations of reactive chlorine compounds, the same compounds known to be present over the Antarctic when ozone depletion happens. Experiments in 1989 all verified the presence of the Arctic ozone hole and provided detailed measurement of the amount and extent of the depletion, and much information about the chemistry of ozone depletion.

Potential Solutions for the Problem of Ozone Reduction

Concern over the possible connection between CFCs and ozone loss led to a ban on the use of these compounds as aerosol propellants in the United States effective in 1978. The United Nations Environment Program called a conference in Montreal, Canada, in September 1987 to discuss the possible effects of CFCs on stratospheric ozone. Representatives of more than 30 countries took part in the conference, which resulted in a treaty restricting the production of CFCs. The agreement, officially termed the Montreal protocol, called for freezing domestic consumption of the chemicals at 1986 levels by July 1990, and limited production to 110% of 1986 levels. Included in the freeze were CFCs-11, 12, 113, 114, and 115. Nations agreeing to the protocol must reduce consumption 20% by 1994 and to 50% of 1986 rates by 1999. Production rates are to be reduced 10%–40% by the same dates. Actual production will continue to provide a supply of the chemicals for developing countries that cannot afford to respond to remedial measures as fast as the developed countries. The protocol permits developing countries to continue production and exceed current levels if necessary for economic development. They may increase consumption up to 0.3 kg (0.6 lb) per capita. The protocol also freezes the consumption of the halons 1211 and 1301, a more destructive but less prevalent class of chlorine compounds used in fire extinguishers.

The protocol does not solve the problem, but it provides time to both examine the problem and seek solutions. Compliance will reduce but not stop the accumulation of CFCs in the stratosphere. Concentrations may grow to as much as 30 times the 1986 levels. Computer models show at least an 85% reduction in CFC and halon use is needed to stabilize the level of the chemicals in the atmosphere. CFCs have an atmospheric lifetime of 100 years or more. The chlorine already released will continue to remove ozone for at least a century. If surface emissions of CFCs stop today, it will be several decades before maximum ozone depletion occurs and the rate of removal begins to drop. The chlorine content may continue to rise until it reaches a level as much as six times the 1986 levels. If the release of the compounds continues unabated, a reduction in ozone of about 10% will take place (with a possible range of 2%–20%).

It is uncertain whether a real long-term decrease in ozone is occurring. It may just be the shortness of the record and the accuracy of the data, or it may be a natural oscillation that will correct itself. The only significant source of chlorine in the atmosphere is human industrial activity, and atmospheric chlorine doubled from 1965 to 1985. If that doubling is due to human activity, the effects of the CFCs are far greater than originally thought.

CFCs and the Greenhouse Effect

CFCs also play an important role in the greenhouse process. Although present in much smaller quantities in the atmosphere than CO_2, the impact of CFCs is large. The high amount of CO_2 already in the atmosphere is effectively blocking Earth radiation. However, additional CFC atoms added to the atmosphere are 10,000 times more effective at absorbing infrared radiation than are additional molecules of CO_2. Chlorofluorocarbons probably account for 25% of the greenhouse warming. A quadrupling of CFCs might produce an increase in temperature of 0.5°–1.0°C (1°–2°F).

■ Applied Study

Ultraviolet Radiation and Skin Cancer

Observers have been concerned about the relationship between health and the possible reduction in the ozone layer which results in an increase in ultraviolet radiation reaching ground level. Exposure to ultraviolet radiation causes prematurely aged skin, skin cancer, and a weakened immune system. The main reason for increased exposure is the popular desire for a suntan. Ultraviolet radiation produces tanning as well as sunburn of the skin. The hazard of skin cancer is much greater from overexposure (as in a sunburn) than from steady, low doses of ultraviolet radiation. A single, blistering sunburn in a person 20–30 years of age triples that person's chance of skin cancer.

One form of skin cancer, melanoma, is the most lethal form. It is almost always fatal if it spreads to other parts of the body. The disease is almost epidemic in the United States: Some 27,000 new cases were expected in 1990 (American Cancer Society 1990). The incidence of melanoma is increasing at the rate of about 4% each year. Other countries where sunbathing and tanning salons are in vogue are seeing about the same growth in melanoma. In addition, the age at which melanoma is diagnosed is dropping. When first regularly reported, it affected persons aged 40 or over. By 1990, it was common in the 20–40 age group. Of skin cancer fatalities in 1990, about 6300 were cases of melanoma and 2500 were other kinds of skin cancer. The fatality rate from melanoma is about 25%. Early treatment results in a survival rate of over 80% (see Plate 4).

In 1990, more than 600,000 new cases of non-melanoma skin cancer were expected to be diagnosed. Each 1%–2% reduction in average ozone concentration may result in several hundred thousand new cases of skin cancer.

The risk of getting skin cancer can be reduced with reasonable care. The first rule is to avoid exposure to the midday sun, when the intensity is greatest. The most dangerous hours are 10 A.M. until 2 P.M. local time (11 A.M. to 3 P.M. during Daylight Savings Time).

If exposure to the sun is necessary, use a sunscreen with a rating of 15 based on ultraviolet B radiation. Ultraviolet A is also hazardous to health, but not nearly as much as ultraviolet B. Lotions with a rating of 15 provide protection from both A and B radiation, and no evidence proves that sunscreens with a higher rating provide additional protection. Avoid tanning parlors, since their radiation is as bad as or worse than natural sunlight. Children should avoid overexposure to sunlight, because skin damaged early in life is more likely to develop into skin cancer later.

Ultraviolet radiation wavelengths of 0.297 μm are the most effective in producing sunburn. Measurements have not yet shown any increase in this radiation at the surface. ■

Summary

The chemistry of the atmosphere is always changing as a result of the slow evolution of the planet through time. It changes on a short-term basis as a result of fluctuations in solar energy, volcanic eruptions, and other natural events. Now the atmosphere is changing due to human activity. The rapid growth of the human population in the past two centuries has led to a steadily increasing change in the chemistry of the atmosphere. Carbon dioxide concentration has grown by nearly 20% since 1850 and will continue to increase. That increase will in turn cause the mean temperature to rise.

It is not possible to predict how much the atmosphere's chemistry will change in the future because too many variables are involved. Variables include the rate of growth of the human population, the rate of economic development, the rate of consumption of fossil fuels, general changes in technology, the rate at which the ocean removes CO_2, and other elements of climatic change.

Upper-atmospheric ozone is being depleted by the addition of CFCs. These compounds accumulate in the stratosphere and, together with ultraviolet radiation, remove the ozone. The reduction in ozone adds to the potential problem of global heating. It is also causing nearly epidemic increases in cancer in humans. Both global warming and ozone depletion have the potential to make major changes in Earth's climate for centuries.

PLATE 1

Planet Earth from an Apollo spacecraft. The distinguishing features of Earth as seen from space are the dominance of the seas, the extensive cloud cover, and the continents. The clouds are the most efficient reflector of solar radiation.

GLOBAL CHANGE—THE CLIMATIC ELEMENT

It has long been known that Earth's climate varies from place to place, producing different climatic regions with resulting major world biomes. In recent years the ability to see or sense Earth from satellites has greatly added to our knowledge of change through space. At the same time we have come to recognize that change through time is a fundamental characteristic of our planet. The color plates in this section provide photographic and computer generated images of this change through time and space. The impact of change on the human species is also illustrated in the plate showing the effects of ultraviolet radiation on the skin.

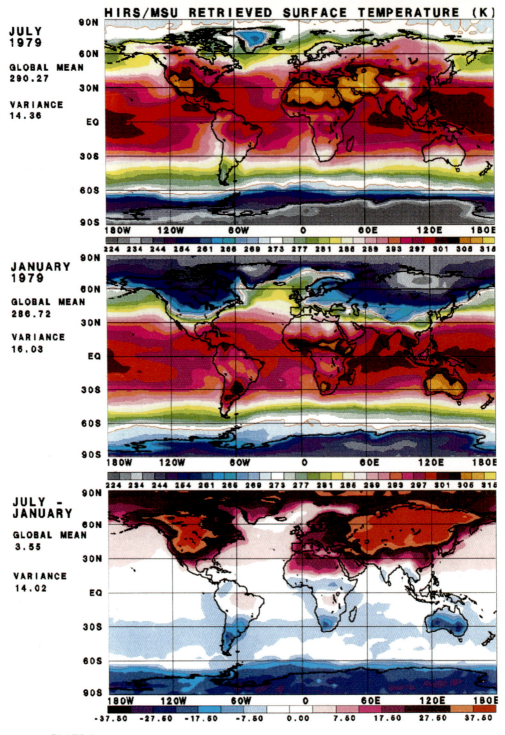

PLATE 2

Global mean temperatures of Earth in July and January of 1979 and the mean difference between them. Note the much larger difference in temperatures in the midlatitudes.

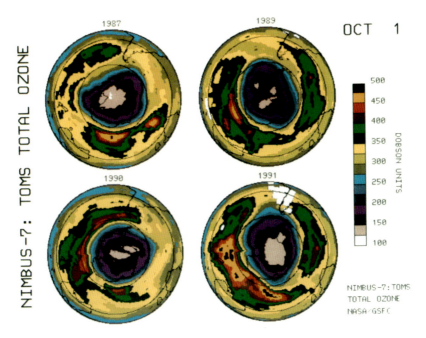

PLATE 3

Total ozone over the Southern Hemisphere as mapped by the Total Ozone Mapping Spectrometer on board the Nimbus-7 polar orbiting satellite. The ozone hole is seen in pink, with the purple colors indicating extremely low ozone values. These four years were the worst up to 1991 with the depleted area nearly the same in each year.

Common Moles

Malignant Melanomas

PLATE 4

Common moles are round and symmetrical, have smooth, even borders, are usually a single shade of brown, and are usually less than 6 mm (1/4 in.) in diameter. Malignant melanomas are generally asymmetrical, have uneven borders (often containing scalloped or notched edges), are usually two or more shades of brown or black, and are larger than 6 mm in diameter.

PLATE 5

Mount Pinatubo, Philippines, erupting in July 1990. A thick cloud of ash and steam rises from the volcano. It is this ash which spread around Earth over the next year, as shown in Plate 6.

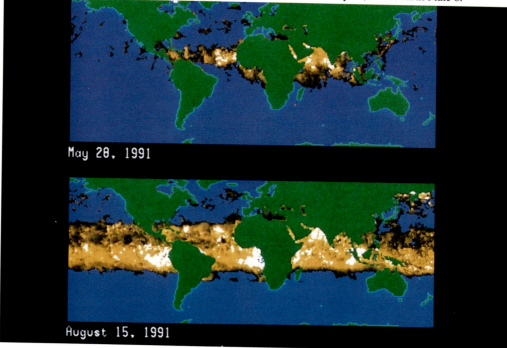

May 28, 1991

August 15, 1991

PLATE 6

Global spread of volcanic aerosols from Mt. Pinatubo detected from the NOAA-11 Meteorological Satellite. Data from the Advanced Very High Resolution Radiometer (AVHRR). Top image obtained May 28, 1991 and the bottom image August 15, 1991.

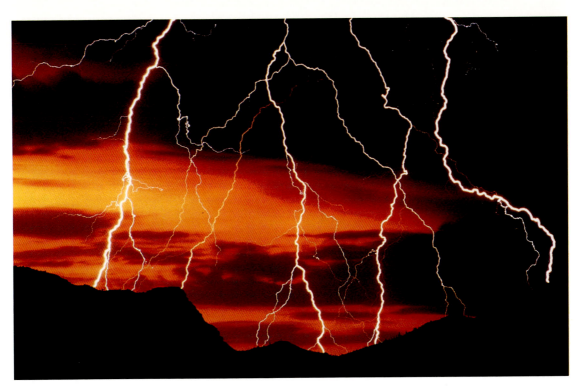

PLATE 7   Multiple lightning strokes near sunset.

Hurricane Bob
NOAA-10 AVHRR 8/19/91

Robert M. Carey
Satellite Research Laboratory
NOAA / NESDIS / ORA

A

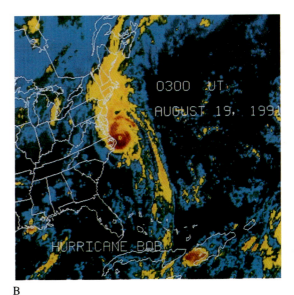

0300 UT
AUGUST 19, 1991

HURRICANE BOB

B

PLATE 8   A) Hurricane Bob from a low-level satellite. The eye, the counterclockwise circulation around the eye, the spiral nature of the clouds, and hence bands of rain are clearly shown. (B) A false color satellite image of Hurricane Bob from high altitude. The red color around the eye corresponds to the highest and coldest clouds. The areas shaded orange and yellow are lower and warmer clouds. The image shows the hurricane traveling north over the Atlantic Ocean just offshore from Cape Hatteras, North Carolina. The major area of precipitation is to the north of the storm center. Note that there are some long bands of clouds extending far behind the storm center.

PLATE 9

A tropical rainforest biome typical of wet climates. This example is in La Amistad Biosphere Preserve in Costa Rica.

PLATE 10

A tropical grassland biome associated with wet-and-dry climates. This is the Samburu Reserve, Kenya, East Africa.

PLATE 11

A tropical desert. In the photo are date palms and camels, both of which require a lot of water. The palm trees, which are in some cases immersed in sand, have their roots in water in an oasis. Camels can store water in their humps and can go for a number of days without drinking.

PLATE 12

A coastal desert in the tropics. Here in the Peruvian desert it very rarely rains even though adjacent to the ocean. Subsidence of dry air in the subtropical high produces a very stable lower atmosphere.

PLATE 13

A midlatitude rainforest in Olympic National Park, Washington. Frequent midlatitude lows from over the Pacific Ocean provide abundant rains to the area.

PLATE 14

A midlatitude grassland in National Bison Refuge, Montana. There is enough rainfall here to support a tall-grass prairie.

PLATE 15

A midlatitude grassland in a low rainfall area. This short-grass prairie is in Wind Cave National Park, South Dakota.

PLATE 16

A midlatitude desert in Organ Pipe National Monument, Arizona. Most desert vegetation blooms in the spring.

PLATE 17 The polar desert of Antarctica. Adelie penguins nest along the coast.

5

MOISTURE IN THE ATMOSPHERE

Chapter Overview

Changes of State

The Hydrologic Cycle

Water Vapor and Its Measurement

Evaporation

Evapotranspiration

Condensation Near the Ground: Dew, Mist, and Fog
- Advection and Radiation Fogs
- Foggy Places

Condensation Above the Surface: Clouds
- Condensation Nuclei
- Cloud Forms and Classification

The significance of water as an atmospheric variable is a result of its unique physical properties. Water is the only substance that exists as a gas, liquid, and solid at temperatures found at Earth's surface. This special property enables water to cycle over Earth's surface. While changing form one form to another, it acts as an important vehicle for the transfer of energy in the atmosphere.

Changes of State

The chemical symbol for water, H_2O, is probably the best-known of all chemical symbols. It tells us that the water molecule is made up of two atoms of hydrogen for one of oxygen. Water in all of its states has the same atomic content, but the molecules are arranged differently (Figure 5.1). At low temperatures, little energy is available and the bonds binding the water molecules are firm. The molecules pack tightly in a fixed geometric pattern in the solid phase. As temperature increases, the available energy weakens the bonds of the ice phase. Because they are not firmly set, bonds form, break, and form again. This permits flow and represents the liquid phase of water. In this liquid stage, bonding still happens, but it is much less compact than in the ice phase. At higher temperatures and with more energy, the water molecules' bonding breaks down and the molecules' movement is disorganized—the gas phase. If the temperature decreases, the molecules revert to a less-energetic phase and reverse the process. Gas changes to liquid and liquid to solid.

Figure 5.2 illustrates these changes of phase. Note that melting, evaporation, and sublimation from the solid to liquid phase all absorb energy. This added energy causes a change in the molecules' bonding pattern. The changes to the water-vapor stage incorporate a great deal of energy, the change from ice to water much less.

The energy absorbed is latent energy; it reverts to the environment when the phase changes reverse. When water vapor changes to liquid, it releases the energy originally absorbed and retained as latent heat. The same is true when water freezes and water vapor sublimates to ice.

The release of latent heat has a number of results. Latent heat provides energy to form thunderstorms, tornadoes and hurricanes. It also plays a critical role in the redistribution of heat energy over Earth's surface. Because of the high evaporation in low latitudes, air transported to higher latitudes carries latent heat with it. This water vapor condenses and releases energy to warm the atmosphere in higher latitudes.

The Hydrologic Cycle

The hydrologic cycle is a conceptual model of the exchange of water over Earth's surface. Figure 5.3 shows one model of the hydrologic cycle. It shows large-scale changes of state with evaporation providing the moisture that condenses and becomes precipitation. Additionally, the diagram shows that transpiration occurs. Transpiration is a special form of evaporation in that moisture does return to the air through evaporation (although via vegetative processes). Climatologists often combine the evaporation and transpiration amounts into a single parameter called **evapotranspiration.**

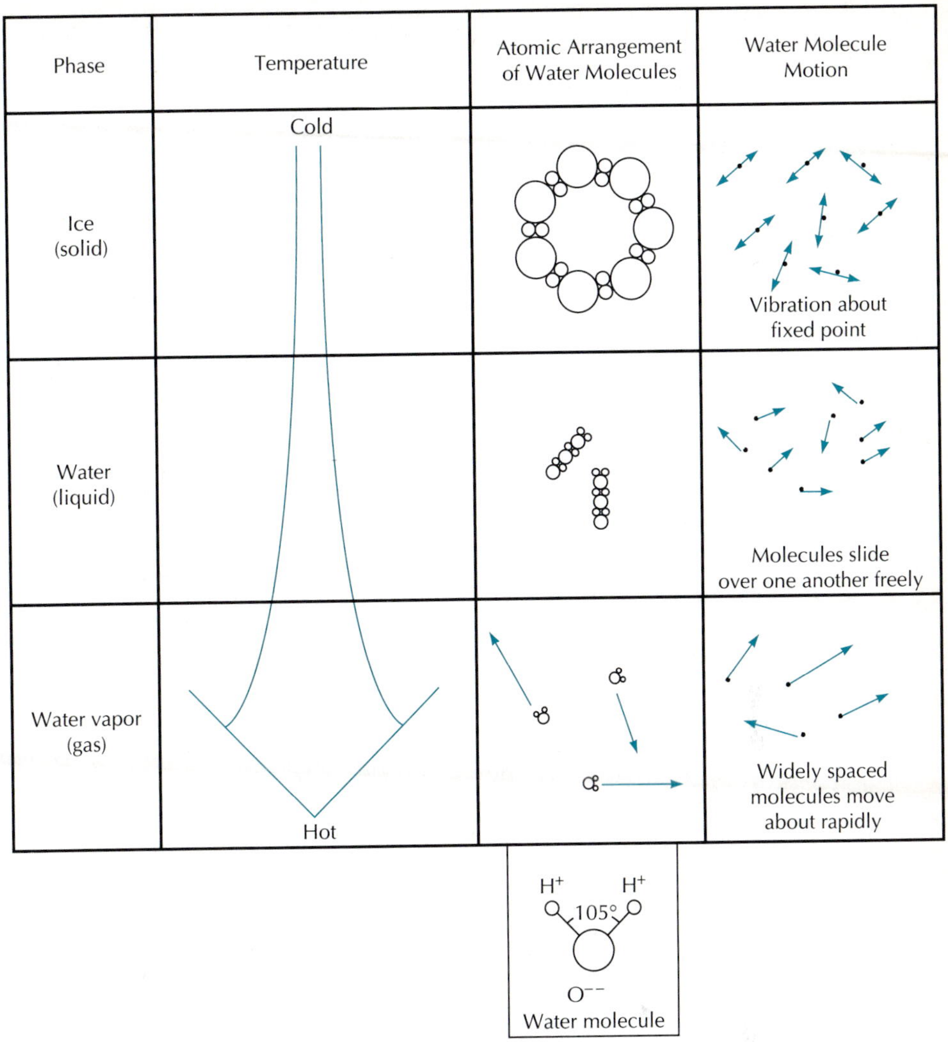

Phase	Temperature	Atomic Arrangement of Water Molecules	Water Molecule Motion
Ice (solid)	Cold		Vibration about fixed point
Water (liquid)			Molecules slide over one another freely
Water vapor (gas)	Hot		Widely spaced molecules move about rapidly

H^+ H^+
$105°$
O^{--}
Water molecule

FIGURE 5.1

Schematic diagram of the arrangement and motion of water molecules in different phases.

Some of the precipitation that falls to the surface passes to the soil, becoming soil moisture that growing plants use, while some passes deeper into the ground to become groundwater. Other precipitation runs off the surface and is collected in ponds, lakes, and reservoirs or flows as surface water in streams and rivers. Eventually water finds its way to the ocean and starts the cycle over again.

How is water distributed over Earth? The divided circle in Figure 5.4 shows that of all the water available on Earth, about 97% occurs in the oceans. About 3% is mostly ice found in the large ice caps. Almost all of the rest is in groundwater. Rivers, lakes, and soil moisture account for less than 1% of the total water. The atmosphere contains only

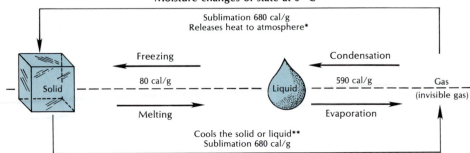

Moisture changes of state at 0 °C

*Applies to changes shown above broken line
**Applies to changes shown below broken line

Heat transfers associated with phase changes of water

Phase Change	Heat Transfer	Type of Heat
Liquid water to water vapor	540-590 cal absorbed	Latent heat of vaporization
Ice to liquid water	80 cal absorbed	Latent heat of fusion
Ice to water vapor	680 cal absorbed	Latent heat of sublimation
Water vapor to liquid water	540-590 cal released	Latent heat of condensation
Liquid water to ice	80 cal released	Latent heat of fusion
Water vapor to ice	680 cal released	Latent heat of sublimation

FIGURE 5.2
Changes of state of water.

0.35% of all the water available. Yet this small amount is the reservoir providing the moisture for clouds and precipitation over Earth's surface.

The actual exchanges of water provide some surprising data. Assume that the average precipitation of the world (the amount that would fall if every place on Earth got the same amount) is 85.7 cm (33.8 in.), and let this amount equal 100%. Most water evaporates from the ocean and the greater part of all precipitation falls over the ocean. As illustrated in Figure 5.5, the amount of water that falls on the land is appreciably less (23%). This is expected, however, for the oceans provide much of the available moisture and occupy a much greater area that the continents.

Water Vapor and Its Measurement

We refer to the gaseous state of water as vapor, but it is a gas like any other. The amount of vapor in the air at any time varies from almost zero over ice deserts to 7% (e.g., 7 g of water vapor for every 100 g of air) in warm, humid areas. The amount present is important, because water vapor plays a major role in the energy budget. It acts as an absorber, as an energy-exchange mechanism through latent heat, and as the source of condensation.

We can express the amount of water vapor in the air in several ways. Water vapor is a gas, and like any gas it exerts pressure on a surface. Consider that when we measure air pressure, the water vapor in the air contributes toward the total pressure. Considered independently of other gases, the amount of pressure exerted by the water vapor is the vapor pressure. The more moisture in the air, the higher the vapor pressure.

Beyond exerting a pressure, water vapor in the air also occupies space and contributes toward the mass of the air. Accordingly, the amount of vapor in the air is expressed as a weight or a volume. The mixing ratio of air is the ratio of the mass of water vapor to a unit mass of dry air (e.g., grams of water vapor per kilogram of air). A similar measure that also expresses water vapor content as a mass is the **specific humidity,** but this relates the mass of water vapor to the mass of air including the water vapor. In the mixing ratio, dry air is the basis for determining the ratio:

mixing ratio = mass of water vapor/mass of dry air
specific humidity = mass of water vapor/total mass of air

Absolute humidity expresses the mass of water vapor present per unit volume of space and thus is expressed in units such as grams of water vapor per cubic meter of air.

One other well-known and widely used measure of water vapor in the atmosphere is **relative humidity.** Relative humidity depends upon the maximum amount of moisture that air can hold, the saturation level.

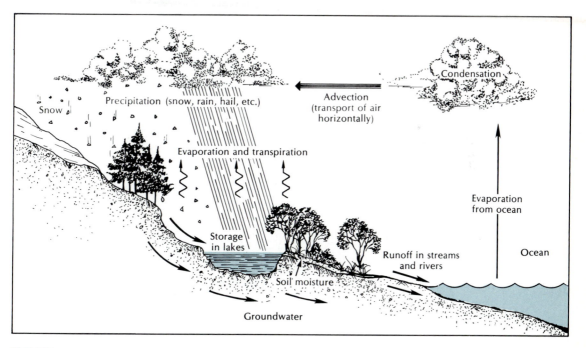

FIGURE 5.3
The hydrologic cycle.

The amount of water vapor that air can hold is a function of air temperature: Warm air can hold more moisture than cold air (see Table 5.1). In fact, the amount of moisture air can hold increases rapidly with increasing temperature (see Figure 5.6). Air over ice

FIGURE 5.4
The proportional distribution of water over Earth's surface.

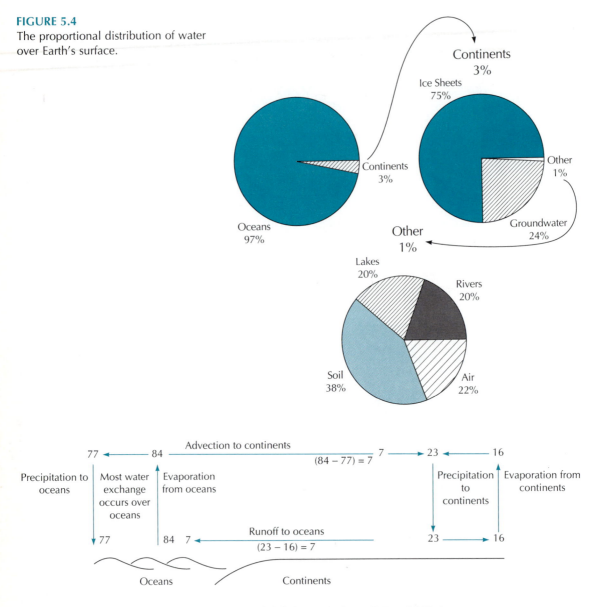

FIGURE 5.5
Estimates of water transfer within the hydrologic cycle. A value of 100 is used to denote average global precipitation.

TABLE 5.1
Saturation vapor pressure over water (in mb).

TEMPERATURE IN °C

TEMPERATURE IN °C	0	1	2	3	4	5	6	7	8	9
40	73.777	77.802	82.015	86.423	91.034	95.885	100.89	106.16	111.66	117.40
30	42.430	44.927	47.551	50.307	53.200	56.236	59.422	62.762	66.264	69.934
20	23.373	24.861	26.430	28.086	29.831	31.671	33.608	35.649	37.796	40.055
10	12.272	13.119	14.017	14.969	15.977	17.044	18.173	19.367	20.630	21.964
+0	6.1078	6.5662	7.0547	7.5753	8.1294	8.7192	9.3465	10.013	10.722	11.474
−0	6.1078	5.623.	5.173	4.757	4.372	4.015	3.685	3.379	3.097	2.837
−10	2.597	2.376	2.172	1.984	1.811	1.652	1.506	1.371	1.248	1.135
−20	1.032	0.9370	0.8502	0.7709	0.6985	0.6323	0.5720	0.5170	0.4669	0.4213
−30	0.3798	0.3421	0.3079	0.2769	0.2488	0.2233	0.2002	0.1794	0.1606	0.1436
−40	0.1283	0.1145	0.1021	0.09098	0.08097	0.07198	0.06393	0.05671	0.05026	0.04449
−50	0.03935	0.03476	0.03067	0.02703	0.02380	0.02092	0.01838	0.01612	0.01413	0.01236

To find the saturation vapor pressure for 13°C, read down the left column to 10°C and across that row to 3°C. The saturation vapor pressure for 13°C is 14.969 mb.

Source: J. J. Hidore and M. C. Roberts, *Physical Geography—A Laboratory Manual.* (New York: Macmillan, 1990), 21.

FIGURE 5.6

Saturation vapor pressure as a function of temperature. Values for below 0°C are shown in the inset.

Source: Reprinted with permission of Merrill, an imprint of Macmillan Publishing Company from *Elements in Meteorology* by Albert Miller and Jack C. Thompson. Copyright © 1979 by Macmillan Publishing Company.

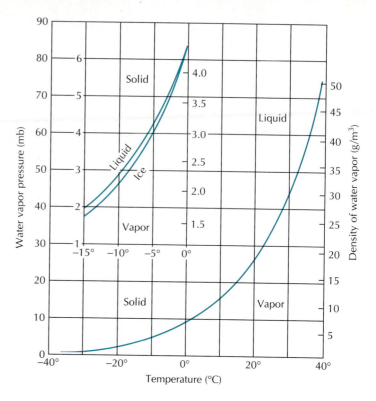

holds less moisture than that over water, a property that has important implications in the growth of rain droplets in clouds.

Of special significance in relation to saturation is the **dew point** (also called dew-point temperature). This is the temperature to which a parcel of air must cool to reach saturation. The name indicates the meaning, for dew provides visible evidence that the air has reached saturation level.

On a cool, calm night, the air near the ground loses heat by radiation to the air above. With little mixing of the air, a thin layer of air next to the ground becomes cold. If the temperature drops to the point at which the air becomes saturated, then moisture condenses in the form of dew. At temperatures above freezing, this is considered the dew-point temperature; below freezing, it is the frost point. (Ice crystals form from water vapor at the frost point.) Ice formed on car windows after a cold night is a clear sign of cooling to the frost point.

The saturation amount forms the basis for relative humidity.

$$\text{Relative humidity} = \frac{\text{amount of water vapor in air}}{\text{amount of water vapor that the air can hold at saturation}} \times 100$$

or

$$= \frac{\text{vapor pressure of air}}{\text{saturation vapor pressure}} \times 100$$

The relative humidity is stated as a percent.

The different measures of water vapor in the atmosphere are used in a variety of ways, depending upon the problem investigated. Because the warmer the air the more water vapor it can hold, some measures of water vapor content change with changing temperature, while others do not. For example, the mixing ratio measures the amount of moisture per weight of air. Even if a parcel of air changes temperature and no moisture is added or subtracted, the mixing ratio remains the same. The mixing ratio is a conservative measure of water vapor. Relative humidity, however, changes as the temperature of air changes (see Table 5.2). When absolute values of water vapor are needed, such as in the analysis of rising air to determine the level of condensation, measures such as the mixing ratio are used. If a value is needed that provides some insight into how close the air is to saturation, then relative humidity is of value. Weather broadcasts have popularized relative humidity, but the nature of relative humidity makes it of limited value in comparing moisture content at different places. A relative humidity of 60% at a temperature of 5°C (41°F) represents a different amount of water vapor in the air when compared with 60% relative humidity at 25°C (77°F).

TABLE 5.2
Relative humidity based on the wet-bulb depression.

DRY-BULB TEMPERATURE IN °C	DEPRESSION OF THE WET BULB IN °C*																			
	1	2	3	4	5	6	7	8	9	10	11	12	13	14	15	16	17	18	19	20
32	93	87	80	74	68	62	57	51	46	41	36	31	27	23	20	16	12			
31	93	87	80	74	68	62	56	50	45	40	34	31	26	22	19	14				
30	93	86	80	74	67	61	54	49	43	38	34	29	25	21	17					
29	93	86	80	73	66	59	53	47	42	37	32	28	24	20						
28	93	86	79	71	65	57	51	45	41	36	32	27	23	18						
27	93	85	78	70	63	56	50	45	40	35	31	26	21							
26	92	84	76	68	61	55	49	44	39	34	30	25								
25	91	82	74	67	60	54	48	43	38	33	28									
24	90	81	72	65	58	52	47	42	37	33										
23	90	81	72	65	58	52	46	41	37											
22	90	81	72	64	58	52	46	41	36											
21	89	80	72	63	57	51	45	40	35											
20	89	80	71	63	57	51	45	39												
19	89	79	71	63	56	50	44	38												
18	89	79	70	62	55	49	44													
17	88	79	70	62	54	48	42													
16	88	79	69	61	54	48														
15	88	78	69	61	53															
14	88	78	69	61																
13	88	78	69																	
12	88	77																		
11	88																			

*The depression of the wet bulb is the difference in temperature between the dry thermometer and the wetted thermometer.
Source: J. J. Hidore and M. C. Roberts, *Physical Geography—A Laboratory Manual.* (New York: Macmillan, 1990), 22.

Moisture content of the air is most commonly measured using a hair hygrometer or a psychrometer. A hygrometer depends upon hair expanding as it absorbs moisture (human hair increases its length by between two and two and a half percent as relative humidity increases from zero to 100%). While a hair hygrometer is used in thermostats, atmospheric scientists use a psychrometer for most weather observations. A psychrometer consists of two thermometers with one bulb covered by a moistened cloth sleeve. Evaporation from the sleeve of this wet bulb results in the temperature it registers being lower than that of the dry bulb. Appropriate tables help determine the moisture content of the air using readings from a psychrometer.

Evaporation

Evaporation is the process by which water changes from a liquid to a gaseous state. Water is sufficiently volatile in solid and liquid states to pass directly into the gaseous state at most environmental temperatures. **Sublimation** is the process of change from ice to water vapor. Since the source of water vapor is at Earth's surface, the amount of water vapor present in the atmosphere decreases with height. Most atmospheric moisture is found below 10,000 m (6.2 mi).

The amount of water that actually evaporates from a given water surface in a given period depends upon the following factors:

1. Vapor pressure of the water surface, which in turn depends upon water temperature. The higher the water temperature, the greater the surface vapor pressure. When water temperature is higher than the air temperature, evaporation will always take place.
2. Vapor pressure of the air. The greater this pressure, the less evaporation will occur. The rate of evaporation varies directly with the difference between vapor pressure of the water surface and vapor pressure of the air.
3. Wind. Air movement is usually turbulent, with moist air removed from near the water surface and replaced by dry air from above. Evaporation thus varies directly with the velocity of the wind. The higher the wind velocity, the more evaporation.

One way to estimate evaporation is Dalton's law:

$$E_o = (e_s - e) f(u)$$

where

E_o = the amount of evaporation
e_s = the vapor pressure of the water surface
e = the vapor pressure of the air
$f(u)$ = a function of the horizontal wind speed

Beyond these basic factors, evaporation rates depend on the nature of the evaporating surface. Salinity of the water will play a role, since dissolved material reduces the rate of evaporation. In fact, the rate of evaporation from sea water is about 5% lower than that from fresh water. Because of its thermal properties, the depth of the water body will also influence evaporation. In deep water in the midlatitudes, the temperature tends to stay near 4°C (39°F) at depths below 6000 m (3.6 mi). Because of the high specific heat of water, temperature changes above 4°C (39°F) involve large amounts of heat absorbed by the water during warming or released by cooling. The temperature lag of deep-water lakes and the sea is greater than that of air. This is an important factor in determining evaporation rates from free water surfaces. In late fall and early winter, the vapor pressure of water is greater than the vapor pressure of air, since the water temperature is greater than the air temperature. In spring and summer, however, the opposite situation exists more often. During the summer months, the air warms more rapidly than water, so vapor pressure of water is less than that of the air, and less evaporation occurs. Visible evidence of this process is the steam seen rising from lakes, streams, and the ocean on very cool mornings and before the rivers and lakes freeze in the winter. On a cold winter day before the lake freezes, more water evaporates from Lake Michigan than the city of Chicago uses in a year.

We can measure the evaporation rate directly using evaporation pans or other instruments. We can also calculate it using measured variables in an evaporation equation. Figure 5.7 shows the world distribution of evaporation.

Evapotranspiration

Analyzing terrestrial water balances is not as simple as analyzing water bodies. For land areas, the net atmospheric transfer of water vapor depends upon the amount of water available and the processes of evaporation and transpiration. Transpiration refers to the loss of water from living plants. The loss cools the plant and keeps its temperature within the tolerable limits. Because transpiration is difficult to measure in the field, climatologists monitor evapotranspiration, which is the combined loss from the surface through evaporation and transpiration. Various factors affect evapotranspiration, including the following:

1. radiation intensity
2. atmospheric temperature
3. atmospheric dew point
4. length of day (photoperiod)
5. wind velocity
6. type of vegetation
7. soil moisture conditions
8. type of precipitation

These influencing factors make evapotranspiration very difficult to measure. It is also difficult to isolate the effects of one or two variables to determine their effect on total evapotranspiration. Methods for estimating evapotranspiration fall into one of three classes: theoretical methods using physical principles of the process, analytical approaches based upon balance-of-energy amounts, and empirical methods based upon historic data. Of the methods drawing upon physical principles, the most widely known is that devised by Penman. His equation for computing evapotranspiration uses vapor pressure, net radiation, and the drying power of air at a given temperature. To derive such a formula, Penman first devised equations to express both net radiation and drying power. The result is a complex set of equations. Their use is simple, however, and researchers for the United Nations use the method as a basic measure.

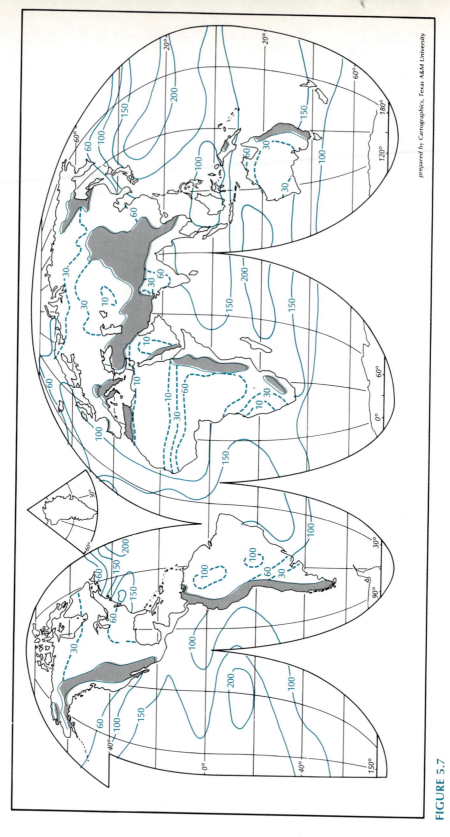

prepared by Cartographics, Texas A&M University

FIGURE 5.7
The world distribution of evaporation.

Thornthwaite's method is also well-known and widely used in the United States. The system depends upon temperature data, for the variables used are mean temperature and daylight hours. Of major significance in the Thornthwaite method is *potential evapotranspiration (PE),* the maximum amount of water that will evaporate from the surface if an unlimited supply of water is available. This is equivalent to evaporation from a water surface. If water is limited, *actual evapotranspiration (AE)* is less than potential.

For the most part, only agricultural research stations directly measure evaporation because of the difficulty of instrumentation. Figure 5.8 shows an evapotranspirometer, which measures evapotranspiration, and a lysimeter, which measures either the actual or potential rates. Using data derived from such instruments, we can test the relative accuracy of estimating equations. The example shown in Table 5.3 shows that the Thornthwaite method, derived for use in a midlatitude rainy climate, underestimates evapotranspiration in a dry climate.

Evapotranspiration from a mixed land-and-water surface is nearly always less than evaporation would be from a similar-sized water surface. On a global scale, evapotranspiration for the continents is some 470 mm (18.5 in.) per year. From the ocean, evaporation is 1300 mm (51.2 in.) per year. Average evapotranspiration from the continents

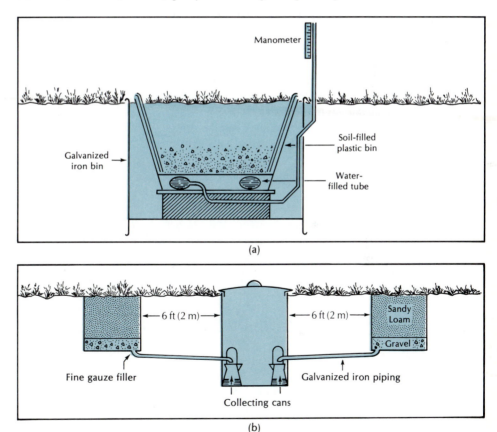

(a)

(b)

FIGURE 5.8
Sections through (*a*) a weighing lysimeter. and (*b*) an evapotranspirometer.

TABLE 5.3
Measured and estimated potential evapotranspiration at an Australian station.

MONTH	T (°C)*	MEASURED EVAPOTRANSPIRATION (MM/DAY)†	ESTIMATED POTENTIAL EVAPOTRANSPIRATION (MM/DAY)‡
Jan.	23.3	7.76	4.47
Feb.	21.1	5.62	3.51
Mar.	19.6	4.21	2.80
Apr.	17.2	2.94	2.04
May	12.6	1.31	1.09
June	10.9	0.99	0.80
July	10.0	0.93	0.74
Aug.	11.0	1.47	0.89
Sept.	13.0	2.30	1.50
Oct.	15.8	4.07	2.09
Nov.	17.8	5.26	2.73
Dec.	10.1	5.99	3.47
Annual total		1296	793

*Temperature.
†Based upon observed values of seven lysimeters.
‡Calculated potential evapotranspiration, using the Thornthwaite method.

varies through time and space, depending mainly on the amounts of water and energy available. In tropical areas where water is ample, evapotranspiration rates are very high. In the lower Amazon Valley and the central Congo River basins, rates of 1200 mm (47.2 in.) a year very nearly approach the evaporation over the open ocean. In parts of the Atlantic and Gulf Plain of the United States, the amount is almost as high.

Evapotranspiration probably reaches a maximum in the Sudd and in the Chad basin in sub-Saharan Africa. Here the rate may reach 2400 mm (94.5 in.) per year, far more than local rainfall. Rivers supply these two extensive areas of swamp and shallow lakes. The White Nile supplies the Sudd, and a series of rivers from near the equator supply Lake Chad. Solar radiation is intense and the air dry—factors that favor evapotranspiration. On the other hand, where temperatures are lower, such as in Northern Europe, evapotranspiration drops to as little as 200 mm (7.9 in.) per year.

Condensation Near the Ground: Dew, Mist, and Fog

Water vapor in the atmosphere eventually condenses to form water droplets, and that condensation leads to the deposit of water at Earth's surface. Most condensation occurs in the formation of clouds, but some does occur close to the surface in the form of dew and fog.

Condensation occurs at the dew point of air at a given temperature and can lead to the creation of dew, mist, or fog. As we have seen, cooling to the condensation level is the main means by which saturation occurs. If water droplets condense in the layers of air immediately above the ground, they form mist or fog. Mist is the suspension of microscopic water droplets that reduces visibility at Earth's surface. It forms a fairly thin, grayish veil that covers the landscape. Fog, in contrast, is the suspension of very small water droplets in the air that reduces visibility to less than 1 km (⅝ mi). Foggy air feels raw and clammy and, given the correct illumination, the fog droplets are often visible to the naked eye. Mist does not provide the same damp, raw feeling and the individual droplets are too small to see.

Advection and Radiation Fogs

While all fog looks the same, its causes vary. The most common types, other than those associated with frontal systems (see Chapter 10), are air mass fogs. These form as either advection or radiation types.

Advection is merely the horizontal transport of air. Advection fogs can form either by the transport of warm air over a cold surface or the transport of cold air over a warm, wet surface. When warm air makes contact with the cold surface, the air cools, and if the dew point is reached, then condensation will occur in the form of a fog (see Figure 5.9). Steam

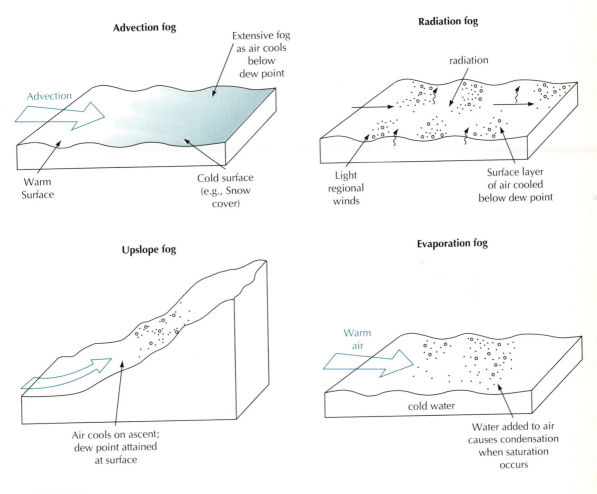

FIGURE 5.9
Schematic diagrams illustrating types of fog formation.

fog (or arctic "sea smoke") forms when cold air blows over a warm sea surface, which results in rapid evaporation from the water. The evaporation saturates the cold air, resulting in steam fogs commonly seen in Arctic regions.

Radiation fog forms as an extension of dew. It occurs under clear skies in relatively still air and forms when nocturnal cooling of the ground chills the layers of air near the surface. If the temperature drops to the dew point, then a layer of fog forms. Compared to advection fog, radiation fog lasts but a short time and "burns off" in the early morning when the sunshine heats the layer of chilled air.

Another type of fog forms when air rises up the side of a mountain. This upslope fog is due to adiabatic expansion, a phenomenon explained below.

Foggy Places

Conditions favorable for fog formation occur in some areas more frequently than in others. Figure 5.10 shows areas of the United States that frequently experience fog and relates the distribution to type of fog.

Advection fogs often are widespread and persistent. The foggiest place in the United States is in the Libby Islands off the coast of Maine. This is part of the extensive fog belt associated with the Newfoundland coast and the offshore Grand Banks. Here, air flowing off the relatively warm waters of the North Atlantic drift crosses the cold water of the

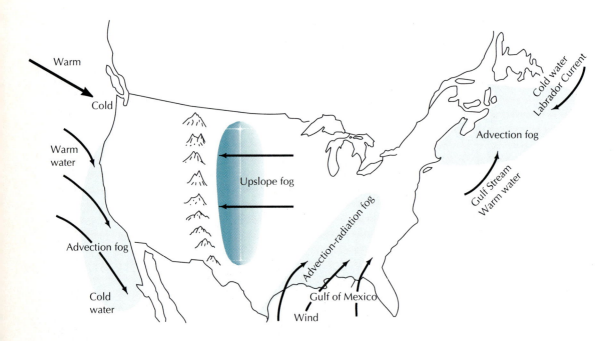

FIGURE 5.10
Winds associated with some of the common fogs.
Source: S.D. Gedzelman, *The Science and Wonders of the Atmosphere.* (New York: Wiley), 1980, 169. Reprinted by permission of John Wiley & Sons, Inc.

Labrador current. The warmer air cools below its dew point, and fog results. A similar situation occurs off the Aleutian Islands in the North Pacific. Warm air from the North Pacific drift comes in contact with the cooler waters of the Bering current, yielding widespread advection fog. Such fogs are most frequent from March to September.

Fogs are common along California's coast. Cold upwelling water occurs close to the coast, where air moving in from the warmer Pacific waters comes in contact with it. Often, the fog starts here as low stratus clouds because wind action causes the air to mix. Air that has come in contact with the cold water cools further. When that cooler air rises, it results in fog forming above the sea surface. When this low layer of moisture reaches the hilly coast, people see it as fog at ground level. When fog actually forms at sea level, it is usually very dense and minimizes visibility.

Like the Newfoundland fogs, the fogs of California also occur during summer. During the dry summer, the moisture that condenses provides some moisture for the coastal red-wood trees and significantly, provides a humid environment that reduces loss of moisture through evaporation. In coastal Chile, a similarly formed fog produces enough moisture to maintain green plants—the garua—in a desert region.

The upslope fog shown in Figure 5.10 is a result of adiabatic expansion. On the eastern side of the Rockies, the land slopes gradually upward from the Great Plains. When an eastern wind blows in the area, warm, moist air slowly ascends the slope, cools, and condenses. Sometimes known as "Cheyenne fog," the right conditions can lead to a thick blanket of fog extending from Amarillo, TX, to Cheyenne, WY.

Condensation Above the Surface: Clouds

Clouds provide the most readily visible weather phenomena. Their beauty is celebrated in poems, songs, and paintings. While atmospheric scientists also see beauty in clouds, they look to the form and appearance of clouds as important keys to understanding and predicting atmospheric conditions.

Many people assume that when air reaches a relative humidity of 100%, condensation occurs. Such, however, is not always the case. It is possible to cool air in a sealed chamber until the relative humidity exceeds 100% and condensation may still not occur. This cooled air is supersaturated, for its vapor pressure exceeds the saturation vapor pressure without condensation occurring.

If the air used in the laboratory experiment just described is not purified and contains dust and other particles that are normally in the air we breathe, then some water droplets form when the relative humidity reaches 100%. Observations show that for condensation of water vapor to take place in the air, condensation nuclei must be present.

Condensation Nuclei

All water droplets begin as microscopic particles, and some of the optical phenomena in the atmosphere are a result of these particles. The spectacularly beautiful rays of the sun seen in Figure 5.11 occur when the sun is low in the sky and partially hidden by clouds. These rays, known as crepuscular rays, appear to fan out as a series of giant searchlight beams. The intensity of the rays is a function of how much light is scattered, which in

FIGURE 5.11
Crepuscular rays.
Source: © 1990 A & J Verkaik/Skyart.

turn indicates the concentration of fine particles suspended in the air. Where air is free from debris, such as in polar areas, crepuscular rays do not occur. We can demonstrate this using a bright flashlight. In most places it is possible to see the beam of a flashlight quite distinctly at night. In clean air the rays are invisible.

Condensation and the role of condensation nuclei are different at temperatures above freezing versus below freezing. We are concerned here with the condensation of air at temperatures above freezing. The particles on which water condenses to form cloud droplets vary in size (Figure 5.12). The smallest particles, called *Aitkin* particles, have diameters of less than 0.4 μm. They originate primarily in the natural decay of vegetation and combustion of fossil fuels. They are so small that they play but a small role in providing nuclei for condensation. On the other hand, the largest particles, called *giant* particles, have diameters greater than 2 μm. These include salt particles from breaking waves and chemical compounds derived from rocks weathering at Earth's surface. Some nuclei are hygroscopic—they have an affinity for water. Tiny salt particles in the air are one such example. You may have noticed that ordinary table salt becomes soggy on warm moist days when it absorbs moisture from the air. (Hygrophobic particles do not have this property of attracting water. They also act as nuclei, but under different moisture conditions.) We have said several times that at 100% relative humidity, air becomes saturated. In fact, with hygroscopic nuclei present, condensation can occur below 100% relative humidity. Without these nuclei, it may not occur until relative humidity exceeds 100%. This happens because we define saturation based on conditions that occur over a flat surface of pure water. The droplets that occur in clouds are neither flat nor pure. As a result, two additional effects—the *solute effect* and the *curvature effect*—partly determine when condensation begins.

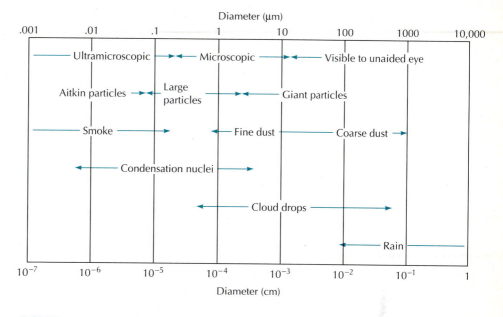

FIGURE 5.12
Particle sizes of selected atmospheric constituents.

The solute effect results when water condenses on a soluble nucleus, forming a solution. For example, a salt nucleus dissolves to form a water-droplet solution that has properties different from pure water. The vapor pressure of any solution is less than that of pure water, so the evaporation rate from the solution is less than that from pure water. In addition, the smaller the droplet, the more concentrated the solution. Because of this, small droplets form at a relative humidity level at which pure water droplets will not form.

The curvature effect rests upon water being a polar molecule with weak charges between the molecules. The intermolecular forces are strongest on a flat water surface and decrease as the water surface becomes more curved. In very small droplets, molecular bonding is lowest and water thus evaporates rapidly. The curvature effect is directly opposite the solute effect in that the former favors growth of small droplets and the latter their evaporation.

Since the solute and curvature effect are directly opposite, their combined influence in clouds is important. The solute effect decreases as the water droplet enlarges. The solution weakens by the addition of water to the droplet. The curvature effect reduces evaporation from the droplet as it grows. Acting together, the two processes result in a cloud droplet whose size is in equilibrium at the existing moisture levels in the cloud. Unless other factors come into play, the droplets will remain suspended in the cloud rather than fall as precipitation from the cloud.

Cloud Forms and Classification

Before 1800, clouds had no formal names and we had little knowledge of cloud mechanics. A young Englishman named Luke Howard (1772–1864) stepped in to provide a new perspective on clouds. In 1803, he presented a classification of clouds into main and secondary types and gave them Latin names. He distinguished three principal cloud forms:

Stratus (from Latin *stratum* = layer) cloud—lying in a level sheet
Cumulus (from Latin *cumulus* = pile) cloud—having flat bases and rounded tops, and being lumpy in appearance
Cirrus (from Latin *cirrus* = hair) cloud—having a fibrous or feathery appearance

The classification was so well-designed that it remains the basis of the system in use.

As knowledge of clouds increased, however, a more comprehensive system was required. To establish worldwide uniformity, in 1895 an International Meteorological Committee published a system for naming and identifying clouds. This system was revised several times. The international standard is now under the auspices of the World Meteorological Organization (WMO), which publishes The International Cloud Atlas. In the international classification system, the main types are genera, which contain species and, in turn, several varieties. Despite the infinite variety of clouds, they fall into one of 10 basic types or genera. (See Table 5.4 and Figure 5.13.)

(a)   (b)   (c)   (d)

FIGURE 5.13
Examples of (a) cirrus, (b) cirrocumulus, (c) cumulus, and (d) cumulonimbus clouds.
Sources: (a) © George Porter, National Audubon Society; (b) © 1989 Roger Appleton, Photo Researchers, Inc., New York; (c) Van Bucher, Photo Researchers, Inc., New York; (d) Robert H. Wright, National Audubon Society, Photo Researchers, Inc., New York.

TABLE 5.4
Cloud genera.

Cirrus (Ci). Detached clouds in the form of white, delicate filaments, or white or mostly white patches or narrow bands. These clouds have a fibrous (hair-like) appearance or a silky sheen or both.

Cirrocumulus (Cc). Thin, white patch, sheet or layer of cloud without shading, composed of very small elements in the form of grains, ripples, etc., merged or separate, and more or less regularly arranged; most of the elements have an apparent width of less than 1° (approximately the width of the little finger at arm's length).

Cirrostratus (Cs). Transparent, whitish cloud veil of fibrous or smooth appearance, totally or partially covering the sky, and generally producing halo phenomena.

Altocumulus (Ac). White or gray, or both white and gray, patch, sheet or layer of cloud, generally with shading, composed of laminae, rounded masses, rolls, etc., sometimes partly fibrous or diffuse, and may or may not be merged; most of the regularly arranged small elements usually have an apparent width of between 1° and 5° (approximately the width of three fingers at arm's length).

Altostratus (As). Grayish or bluish cloud sheet or layer of striated, fibrous or uniform appearance, totally or partly covering the sky, and having parts thin enough to reveal the sun at least vaguely, as through ground glass. Altostratus does not show halo phenomena.

Nimbostratus (Ns). Gray cloud layer, often dark, the appearance of which is rendered diffuse by more or less continually falling rain or snow which in most cases reaches the ground. It is thick enough throughout to blot out the sun. Low, ragged clouds frequently occur below the layer with which they may or may not merge.

Stratocumulus (Sc). Gray or whitish, or both gray and whitish, patch, sheet or layer of cloud that almost always has dark parts, composed of tessellations, rounded masses, rolls, etc., that are nonfibrous (except for virga) and may or may not be merged; most of the regularly arranged small elements have an apparent width of more than 5°.

Stratus (St). Generally gray cloud layer with a fairly uniform base, which may give drizzle, ice prisms or snow grains. When the sun is visible through the cloud its outline is clearly discernible. Stratus does not produce halo phenomena (except possibly at very low temperatures). Sometimes stratus appears in the form of ragged patches.

Cumulus (Cu). Detached clouds, generally dense and with sharp outlines, developing vertically in the form of rising mounds, domes or towers, of which the bulging upper part often resembles a cauliflower. The sunlit parts of these clouds are mostly brilliant white; their bases are relatively dark and nearly horizontal. Sometimes cumulus is ragged.

Cumulonimbus (Cb). Heavy and dense cloud, with a considerable vertical extent, in the form of a mountain or huge towers. At least part of its upper portion is usually smooth, or fibrous or striated, and nearly always flattened; this part often spreads out in the shape of an anvil or vast plume. Under the base of this cloud, which is often very dark, there are frequently low ragged clouds either merged with it or not, and precipitation, sometimes in the form of virga.

TABLE 5.5

Approximate height range of cloud bases.

LEVEL	RANGES IN POLAR REGIONS (KM)	RANGES IN TEMPERATE REGIONS (KM)	RANGES IN TROPICAL REGIONS (KM)
High	3–8	5–13	5–18
Middle	2–4	2–7	2–8
Low		from surface to 2 km	

A second aspect of cloud classification is their altitude. Clouds of similar shapes occur at different levels in the troposphere: high, middle, and low (see Table 5.5). The approximate height for these levels varies with latitude. As already noted, the structure and thickness of the atmosphere varies from equator to pole. To some extent, the 10 genera of clouds depend on height. Those described as cirrus (or with the prefix *cirro-*) are high clouds. Those with the prefix *alto-* are middle clouds. Names for low clouds lack prefixes. An exception to this is the nimbostratus cloud, which is classified as both a middle and a low cloud. The word *nimbus* (or prefix *nimbo-*) applies to a cloud from which rain is falling. It derives from the Latin word for "violent rain."

Figure 5.14 is a generalized diagram illustrating the classification of clouds according to genera and height. Note that cumulonimbus, the thunder cloud, extends from low, through middle, to high altitudes.

Differences in the structure and shape of clouds permit identification of cloud species. We add a species name to cloud genera to add further information about the cloud. A cloud is classified within only one genus. If it is a cumulus cloud, it cannot be a stratus or cirrus. However, the species name can apply to any genera. For example, clouds of the species *castellanus* appear to have turrets like a castle. Such a shape applies to many clouds, including cirrus and stratocumulus. If the clouds have no distinct structure or shape, the species name is unnecessary.

Table 5.6 lists cloud species and their descriptions. The clouds listed here are secondary in importance to the 10 genera, and few persons commit them all to memory. The important factor is the potential for generating precipitation from the clouds.

Summary

Water molecules can exist in gaseous, liquid, and solid states at standard atmospheric temperatures. Changes from one state to another involve energy transfer through absorption and release of energy. The hydrologic cycle conceptualizes the movement of water over the globe.

The ability of air to contain water vapor is a function of temperature. We can designate the quantity of water in the atmosphere in several ways, including vapor pressure, mixing ratio, and relative humidity.

Water changes state from liquid to gas through the process of evaporation and transpiration and the combined process of evapotranspiration. Rates of transfer depend upon vapor pressure of the water surface and the air as well as wind speed, together with other

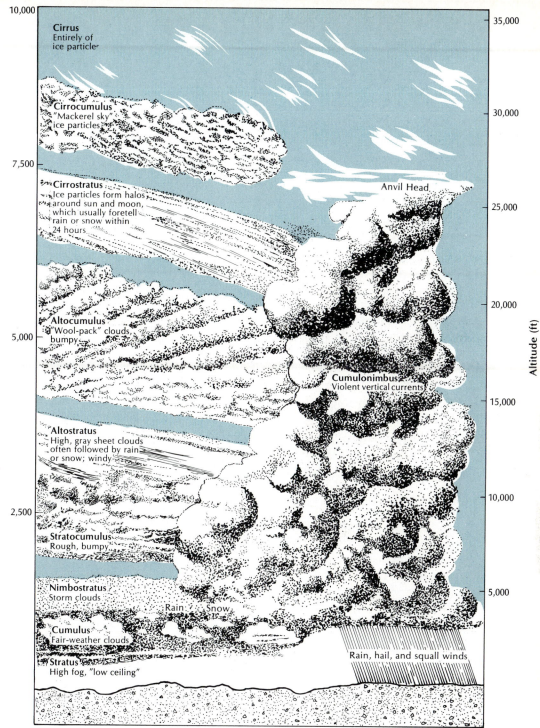

Altitude (mi)

10,000

Cirrus
Entirely of
ice particles

Cirrocumulus
"Mackerel sky"
ice particles

7,500

Cirrostratus
Ice particles form halos
around sun and moon,
which usually foretell
rain or snow within
24 hours

Anvil Head

Altocumulus
"Wool-pack" clouds
bumpy

5,000

Cumulonimbus
Violent vertical currents

Altostratus
High, gray sheet clouds
often followed by rain
or snow; windy

2,500

Stratocumulus
Rough, bumpy

Nimbostratus
Storm clouds

Rain   Snow

Cumulus
Fair-weather clouds

Rain, hail, and squall winds

Stratus
High fog, "low ceiling"

Altitude (ft)

35,000

30,000

25,000

20,000

15,000

10,000

5,000

FIGURE 5.14
Cloud forms.

103

TABLE 5.6
Cloud species.

Arcus: shaped like an arc; refers to the lower (dark and threatening) part of a cumulonimbus cloud, particularly of the line-squall type.

Calvus: bald; refers to the absence of sprouting (cauliflower) structure as well as cirriform appendages in the upper part of a cumulonimbus cloud (see also *congestus* and *incus*).

Capillatus: hairy; fibrous or striated structure of upper part of cumulonimbus.

Castellanus: turreted; heap-shaped towers, resembling miniature cumulus protruding from clouds in the middle and upper troposphere, especially altocumulus.

Congestus: congested, heaped; sprouting, towering structures in the upper portion of a developing cumulus.

Fractus: fractured, torn; ragged fragments of clouds, notably stratus, cumulus, and nimbostratus.

Humilis: humble, small, flat; used to characterize nondeveloping cumulus clouds.

Incus: anvil-shaped; cirriform mass of cloud in the upper part of a developed cumulonimbus.

Intortus: twisted, entangled; used to describe a type of cirrus.

Mamma: shaped like udders; protuberances hanging down from the undersurface of a cloud; most pronounced in connection with thundery cumulonimbus.

Pileus: cap- or hood-shaped; accessory cloud of small horizontal extent above, or attached to, the upper part of a cumulus cloud, particularly during its developing phase.

Spissatus: spiss, compact; describes cirrus that is sufficiently dense to appear grayish when viewed in a direction toward the sun.

Tuba: tube-shaped; typical of clouds associated with tornadoes, water spouts, etc.

Uncinus: hooked; typical of streaky cirrus drifting in strong winds in the upper troposphere.

Velum: flap; an accessory cloud of considerable horizontal extent, sometimes connecting the upper parts of several cumulus clouds.

Virga: twig; trails or streaks of precipitation, hanging from the undersurface of a cloud but not reaching the ground.

environmental variables. Estimates of the amount transferred are determined by direct measurement or by mathematical models using surrogate data.

Condensation occurs near the ground as dew, mist, or fog and at higher levels as clouds. Fog forms as a result of advection, radiation, and adiabatic processes. A high incidence of fog occurs in specific locations. Condensation nuclei play an important role in the condensation process. In clouds, both the solute and curvature effects are significant.

Clouds fall into 10 genera, with form (cirriform, stratiform, cumuliform) and elevation providing the basic classification variables. Variations within genera are classified as species.

6

PRECIPITATION

Chapter Overview

The Ascent of Air
- Vertical Motion in the Atmosphere
- Adiabatic Warming and Cooling
- Lapse Rates

Water in Clouds

Cold-Cloud Precipitation

Warm-Cloud Precipitation

Spatial Distribution of Precipitation

Temporal Distribution of Precipitation

Floods and Droughts
- Floods
- Droughts

The word *precipitation* is used to describe any of the various forms of water particles that fall from the atmosphere to reach to ground. It is a useful word, for it describes water forms ranging from snowflakes and drizzle to hail and raindrops (Table 6.1). As discussed later in this chapter, the form that precipitation takes determines the nature of its impact upon people and the environment. The precipitation process—the mechanisms in clouds that produce precipitation—is the end product of a whole set of events. The ascent of air, its cooling to form clouds, and the characteristics of the various clouds that form all act as building blocks in the process by which precipitation is produced. Our concern is now how and why air ascends, the behavior of water droplets (or ice crystals) in clouds, and the theories that explain the cause of precipitation. Once we understand these phenomena, we can better comprehend the distribution of precipitation over the globe.

TABLE 6.1
Types of precipitation.

Rain is precipitation of liquid water particles with diameters over 0.5 mm, although smaller drops are still called rain if they are widely scattered.

Drizzle is a fairly uniform precipitation composed exclusively of fine drops of water with diameters less than 0.5 mm. Only when droplets of this size are widely spaced are they called rain.

Freezing rain or *Freezing drizzle* is rain/drizzle that freezes on impact with the ground, with objects at Earth's surface, or with aircraft in flight.

Snow is precipitation of ice crystals most of which are branched. At temperatures higher than about −5°C (23°F) the crystals are generally agglomerated into snowflakes.

Snow pellets are composed of white and opaque grains of ice. The grains are mostly spherical and have a diameter of 2–5 mm. The grains are brittle and when falling on a hard surface they bounce and break up. Snow pellets are also known as soft hail and graupel.

Snow grains are very small (less than 1 mm in diameter) grains of white, opaque ice. Snow grains are also called graupel.

Ice pellets are comprised of transparent or translucent pieces of ice that are spherical or irregular and have a diameter of 5 mm or less. They are composed of frozen raindrops or largely melted and refrozen snowflakes.

Hail is precipitation of small balls or pieces of ice (hailstones), with diameters ranging from 5 to 50 mm, falling either separately or agglomerated into irregular lumps. Hailstones are composed of a series of alternating layers of transparent and translucent ice.

Ice prisms (diamond dust) are ice crystals often so tiny that they seem suspended in air. Such crystals may fall from a cloudy or cloudless sky. Mostly visible when they glitter in sunshine (hence diamond dust), they occur at very low temperatures.

Fog, ice fog and *mist* are also considered forms of precipitation.

The Ascent of Air

Water vapor condenses to form a mist and fog largely as the result of advection and radiational cooling. Clouds formed by water-vapor condensation rely upon another process that depends upon vertical motion in the atmosphere. The significance of this process cannot be overstated, for the presence or absence of clouds and the occurrence of precipitation ultimately depend upon the upward motion of air.

Vertical Motion in the Atmosphere

To help explain many of the processes that occur within the atmosphere, meteorologists often use the concept of a parcel of air. A parcel is considered a volume of air small enough to have uniform properties (such as temperature and water-vapor content), yet large enough to respond to the meteorological processes that influence it. While it can be any size, visualize it as being about 1 m³ (35.3 ft³) in volume.

The gas laws describe relationships between pressure, density, and temperature. One of the important results of the laws is that warm air is less dense than cold air. This difference in density is demonstrated every time a hot-air balloon rises (see Figure 6.1). Successful ascent of a balloon depends upon inflating it with hot air. As long as the temperature inside the balloon exceeds that of the surrounding

FIGURE 6.1
An ascending hot-air balloon.
Source: Snowmass Resort Association.

air, the balloon will continue to rise. In many ways, a hot-air balloon acts like a parcel of air.

A parcel of warm air surrounded by cooler air tends to rise and the parcel's pressure drops. Following the gas laws, the parcel also expands and cools. The air's rising, expanding, and cooling help illustrate the effect of vertical motion in the atmosphere.

Various mechanisms cause air to move upward. Moving air rises when it encounters a physical barrier such as a mountain, for example. The moving air cannot pass through the barrier and is forced to rise over it, as Figure 6.2a shows. Such lifting is termed oro-

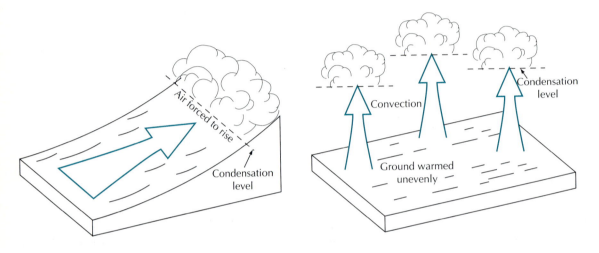

(a)   (b)

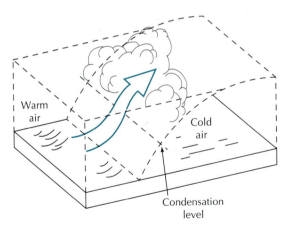

(c)

FIGURE 6.2
Processes responsible for the uplift of air: (a) The orographic effect. (b) Convection. (c) Uplift along fronts.

graphic uplift (the term is derived from the word *orography*, which refers to the uneven shape of Earth's surface). Orographic lifting is a mechanical process. Lifting also results from the properties of air parcels interacting. Such dynamic lifting occurs as a result of convection and frontal activity.

Convective lifting occurs in warm, moist air and begins with heating from the ground surface. When the surface is very warm, it heats the air in contact with the ground. It expands in response to the heating and, having a lower density than the surrounding air, it begins to rise. (See Figure 6.2*b*.)

Cyclonic lifting occurs when air masses with different temperature and moisture characteristics come together. The boundary between the air masses is termed a front, and uplift of air at these fronts is of major meteorological importance. The basic process involved at a cold front, illustrated in Figure 6.2*c*, is considered in greater detail in Chapter 10.

Adiabatic Warming and Cooling

Once mechanical or dynamic lifting causes air to rise, it undergoes physical changes. Recall that atmospheric pressure decreases with altitude, subjecting a rising parcel of air to decreasing pressure around it and causing the parcel of air to expand. As the parcel expands, the distance between the individual gas molecules within the parcel increases. Thus, when the air expands, its molecules must do work (i.e., "push" other molecules out of the way). That work requires energy; the gas molecules themselves supply the energy. Use of this energy reduces the molecules' kinetic energy, and the temperature of the parcel decreases. Note that the decrease of temperature is an internal response to expansion of the air parcel; no heat is lost to or gained from the environment surrounding the parcel. Because of this, the process is called **adiabatic,** defined as the internal changes within a gas during expansion and contraction when no energy is removed from or added to the gas. Note that the adiabatic process applies to both expansion and contraction and hence it applies to both rising and sinking air parcels.

The first law of thermodynamics states that the temperature of a gas may be changed by the addition (or subtraction) of heat, a change in pressure, or a combination of both. The adiabatic processes disregard the addition and subtraction of heat, instead explaining temperature changes as

$$\text{Change in temperature} = \text{constant} \times \text{change in pressure}$$

This adiabatic form of the first law of thermodynamics is of extreme importance in understanding many atmospheric processes, especially those concerning lapse rates.

Lapse Rates

In the standard atmosphere, temperature decreases with height in the troposphere at an average rate of 6.5°C per km (3.5°F/1000 ft). This standard lapse rate reflects the difference between the average surface temperature (15°C) and the average temperature at the tropopause (−59°C at 11 km). At any given time, the lapse rate may vary substantially from this mean value. The standard lapse rate is a measured value. On days when a dense, cold air mass occurs, the lapse rate measured will differ from that on another day when different conditions exist. This lapse rate may be determined, for example, when temperatures are taken at various elevations on a mountain. Since it measures the temperature of the environment, it is called the *environmental lapse rate*.

When a parcel of air is forced to rise over a mountain, it cools at a rate independent of the environment. The cooling of the parcel will occur as a result of the adiabatic cooling process. Because the rate of cooling is based upon a physical change within the parcel (a response to decreasing pressure), the rising parcel will cool at the adiabatic lapse rate of 10°C for every 1000 m of ascent (5.5°F/1000 ft). Unless condensation occurs within the rising air, this rate of change is a constant value and is called the *dry adiabatic lapse rate*.

Suppose that an air parcel rises and cools sufficiently to attain its dew point. At the saturation temperature, water vapor will condense to form water droplets. This is a change of state and, as previously noted, energy is released as part of the process. This available energy modifies the rate at which the rising air cools. That rate is known as the *moist adiabatic lapse rate*. Unlike the dry adiabatic lapse rate, the moist rate varies, but a rate of 6°C per 1000 m (3°F/1000 ft) is a good estimate of the moist adiabatic lapse rate.

An example of the determination of the upward displacement of air can be demonstrated using the flow of air up the side of a mountain (see Figure 6.3). Condensation requires that the air be lifted high enough for saturation to occur—it must attain its *lifted*

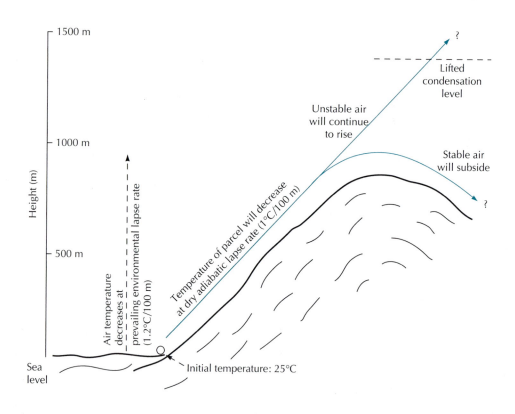

FIGURE 6.3
Air cools adiabatically as it rises. Continued ascent depends on the relative stability of the air.

FIGURE 6.4

A graphic representation of the adiabatic conditions illustrated in Figure 6.3.

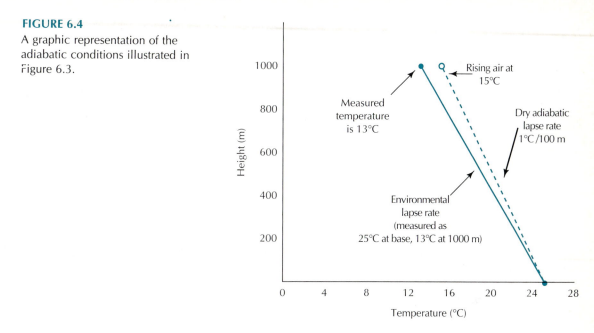

condensation level (LCL). This means that as the air is forced to the top of the mountain, it must continue to rise if condensation is to occur; the alternative is for the air to flow down the other side of the mountain, be compressed and warmed. Whether the air rises or flows down depends on how stable the air at the top of the mountain is.

Figure 6.4 provides a graphic solution to the problem. The decrease of temperature from the bottom to the top of the mountain (here considered to range from sea level to 1000 m) is shown by the environmental lapse rate (which in this case is 1.2°C per 100 m). At the top of the mountain, the environmental temperature is thus the initial temperature, 25°C, less 12°C (1.2 × 10) = 13°C. A parcel of air rising up the side of the mountain will cool at the dry adiabatic lapse rate of 1° C per 100 m. At the top of the mountain its temperature will be 25°C minus 10° C (1.0 × 10), or 15°C. The rising parcel of air is thus warmer than the air into which it is rising and, because warm air will rise above cold, the parcel will continue to rise. This rising parcel is in unstable equilibrium and could rise to the condensation level, where clouds may form.

This example shows that the key to assessing stability is a comparison of the environmental and adiabatic lapse rates. For example, if the environmental lapse rate had been 0.9°C per 100 m, then the environmental temperature at the top of the mountain would be

initial temperature − measured cooling rate × height of mountain

= 25°C − 0.9°C/100m (for 1000m)

= 25°C − 9°C

= 16°C

FIGURE 6.5
The potential temperature of an air parcel is given by its temperature when brought to the surface at the dry adiabatic rate.

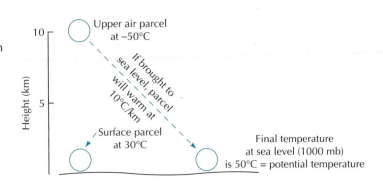

The rising parcel would cool at the dry adiabatic lapse rate as before and would have a temperature of 15°C. The rising air would then be cooler than the air into which it was rising and thus would have a tendency to subside down the other side of the mountain.

The preceding example illustrates that we can determine the type of equilibrium that occurs by comparing the lapse rates with the existing environmental lapse rate. The relationship between these two lapse rates provides a guide to the stability of air: Stable air tends to return to its initial position once the mechanism causing uplift is removed; unstable air continues to rise.

Air's adiabatic lapse rate helps explain an apparent paradox about the atmosphere. We know that warm air rises while, at the same time, air is colder aloft than at the surface. Why, then, does the cold air not sink to replace the warm? The answer lies in an understanding of *potential temperature*. Figure 6.5 compares a parcel of air at the surface with a temperature 30°C and a parcel at 10 km with a temperature of −50°C. Although the surface air has a higher temperature than the air aloft, its warmth is a result of its internal energy and the pressure level at which it occurs. To compare the two parcels on an equal basis, we can bring the parcel at 10 km down to the surface. Since the adiabatic lapse rate applies to both ascending and descending air, lowering the parcel warms it adiabatically at a rate of 10°C per km, so its temperature at the surface would be 50°C. The parcel aloft is thus potentially warmer than the surface air. Meteorologists call this concept the potential temperature—the temperature that an air parcel would have if it were moved adiabatically to a pressure of 1000 mb, the approximate surface level.

Water in Clouds

The water droplets or ice crystals that make up a cloud are heavier than the air surrounding them, but as we all know, they do not always fall from the clouds. This phenomenon rests upon the fact that the drops in fog and clouds are so small that they fall very slowly through the air. Usually, upward motion occurs within a cloud because, as we have found, most clouds form through the forced ascent of air. This upward motion more than compensates for the fall of the small cloud drops that remain suspended in the cloud. For the water to fall from the cloud, a droplet must attain enough velocity to overcome the resistance of the air and the upward motion.

Consider the forces acting upon a cloud droplet. First, it is subjected to gravity's downward force. Countering gravity are the air's buoyancy and the falling droplet's frictional retarding force. As might be expected, friction's effect is directly proportional to the rate at which the droplet is falling (i.e., the faster the droplet falls, the greater the frictional drag). It follows that, at some point, the forces of resistance (buoyancy and friction) will equal the force of gravity; the droplet will then fall at a constant speed. The speed at which a droplet must fall to attain this point is called the **terminal fall velocity**.

The terminal velocities of water drops depend upon their size, expressed as the radius of the droplet. The terminal fall velocity for small drops is very low. Keeping the droplets suspended at the same level does not require that much updraft in the cloud. If a cloud droplet with a diameter of 0.04 mm (.0016 in.) occurred in air with no vertical motion, it would fall at a rate of 5.4 cm (.213 in.) per sec; to fall from a cloud base of 500 m (1640.5 ft) to the ground would take 30 hr! A droplet this size moving at such a low velocity evaporates long before it reaches the ground. In comparison, the terminal velocity of larger drops is much greater. A raindrop 4 mm (.157 in.) in diameter falls the same 500 m from cloud to ground in less than 10 min.

It becomes clear that for precipitation to fall from a cloud requires a process by which the small droplets, with their low terminal velocities, become larger. Consider, however, what must happen for a cloud droplet (radius 10 μm) to become a small droplet of the drizzle variety (radius 150 μm). The cloud droplet must increase its volume some 3000 times! To become a raindrop 1000 μm in diameter, it must increase its volume almost a million times (see Figure 6.6).

By introducing a salt nucleus (0.1 μm radius) into slightly supersaturated air at 0°C, a water droplet of 1 μm radius would form in 1 sec. Thereafter, it would take 10 min to form a droplet of 8 μm and 1 hr to produce a droplet of radius 20 μm. Further growth is even slower and the large droplets that make up rainfall could never grow in this way. So,

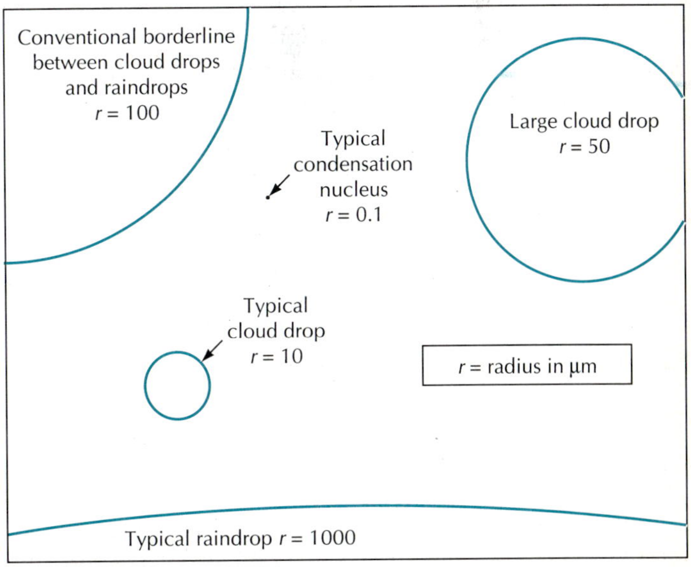

FIGURE 6.6

Comparative diameter of nuclei, cloud droplets, and rain drops (in μm).

Conventional borderline between cloud drops and raindrops
r = 100

Typical condensation nucleus
r = 0.1

Large cloud drop
r = 50

Typical cloud drop
r = 10

r = radius in μm

Typical raindrop r = 1000

while a hygroscopic nucleus aids in the formation of cloud droplets, continued growth to produce drops large enough to fall from the cloud requires other explanations. Two major theories explain what happens: One applies to cold clouds (with temperatures below freezing), and the other to warm clouds (temperature above freezing).

Cold-Cloud Precipitation

Figure 6.7 shows a diagram of a cloud in which the freezing level occurs above the base of the cloud, but well below its top. At lower altitudes where the temperature is above freezing, water droplets form. At altitudes where the temperature is below freezing, we would expect condensation to occur as ice. This is not necessarily the case, however. A particle of ice comes into being either by the freezing of a tiny water droplet or by formation directly from water vapor through sublimation. But while water never melts until its temperature is 0°C, pure water does not always freeze at 0°C. Water that continues to exist as water below the freezing point is called *supercooled water*. Clouds with small quantities of water contain a large proportion of supercooled water. Such water can exist at temperatures down to −40°C, when spontaneous freezing occurs.

In the example shown in Figure 6.7, at altitudes above freezing level the cloud will consist of both ice crystals and supercooled water. The number of supercooled droplets will exceed the ice crystals, since condensation nuclei are much more common than ice

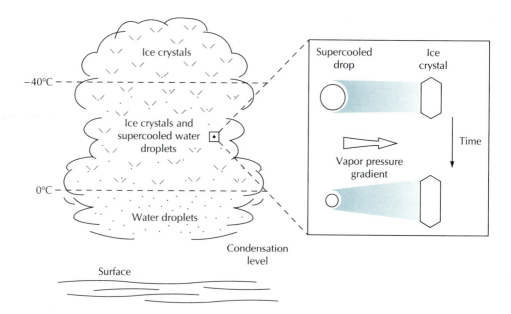

FIGURE 6.7

In a cold cloud, supercooled water droplets and ice crystals occur in the same area. The ice crystals grow at the expense of the supercooled water droplets.

nuclei. At higher levels and at lower temperatures, ice is found. Precipitation originates in the zone in which the supercooled water and ice occur together.

As early as 1911, the famed scientist Alfred Wegener (who is best known for his theory of drifting continents) noted that the coexistence of supercooled water and ice crystals could play a significant role in precipitation formation. Swedish meteorologist Tor Bergeron later incorporated Wegener's ideas into a complete theory of precipitation. That theory was confirmed by the German scientist W. Findeisen in the late 1930s. A comprehensive precipitation theory therefore should be correctly labeled the Wegener-Bergeron-Findeisen process for precipitation from cold clouds. We will refer to it as the Bergeron process.

The Bergeron process rests upon the fact that the saturation vapor pressure over ice is less than that over water; thus, air that is saturated with respect to water will be supersaturated with respect to ice. This means that more water vapor will condense onto the ice and the crystal will grow. At the same time, the water droplet will evaporate and, over time, the ice crystal will grow at the expense of the water droplet. The ice crystal will become large enough to fall from the cloud. On passing through warmer air, the ice will melt and rain will fall at the surface.

The Bergeron process implies that for rain to occur, clouds must extend to a height above freezing. Such an interpretation was assumed to be true until the late 1930s, when new data became available. It was found that large quantities of rain did fall from warm clouds, those clouds whose tops did not attain the freezing level. An alternative theory was required to explain this phenomenon.

Warm-Cloud Precipitation

Water droplets within a cloud vary in size. Large or giant water droplets may well occur around hygroscopic condensation nuclei. Because the terminal velocity of droplets is a function of their radius, within any given cloud droplets fall at different rates (with the larger falling more rapidly). We can thus visualize a large drop passing through the cloud and overtaking smaller, slowly moving droplets. When this occurs, the *collision-coalescence process* takes place.

Figure 6.8 shows an idealized view of the process. Subsequent to the formation of droplets on hygroscopic nuclei, larger drops will fall, collide, and coalesce with smaller droplets. The figure illustrates how some droplets will follow a streamline around the larger drop, while others collide with it. Clearly, droplets beyond the distance x will not make contact with the larger drops. The radii (r) of the drops will determine the amount of collision that occurs, and we can identify a *collision efficiency* factor. The greatest efficiency occurs when the ratio of the radius of the small droplet to that of the large drop is between 0.6 and 0.7.

When the drop and the droplet do collide, they will either coalesce or bounce apart. Generally, if the two drops are a similar size, they will coalesce temporarily, oscillate, then break up into smaller fragments. As the difference in size increases, the *coalescence efficiency* changes. The efficiency is a function of the size of the droplet, its velocity, and the angle at which it strikes the larger falling drop. High collision and coalescence efficiencies will produce a large number of drops, and the warm cloud produces plentiful rainfall.

FIGURE 6.8
The collision-coalescence process requires that small droplets collide and coalesce with larger, falling drops. Streamline flow around a large droplet will occur if the droplet is more than distance *x* from the falling drop. Radii (*r*) of the drops determine the rate at which collision occurs.
Source: After J. A. Day, *The Science of Weather* (New York: Addison-Wesley, 1966), 143.

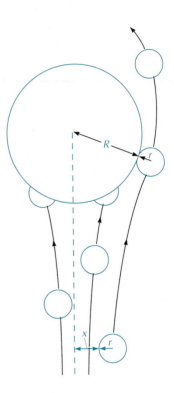

   Experiments have shown that atmospheric electricity aids in coalescence. If a droplet and a drop have different electrical charges, electrical attraction enhances coalescence. The time a droplet has to grow also enhances the coalescence process. A cloud that lasts less than 20 min, for example, will be incapable of producing rainfall in this way. Both thick cloud layers and long-lasting clouds will enhance raindrop formation.
   The most favorable conditions for rain from a warm cloud include:

1. a wide variety of drop sizes in the cloud initially
2. air with a high liquid content
3. deep clouds (e.g., several thousand meters)
4. persistent clouds
5. some updrafts should occur in the clouds

Such conditions occur frequently in moist tropical areas, resulting in clouds that provide abundant rainfall.

Spatial Distribution of Precipitation

   The amount of precipitation received at any place on Earth's surface depends on a variety of factors. The basic element is the amount of water vapor in the air, which varies geographically and seasonally. The mean precipitable water content of the atmosphere at a given moment is 25 mm (.984 in.), with a maximum near the equator of 44 mm (1.73 in.) and a minimum in the polar regions of 2–8 mm (.079–.315 in.), depending on the season. In latitudes of 40°–50°, precipitable water ranges to above 20 mm (.789 in.) in the summer and drops to around 10 mm (.394 in.) in the winter. The presence of water vapor is a necessary but not a sufficient condition for precipitation. The amount of atmospheric water vapor over an area has no bearing on the resulting precipitation. To illustrate this, we can compare conditions over El Paso, TX, and St. Paul, MN. The average moisture content above these cities is about the same, yet the mean annual precipitation is more than three times greater at St. Paul. Other factors must induce precipitation.

If the total amount of precipitation over the surface of Earth were evenly distributed, it would average 880 mm (31 in.) per year. In reality, it varies from near zero to almost 12 m (393 ft) (see Figure 6.9). The total amount of precipitation received depends upon several factors:

1. whether air converges (to give uplift) or diverges (spreads out) in the area
2. air-mass origin, an indication of the temperature and moisture conditions of the air
3. topographic conditions
4. distance from the moisture source—the greater the distance from the source of the moisture, the less water vapor will be present in the air because of prior precipitation loss

Combining these factors, the areas of greatest precipitation are mountain areas of the tropics, where air from the ocean frequently converges. Where such conditions exist, rainfall may reach 12 m (393 ft) per year (Tables 6.2 and 6.3).

Two general locations where precipitation totals tend to exceed the average for Earth lie relatively far apart. One is near the equator, which experiences convergence of moist tropical air most of the year. The equator lies beneath the low-pressure convergence area caused by the intertropical convergence (ITC). The trade winds moving toward the equator pick up moisture over the oceans and, when lifted in the ITC, yield abundant moisture. Precipitation is increased over coastal areas by orographic lifting and increased convection begun by surface heating. Average precipitation ranges from 1.5 m to 2.0 m annually but in some cases rises much higher.

The second situation that gives rise to above-average precipitation is found on the west side of the continents in the midlatitudes. Precipitation there is due to convergence of maritime air and orographic intensification. The zone is most well-defined in latitudes of 50°–60°. The totals run above 1.5 m along the coasts. The amounts are not as high as in the tropics, because the moisture capacity of the air is much lower than that of the maritime tropical air of the tropics.

The arid regions of the world occur in three sections:

1. Extending in a discontinuous belt between about 20° and 30° north and south of the equator. These areas constitute the great tropical deserts, which owe their aridity to large-scale atmospheric subsidence.
2. The continental deserts in the interiors of continents are arid as a result of their distance from the sea, the major source of the water vapor.
3. The polar deserts of the Arctic and Antarctic constitute the third great area of low precipitation. The low temperatures of these regions, along with subsidence of air from aloft, are probably the most important factors contributing to the low totals.

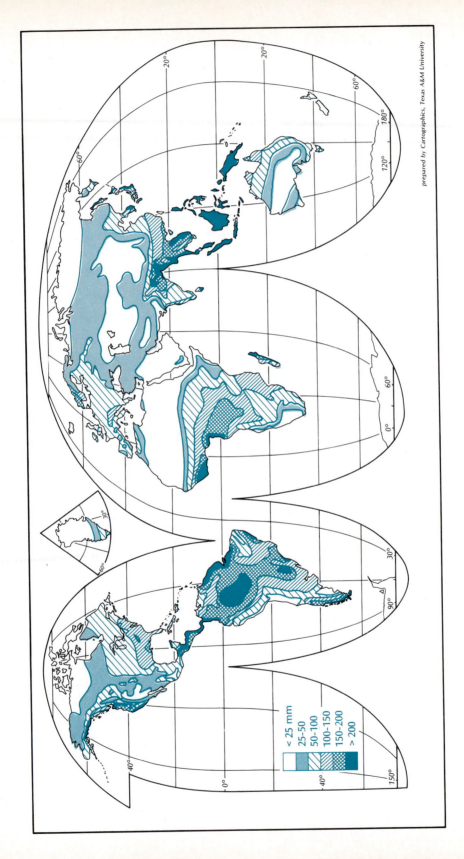

FIGURE 6.9
Mean precipitation over Earth (in mm).

prepared by Cartographics, Texas A&M University

Legend:

< 25 mm
25–50
50–100
100–150
150–200
> 200

Mountain ranges play a significant role in the spatial distribution of precipitation. The windward slopes of mountains receive the greatest amount of precipitation. In the lee of the mountain ranges the precipitation decreases markedly, producing a rain-shadow effect. The predominant flow of air along the West Coast of North America is from west to east. The mountains produce alternate zones of high precipitation and low precipitation—high on the mountain slopes and low in the intervening basins and valleys. In California, the Coast Range is the first barrier to the onshore winds. Precipitation is substantial, and forests grow on the windward slopes of these mountains. Precipitation increases with height to the crest at about 760 m (2493 ft). In the Great Valley, precipitation is much lower; there, a grassland environment exists. Eastward over the Sierra Nevada Mountains, precipitation increases to two or three times that of the Great Valley (Figure 6.10). Forests cover the slopes of the mountains near the summit. Crossing the crestline (2600 m) (8531 ft), air descends slightly and warms adiabatically, and relative humidity drops. Precipitation declines rapidly, and the clouds begin to evaporate. Annual precipitation drops from around 1300 mm (51 in.) at the crest to about 150 mm (5.9 in.) at Reno, NV, giving Reno a desert climate. The arid zone extends from the Mexican border north into Canada between the Sierra Nevada/Cascade Range and the Rocky Mountains because of the drying of the westerlies as they cross the mountains. The process is repeated as the currents continue to flow eastward over the Wasatch and Rocky Mountains. The Rockies, with a crestline in excess of 3300 m (10,827 ft), cause still further drying of air from the Pacific. Little precipitation falls east of the Rockies from air origination over the Pacific Ocean. The chief moisture sources for the continent east of the Rockies are the Gulf of Mexico and the Atlantic Ocean.

Temporal Distribution of Precipitation

Total annual precipitation alone is an insufficient measure of moisture availability, for it does not take into account how precipitation is distributed throughout the year. This temporal distribution is the dominant factor in determining use of precipitation. There are major discrepancies between the *frequency* of precipitation, measured by the number of days per year on which measurable precipitation falls, and *mean annual* precipitation. In some parts of the world, precipitation falls nearly every day of the year. Bahia Felix, Chile, averages 325 days per year with measurable precipitation; thus there is an 89% chance that it will rain or snow on any given day of the year. Buitenzorg, Java, in the tropics, averages 322 days per year with thunderstorms. At the other end of the scale are desert areas such as Arica, Chile, which averages only about one rain day annually; at nearby Iquique, Chile, 14 years passed with no measurable rainfall (1899–1913). It may seem strange that one of the areas with the least-frequent rainfall lies in the same country as one of the rainiest places in the world. The size and shape of the country, topography of the region, and general circulation of the atmosphere all play a part in the explanation.

Over most of Earth's surface, precipitation is seasonal, with precipitation more likely during certain periods and less likely during others. Surprisingly, perhaps, sites with the highest annual total rainfall occur in areas with a pronounced dry season. Cherrapunji,

TABLE 6.2
Record intense rainstorms.

TIME	AMOUNT		PLACE	DATE
	MM	IN.		
1 min	38	1.50	Barot, Guadeloupe, West Indies	Nov. 26, 1970
5 min	63	2.48	Panama	1911
8 min	126	4.96	Fussen, Bavaria, Germany	May 25, 1920
15 min	198	7.79	Plumb Point, Jamaica	May 12, 1916
20 min	206	8.10	Cuerta-de-'Arges, Romania	Date unknown
30 min	235	9.25	Guiana, VA	Aug. 24, 1906
42 min	305	12.00	Holt, MO	June 22, 1947
1 hr	401	15.78	Muduocaidong, Nei Monggol, China	Aug. 1, 1977
2 hr 10 min	483	19.02	Rockport, WV	July 18, 1889
2 hr 45 min	555	22.00	D'Hanis, TX	May 31, 1935
4 hr 30 min	782	30. 80	Smethport, PA	July 18, 1942
6 hr	840	33.07	Muduocaidong, Nei Monggol, China	Aug. 1, 1977
9 hr	1087	42.79	Belouve, La Reunion, Indian Ocean	Feb. 28, 1964
10 hr	1400	55.12	Muduocaidong, Nei Monggol, China	Aug. 1, 1977
24 hr	1870	73.62	Cilaos, Reunion Island	Mar. 15–16, 1952
5 days	3810	150	Cherrapunji, India	Aug. 1841
31 days	9300	366	Cherrapunji, India	July 1861
1 yr	26460	1041	Cherrapunji, India	Aug. 1860–July 1861

India is an example. An average of more than 10 m (33 ft) of rain falls there each year, but rain is infrequent in several months.

A number of index values have been introduced to evaluate the seasonality of precipitation. A precipitation concentration index, for example, expresses the monthly concentration of rainfall over a year using a dimensionless value. Figure 6.11 provides a guide to global seasonality.

TABLE 6.3
World record extremes of precipitation.

Longest period without rain	14 years	Iquique, Chile
Lowest average annual precipitation	0.5 mm	Arica, Chile
Greatest number of days per year with precipitation	325	Bahia Felix, Chile
Greatest number of days per year with thunderstorms	322	Buitenzorg, Java
Highest annual average precipitation	11.98 m	Mt. Waialeale, Kauai, HI
Greatest 24-hr snowfall (USA)	1.93 m	Silver Lake, CO, April 14–15, 1921
Highest one-season snowfall (USA)	28.5 m	Paradise Ranger Station, Mt. Rainier, WA 1971–1972

In most parts of the world, agricultural activities are adjusted according to the rainy season. Usually planting takes place near the beginning of the rainy season and harvesting after the rainy season has ended. If the precipitation occurs during the winter, much of it may be in the form of snow. In some locations snow does not become effective

FIGURE 6.10
Schematic cross-section showing the relationship between elevation and precipitation in the western United States.

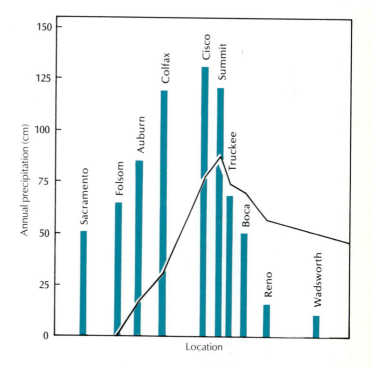

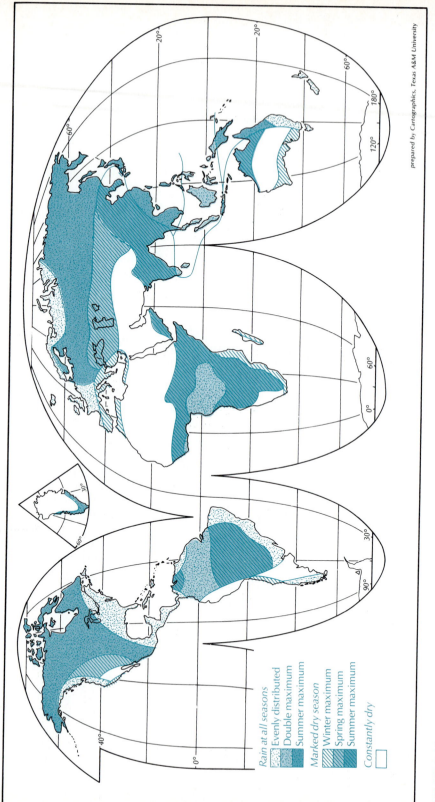

FIGURE 6.11

Generalized map showing the seasonal distribution of precipitation.
Source: After A. A. Miller, *Climatology* (London: Methuen, 1965).

prepared by Cartographics, Texas A&M University

Rain at all seasons
Evenly distributed
Double maximum
Summer maximum

Marked dry season
Winter maximum
Spring maximum
Summer maximum

Constantly dry

moisture until melting takes place in the spring. Much of the precipitation in far western North America occurs in the winter in the form of snowfall. As a result, massive schemes for storing the runoff in the spring and irrigation distribution systems for delivering the water to the fields in the summer have been constructed throughout the western states.

Floods and Droughts

Precipitation is distributed very unevenly in both time and space over Earth's surface. Some places are perpetually dry and others wet, and precipitation normally expected at a location sometimes varies. Floods and droughts are two examples of poor distribution that can create major hazards for the human population.

Floods

Technically, a flood may be defined as the condition when any stream or lake rises above its banks. Although arbitrary, this is a useful definition of a flood. All natural-stream floods are due primarily to surface runoff, which may result from heavy rainfall, snow melting, or a combination of both. Floods caused by rain can result from either short periods of high-intensity rainfall (such as rates of 2.5 cm/hr or 25–45 cm/day) or from prolonged periods of steady rains lasting for several days or weeks. Flood runoff from small watersheds usually has different causes than floods on large drainage basins. Small watersheds are defined somewhat arbitrarily; the average would be 25 km^2 (10 mi^2) or less. Watersheds of this size can be completely covered by a single convective storm, and most floods on small drainage basins are caused by cloudbursts. (The rainfall is so intense that the stream channels cannot carry off the water as fast as it falls.) The floods that isolated New Orleans and created havoc in other parts of Louisiana and Mississippi in April 1983 were caused by a series of thunderstorms over several days. The flood that destroyed parts of Rapid City, SD, in 1972 was caused by an extremely intense storm that covered a very restricted area. The rainstorm lasted less than 8 hr and produced flooding on only a few small watersheds. Figure 6.12 shows the effects of such a local storm on Kansas City.

Large drainage basins require extended precipitation from cyclonic storms or massive snowmelt to produce flooding. The Mississippi River floods of the spring of 1973 are a good example. The winter in the upper valley was very wet, the ground was saturated, and when the spring melt occurred, almost all of the water ran off the land surface. To add to the problem, a series of cyclonic storms moved across the drainage basin. The result was one of the highest floods in recent history on the lower Mississippi River.

The timing of floods varies over Earth's surface. Surrounding the Red Sea is a complete network of streambeds that are dry most of the time. During short rainy periods they often flood. The entire Amazon River Basin receives large amounts of rain at some time during the year, and some place in the large basin is usually flooded at any given time. The Ganges River of India is frequently flooded during the monsoon season, from April onward, because snowmelt in the Himalaya Mountains and excessive rain combine to produce overflow.

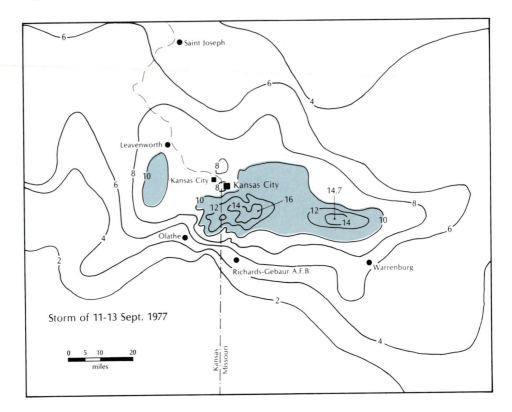

FIGURE 6.12
Intense local rainstorm in Kansas City, KS. (Data in in.)

Droughts

Droughts imply a shortage of water, yet most areas of the world are accustomed to regular periods of dryness. Dryness follows three temporal patterns: perennial, seasonal, and intermittent. Perennial dryness characterizes the great deserts of the world, where water is available only incidentally in the form of occasional rain or in the form of an oasis fed by an exotic river or groundwater. An exotic river is one like the Nile that originates in a humid area and flows into or through a desert. The oasis of Hofuf in Saudi Arabia is an example of an oasis fed by groundwater. Seasonal dryness occurs wherever the climate includes a distinct period of dry weather each year. Intermittent dryness occurs whenever the precipitation level falls in humid areas or when in areas of seasonal dryness the rainy season does not materialize or is very short.

Droughts that create a major problem for humans today produce not just dry conditions, but abnormally dry conditions; the absence of precipitation when it normally can be

TABLE 6.4
The Palmer meteorological drought index.

VALUE	CLASS (RELATIVE TO THE PARTICULAR LOCATION)
≥4.00	Extremely wet
3.00 to 3.99	Very wet
2.00 to 2.99	Moderately wet
1.00 to 1.99	Slightly wet
0.50 to 0.99	Incipient wet spell
0.49 to −0.49	Near normal
−0.50 to −0.99	Incipient drought
−1.00 to −1.99	Mild drought
−2.00 to −2.99	Moderate drought
−3.00 to −3.99	Severe drought
≤−4.00	Extreme drought

Source: After Palmer (1965).

expected and a demand exists for it; or much less precipitation than would be expected at a given time. In some locations where daily rainfall is the usual condition, a week without rain would be considered a drought. In parts of Libya, only a period of two or more years without rain would be considered a drought. Along the floodplain of the Nile River, rainfall is unimportant in determining drought. Prior to the construction of the Aswan High Dam, a drought was any year when the Nile failed to flood. The annual flood provided the soil moisture for agriculture along the river all the way from Khartoum, Sudan, to Alexandria, Egypt. In the monsoon lands of the world, a rainy season producing half the normal precipitation may bring drought; in areas that normally have two rainy seasons, the failure of one would be considered a drought. Drought is thus a relative term, and the total rainfall in a place is not a suitable indication of drought. For this reason it is also difficult to devise a quantitative index for drought that is widely applicable. Nearly every location has its own criteria for drought conditions.

Drought also has different meanings depending on user demands: We can identify meteorologic drought, agricultural drought, and hydrologic drought. Meteorologic droughts are irregular intervals of time, most often months or years in duration, when the water supply falls unusually below that expected on the basis of the prevailing climate. Agricultural drought exists when soil moisture is so depleted as to affect plant growth. Since agricultural systems vary widely, drought must be related to the water needs of the animals or crops in that particular system. There are also degrees of agricultural drought that depend on whether the drought affects only shallow-rooted plants or deep-rooted plants. Both growing-season and dry-season precipitation can affect crop yields. The Thornthwaite water budget method is often used to assess relative drought. An alternate measure is the Palmer meteorologic drought index. The value ranges from +4 to −4, as shown in Table 6.4. A slightly modified Palmer index is used to develop a generalized map of abnormally dry conditions for crops that is published as part of the *Weekly Weather and Crop Bulletin*.

The temporal and spatial scales of drought vary widely. The duration of droughts varies as much as their timing. We cannot predict how long a drought will last for the same reasons that we cannot predict other irregular oscillations with accuracy. Drought simply ends when the rains begin and the streams rise. We are unable to predict either when they will occur or how long they will last, much less prevent them from occurring. We know only that they are a part of the natural system and that they will occur again, perhaps with even greater duration and intensity than in the past. They may be very local in extent, covering only a few square miles; or they may be widespread, covering major sections of continents. Even in large-scale droughts, the intensity is likely to vary considerably.

Summary

As air rises, it cools at the dry adiabatic lapse rate until the condensation level is attained. Thereafter, cooling occurs at the slower moist adiabatic lapse rate. Whether air will rise sufficiently to form clouds and provide the potential for rainfall depends upon the air's stability. We determine that stability by comparing the adiabatic rates with the prevailing environmental lapse rate.

For rain to fall from clouds, individual cloud droplets must overcome buoyancy and friction and attain their terminal velocities. Small droplets have very small terminal velocities and evaporate long before they reach the surface. Precipitation thus depends on the growth of cloud droplets or ice crystals. In cold clouds, growth occurs through the Bergeron process whereby ice crystals grow at the expense of supercooled water. In warm clouds, the growth of water droplets depends upon the collision-coalescence theory.

The spatial distribution of precipitation shows that two global areas receive inordinately high rainfall, while three vast regions are classified as arid deserts. Total annual precipitation is an insufficient measure of moisture availability, disregarding how precipitation is distributed over time. Precipitation seasonality is significant to the climate of a location.

Distribution of precipitation fluctuates widely. While floods have always occurred, evidence indicates that human activities may be increasing their intensity. Droughts may be defined in terms of user demands: meteorologic, agricultural and hydrologic. The Palmer index is one quantitative measure of drought intensity.

7

MOTION IN THE ATMOSPHERE

Chapter Overview

Mechanics of Motion
- Atmospheric Pressure
- Pressure Systems
- Pressure Gradient
- Atmospheric Vorticity
- Coriolis Force
- Friction

Vertical Motion in the Atmosphere

Daily Wind Systems
- Land and Sea Breeze
- Mountain and Valley Breeze

Chaotic Behavior in the Atmosphere
- Turbulent Flow
- The Butterfly Effect
- Sensitivity Dependence on Initial Conditions
- Initial Conditions and Climate Forecasting
- Persistence (the Joseph Effect)

The atmosphere is seldom still. Air moves from place to place in an unending stream. Sometimes the air moves slowly and silently. At other times it howls and shrieks with velocities of hundreds of kilometers per hour. The wind is sometimes caressing and at other times devastating.

Winds are extremely important to Earth, since they transfer heat from near the equator toward the poles. This transfer cools tropical regions and warms polar regions. Winds also carry water vapor from oceans to the continents. Without wind systems there would be no life as we know it on the land masses.

Mechanics of Motion

Wind is the name given to air moving horizontally over Earth, and air currents refer to air moving vertically. This chapter deals mainly with wind. Wind has two main characteristics: direction and velocity. A wind's direction is that from which it blows. They are named for the main points on the compass rose or assigned a bearing in degrees from the north. A wind blowing from east to west is an east wind or a 90° wind. When the direction of the wind is changing, it is veering or backing. A veering wind is changing in a clockwise direction; a backing wind is changing in a counterclockwise direction.

Atmospheric Pressure

Atmospheric pressure is the force per unit area, or the weight exerted by the atmosphere. We measure atmospheric pressure in two primary ways. The one used internationally is the millibar (mb). A bar is a force of 1 million dynes per cm^2. A dyne is the force needed to accelerate a mass of 1g 1 cm per sec per sec. One mb equals $\frac{1}{1000}$ of a bar. In the standard atmosphere, pressure at sea level at 15°C is 1013.2 mb. In the United States, we often measure atmospheric pressure in inches of mercury. A column of mercury about 30 in. in height just offsets the weight of a column of atmosphere of the same area. A column of mercury 1 in. on each side and 30 in. high weighs 14.7 lb. This is the weight of a column of the standard atmosphere 1 in.^{2} in area. The equivalent of 14.7 lb. per in.^{2} or 1013.2 mb is 29.92 linear in. of mercury. In the real world, surface pressures vary routinely from about 950 mb (28 in.) to 1050 mb (31 in.). Table 7.1 provides some extremes of pressure.

Chapter 1 described the variation in atmospheric pressure and density that occurs with height in the standard atmosphere. Like many other atmospheric variables, pressure changes with time and space. For example, at the two poles it is higher than average. This is because of the cold temperatures and subsidence of air. In other zones, atmospheric pressure averages less than the global mean.

Some areas have seasonal changes in pressure. Central Asia experiences marked differences from summer to winter, largely due to wide swings in temperature. In the mid-latitudes, day-to-day changes in pressure often result from traveling storms. We also see regular daily variations: maxima around 10:00 A.M. and 10:00 P.M. and minima around 4:00 A.M. and 4:00 P.M. We don't know the exact cause for these daily changes, but they relate to the daily flood of solar energy that moves around Earth.

TABLE 7.1
Extremes of sea-level barometric pressure.

PRESSURE		
MB	**IN.**	**EVENT**
1083.8	32.00	Highest recorded. Agata, Siberia, Dec. 31, 1968
1078.2	31.85	Highest recorded in North America Northway, AK, Jan. 31, 1989
1063.9	31.42	Highest in the lower 48 states. Miles City, MT, Dec. 24, 1983
870	25.7	Lowest recorded. Typhoon Tip, Pacific Ocean near Guam, Oct. 12, 1979

Humans are normally not sensitive to small variations in pressure. We are most likely to be aware of them when going up or down in an airplane or driving up or down a long mountain road. During take-off and landing, the pressure changes in a modern jet aircraft are large. Cabins in commercial jets are now pressurized so the pressure normally stays at about 50% of that at sea level. We are most sensitive to pressure changes when we have a head cold and congestion in the eustachian tube. Air cannot flow in or out of the inner ear to keep even with the outside pressure. When the difference is very great, it may cause severe earache.

Pressure Systems

Areas of below- or above-normal pressure develop in the atmosphere. Some specific terms refer to these areas:

1. A *low* is an area in the atmosphere where pressure is less than the surrounding area.
2. A *trough* is an elongated area of low pressure.
3. A *high* is an area of the atmosphere where barometric pressure is higher than the surrounding area.
4. A *ridge* is an elongated area of high pressure.

In all cases the pressures are relative and do not have fixed values.

Pressure Gradient

Differences in heating and internal motion in the atmosphere produce differences in atmospheric pressure. This difference in pressure over space is a **pressure gradient**, and air will move down this gradient from high to lower pressure. Figure 7.1 shows how high and lows appear on a weather map using isobars to show pressure patterns. Figure 7.2*a* shows a simplified map of pressure systems over North America. The rate at which air moves depends on the steepness of the gradient. When small differences in pressure occur over large areas, a weak gradient gives rise to weak winds. A steep gradient causes rapid motion or high winds (Figure 7.3).

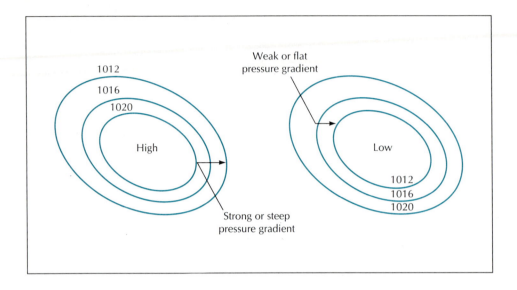

FIGURE 7.1
A map of a simple high- and low-pressure system. In a high-pressure system, pressure decreases outward from the center. In a low-pressure system, pressure decreases towards the center. Closely spaced isobars show a steep pressure gradient, and widely spaced isobars show a flatter pressure gradient.

Atmospheric Vorticity

Air moves wherever a pressure gradient exists. Once air is in motion, other forces come into play to influence its direction. If Earth did not rotate, the air would move directly down the pressure gradient. Earth *is* rotating, however, and this rotation plays a very important part in determining atmospheric circulation.

Rotating bodies are subject to the law of conservation of momentum. This law states that if no external force acts on a system, the total momentum of the system stays unchanged. Angular momentum is a function of the mass of the rotating body, the angular velocity (degrees per unit time), and the radius of curvature. The product of these elements stays constant without external interference. A tetherball serves as a good illustration. As the ball goes around the pole and the radius shortens, the angular velocity increases to keep the momentum the same as it was initially. Another example of the conservation of momentum is that of a figure skater going into a spin. The moving skater turns into a slow spin with arms and perhaps a leg extended. When the skater pulls in his or her limbs while stretching upward, this movement reduces the radius of the spin and the skater turns faster and faster.

Since air has mass, it also has momentum. As it moves north or south, the rotational velocity changes as its distance to Earth's axis changes. In actual practice, the increase in angular velocity is much less than it might be. Friction dissipates the energy as does

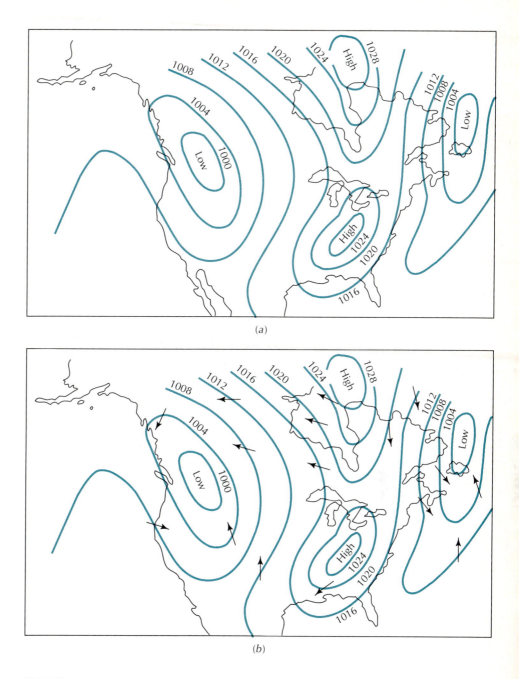

FIGURE 7.2
(a) Pressure systems over North America. (b) Surface wind directions across isobars.

FIGURE 7.3

Horizontal and vertical pressure patterns and wind direction: (*a*) Pressure gradient and wind direction. Air flows from high to low pressure, or down the pressure gradient (data in mb). Wind speed is determined by the steepness of the gradient. Wind speed at (i) will be greater than at (ii). (*b*) Model cross-section of convergence and divergence.

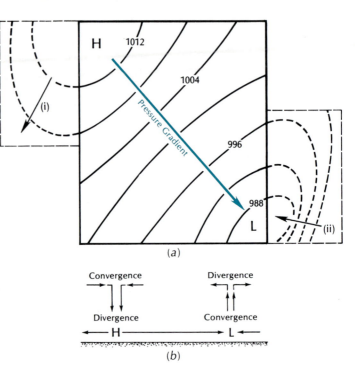

(*a*)

(*b*)

diffusion in large midlatitude eddies or cyclones. Conservation of momentum is a part of the acceleration process in mesocyclones and tornadoes. As the converging air travels a shorter and shorter path, the velocity increases accordingly. Figure 7.4 provides a schematic model of momentum conservation.

FIGURE 7.4

In a converging system such as a mesocyclone, tornado, or hurricane, vertical stretching reduces the radius of the rotating air and leads to an increase in wind velocity. In this example, the mass of rotating air *M* is set at 1 for the sake of simplicity. It is the same for the initial rotating mass and for the stretched mass.

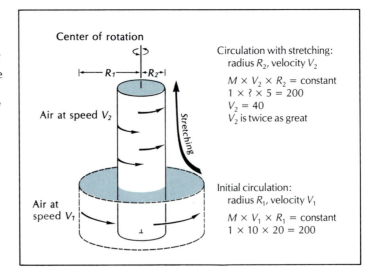

Center of rotation

Air at speed V_2

Air at speed V_1

Stretching

Circulation with stretching:
  radius R_2, velocity V_2

$M \times V_2 \times R_2 = $ constant
$1 \times ? \times 5 = 200$
$V_2 = 40$
V_2 is twice as great

Initial circulation:
  radius R_1, velocity V_1

$M \times V_1 \times R_1 = $ constant
$1 \times 10 \times 20 = 200$

TABLE 7.2
Examples of drift due to the Coriolis force.

1. A person walking at a rate of 10 m per min (0.4 mph) toward a point in space drifts 40 m per km (250 ft/mi).
2. An auto traveling 100 km per hr (60 mph) drifts 4.5 m (15 ft) in 1.6 km (1 mi). The tendency to drift is offset by friction with the road.
3. A bullet fired at 120 m per sec (400 ft/sec) drifts 2.5 mm in 1 sec (0.1 in. in 400 ft).
4. An artillery shell fired at 750 m per sec (2500 ft/sec) drifts 60 m (200 ft) in 33 km (20 mi).
5. A missile launched from the North Pole toward New York City at a rate of 1.6 km per sec (1 mi/sec) lands near Chicago.
6. An airliner leaving Seattle for Washington, DC along a great circle route and not corrected passes over South America near Lima, Peru.

Coriolis Force

Earth's rotation results in a process called the Coriolis force. The Coriolis force acts on any object moving free of Earth's surface, including aircraft, ballistic missiles, and long-range artillery (see Table 7.2). The rotation of Earth causes winds in the Northern Hemisphere to turn in a clockwise direction or to the right. In the Southern Hemisphere, the winds turn counterclockwise (see Figure 7.5). Mariners have long known and recorded the process, which was first explained by French mathematician G. Coriolis (1792–1843). The rate of curvature imparted to the moving air is a function of the velocity of the wind and the latitude.

At the equator, the Coriolis force is zero, and it is greatest at the two poles. Although the Coriolis force is relatively small in itself, it is significant because air streams travel long distances over Earth.

Friction

Winds blowing over the surface are subject to three variable forces: the pressure gradient, the Coriolis force, and friction. Friction works in opposition to the pressure gradient to

FIGURE 7.5
The Coriolis effect acts on the wind to deflect it to the right in the Northern Hemisphere and to the left in the Southern Hemisphere. Deflection in each hemisphere is always in the same direction, regardless of wind direction.

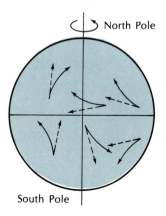

FIGURE 7.6

(a) The relative direction of the three forces that affect surface wind direction and velocity. (b) The mean angle of wind direction relative to the pressure gradient over land and the sea.

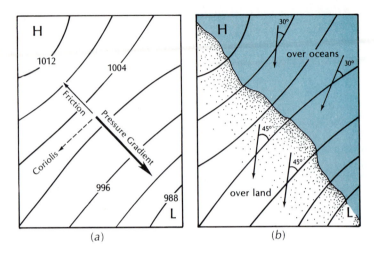

reduce wind velocity and the Coriolis force. With the rotational effect reduced, the pressure gradient dominates. At the surface, then, winds blow from high to low pressure at an angle across the isobars. Over the oceans, where friction is less than over land, the winds cross the isobars at angles of 10°–20° at 6 m above the surface. Oceanic winds blow at or about 65% of the velocity dictated by the pressure gradient. Over the land masses, where the surface is rougher and friction greater, the wind crosses the isobars at an average angle of 30°. Because of greater friction, winds blow with a velocity of about 40% of gradient velocity (see Figure 7.6).

Atmospheric pressure changes slowly in a horizontal direction. The change may be as little as 2 mb in 160 km (100 mi). In the summer over North America, the temperature difference between the Gulf of Mexico and the Canadian border is small. As a result, the density of the air is about the same and the north-south pressure gradient is weak. Since the pressure gradient is weak, mean wind velocities are low. In the winter, temperatures differ more. It is still warm over the Gulf of Mexico but relatively cold over Canada. The cold air is denser, and a pressure gradient slopes downward toward the south. This transports cold air south and produces higher wind velocities.

Friction will be most effective in the air close to the ground, appropriately called the friction layer. Above that layer, the role of friction decreases and winds are influenced only by pressure gradient and coriolis force. In upper airflow, these two ultimately become balanced to produce a geostrophic wind, a wind that blows parallel to the isobars aloft. The resulting upper air circulation is dealt with in Chapter 8, as part of the general circulation of the atmosphere, and in Chapter 10 where synoptic climatology is examined.

Vertical Motion in the Atmosphere

Vertical motion occurs at a broad range of scales from small systems such as dust devils, to very large systems such as midlatitude cyclones that are hundreds of miles across. Near the surface, differences in temperature cause vertical motion. Temperature differences result in pressure differences, and heated air rises because warm air is less dense than cold air. This phenomenon explains how hot-air balloons stay aloft: When a parcel of air warms, it becomes buoyant and rises. These rising currents of air also provide the lift for soaring hawks and hang gliders. On a sunny day, radiation and conduction from the ground heat the air above. Eventually the air begins to move and something pushes it upward. It may flow up the side of a hill, or it may become warm enough to start upward of its own accord. This starts a column of air rising over a plowed field or along a mountainside. The upward velocity may be enough to lift a large bird or hang glider. This is why hawks can stay aloft without flapping their wings—all they need to do is stay in the rising column. On one occasion more than 40 hawks were observed soaring on a single updraft along the side of Stone Mountain in North Carolina.

Once circulation near the ground begins, it may start a steady stream of upward-moving parcels. Circling hawks tend to stay in an area of relatively small diameter. This is because the central rising core of the column, where the velocity is highest, is fairly small—in most cases, only 100 m (330 ft) or so. The whole vertical system is relatively small; most are less than 500 m (1650 ft) across. Near the outside of the cell, upward velocities are lower because of friction with the surrounding air.

Daily Wind Systems

Daily variations in atmospheric pressure develop in many parts of the world and lead to distinctive wind patterns. These variations in pressure, which result from moving air caused by changing surface temperatures through the day, themselves cause daily changes in wind direction and velocity.

Land and Sea Breeze

One of the daily winds associated with temperature differences on the surface is the land and sea breeze occurring along coastal areas and shorelines of large lakes. The land and sea breeze is a function of the change in temperature and pressure of the air over land in contrast to that over water. In the hours after sunrise, land temperatures rise rapidly while water temperatures rise slowly. When the air above the shoreline heats, it expands and becomes denser. The air rises over the land surface, and cold air flows in from the water to replace it. This flow of air landward is the sea breeze, which usually begins several hours after sunrise and peaks in the afternoon when temperatures are highest. This air will penetrate inland 30 km or more along an ocean coast and to a lesser extent around lakes. The cell is quite shallow, ranging up to several kilometers deep under the best conditions. The gusty sea breeze causes air temperature to decrease and humidity to increase. The complementary land breeze is less well-developed than the sea breeze. It occurs when the land surface cools at night and the water surface remains warm. The temperature and pressure relationships reverse with the higher pressure over the land. At night, however, the contrast between water and land is not as large as it is in the daytime. The water surface does not actually gain heat at night but retains the heat acquired during the daylight period and thus remains relatively warm compared to the land. Usually a lag in the development of the land breeze is due to the system's general inertia. The sea breeze does not set in until well after the temperature differences develop. In Figure 7.7, for example, the lake breeze doesn't begin to blow until 3:00 P.M.

Richard Henry Dana, Jr., described the effects of the land and sea breeze on commercial shipping in 1840 in the days of sail in the following passage from *Two Years Before the Mast*:

> The brig Catalina came in from San Diego, and being bound up to windward, we both got under weigh at the same time, for a trial of speed up to Santa Barbara, a distance of about eighty miles. We hove up and got under sail about eleven o'clock at night, with a light land breeze, which died away toward morning, leaving us becalmed only a few miles from our anchoring-place. The Catalina, being a small vessel, of less than half our size, put out sweeps and got a boat ahead, and pulled out to sea, during the night, so that she had the sea-breeze earlier and stronger than we did, and we had the mortification of seeing her standing up the coast, with a fine breeze, the sea all ruffled around her, while we were becalmed, inshore. (p. 187)

Figure 7.7 shows the effect of a lake breeze on the temperature at Chicago compared to that of nearby Joliet. Up until about 3:00 P.M., temperatures were increasing in a similar manner at both cities. The lake breeze set in at Chicago at about 3:00 P.M., and the temperature there dropped about 4°C (7°F) in a little more than an hour. The temperature at Joliet, 56 km (30 mi) southwest of Chicago, does not show the decline. The lake breeze was not strong enough to reach as far inland as Joliet. The wind shift producing the drop in temperature was nearly 180°, switching from west to east. Later in the evening, when the lake breeze died, the temperature in Chicago jumped sharply.

FIGURE 7.7
The effect of a lake breeze is shown by the difference in temperature over a 24-hr period between Chicago, on the lake, and Joliet, an inland location.

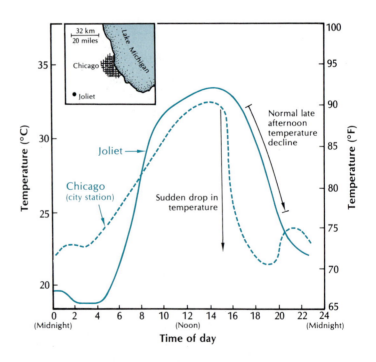

FIGURE 7.8
The mountain and valley breeze. The mountain breeze is a night-time breeze; the valley breeze is a daytime event.

Mountain and Valley Breeze

Similar in formation to the land and sea breeze is the mountain and valley breeze characteristic of some highland areas (see Figure 7.8). In the daytime, the sun heats the valleys and slopes of mountains. The air near the surface heats, expands, and rises up the sides of the mountains. This breeze, called the *valley breeze* (for its place of origin), is a warm wind that occurs in the daytime or late afternoon. The clouds often seen forming over the hills in the afternoon are the result of condensation taking place as the air rises to cooler heights over the mountains.

At night, the valley walls cool. As the air at the surface cools, it flows down the slope due to greater density. This is the night breeze. In some mountain regions subject to frost, the valley slopes are the preferred places for orchards, since the air moving up and down the valley slopes reduces the chance of the stagnant conditions conducive to frost formation.

■ Applied Study

Desert Winds

Severe storms are infrequent in the deserts, primarily because of the lack of moisture to supply the energy. Deserts are very windy, however, particularly in the afternoon and during the hottest months. Air over deserts commonly has a steep lapse rate and is unstable near the surface. **Dust devils** are the most visible example of this instability (see Figure 1). Developing in clear air with low humidity, they become visible as they pick up debris. They reach a height of nearly 2 km (1.2 mi). On occasion they will sustain velocities high enough to blow down shacks or blow screen doors off their hinges. Dust devils are very common in deserts; it is not unusual to see a multitude of them at once. Dust devils result from an intense thermal at or near the ground surface. Surrounding air moves toward the base of the thermal to replace the rising air. As the air moves toward the thermal, the radius of curvature decreases and velocity increases.

Other storms of the tropical deserts are dust storms and sandstorms. These two storms differ according to wind velocity and size of particles carried. A wind's velocity deter-

FIGURE 1
Dust devil in Amboseli National Park, Kenya, East Africa.
Source: Buff Corsi/Tom Stack & Associates.

mines the size of soil particles it can move. Small particles are picked up at lower velocities than are sand particles. When wind velocity is high, wind picks up dust and carries it
relatively high in the atmosphere, sometimes for long distances. One of the readily visible aspects of much desert surface is the lack of fine sediments. **Deflation,** the process of
the wind picking up and carrying away small soil particles, keeps the desert surface swept
clean. Deflation also means that winds blowing out of the deserts are often dust-laden. In
recent years, dust from the Sahara Desert in Africa has been picked up by the wind and
carried across the Atlantic Ocean to Florida and beyond.

Sandstorms occur only when wind velocities are high enough to move the sand. The
minimum velocity needed to move sand is the *threshold velocity*; this velocity depends
upon the sand-particle size and the sand wetness. For medium-sized sand particles (0.25
mm), the threshold velocity is 5.4 m per sec (12 mph). Wind velocities exceed the threshold velocity about 30% of the time. Sand drift increases rapidly as wind velocities
increase above the threshold velocity. An average rate of sand drift of 12 liters per m
width was measured for winds 5.5–6.4 m per sec (12–14 mph) and a rate of nearly 10
times that for wind velocities of 10 m per sec (22 mph). The increase is rapid above the
threshold velocity because the power of the wind increases as the cube of the wind velocity. However, most of the time the wind blows at lower wind velocities (Figure 2).

Sandstorms develop under two different sets of conditions in the deserts of Africa and
the Middle East. The first set of conditions occurs daily: Surface heating during the day
results in turbulence. In the Nafud Desert in Saudi Arabia, for example, sand temperatures at the surface can reach 85°C (180°F). The high rate of heat exchange between the
sand and the atmosphere produce very gusty winds. The sand begins to move soon after
the sun comes up and continues to move until shortly after sunset. This type of sandstorm
is chiefly a daytime event and is most pronounced in the summer. The storms are so regular that one can almost set a clock by the beginning of sand movement.

FIGURE 2

Wind velocities and sand drift during normal daily winds and during storm systems. The two different wind systems move almost the same amount of sand during the course of a year even though the high-velocity winds are infrequent.

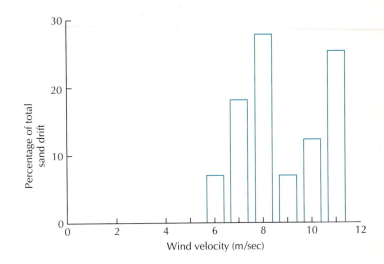

At Al Hofuf, in the Nafud Desert of Saudi Arabia, the number of hours during which sand drift takes place between 6:00 A.M. and 6:00 P.M. ranges from 76% in February to 91% in June. The total number of hours per day when winds exceed the threshold velocity also follows a seasonal pattern. It increases from 3 hr per day in February to 9.4 hr in June. Because of the increasing hours of high-velocity winds in summer, sand movement peaks in summer as well. It increases from 116 liters per m width per day in February to 406 liters in June. In summer, the high frequency of afternoon sandstorms makes travel and other outdoor activities particularly difficult.

The second type of sandstorm results from low-pressure systems passing through a region. These storms often generate winds of higher velocity that last longer than the daily thermal winds. One such storm in the Nafud Desert lasted for a period of 43 hr, with sand blowing constantly and mainly from a single direction. In this storm the wind averaged 15.6 m per sec (35 mph) for hours at a time. This particular storm thwarted the U.S. attempt to rescue the hostages from Tehran in 1980. During a stopover at night in the south of Iran, blowing sand damaged some equipment. The damage, along with an accident between two aircraft, forced the team to turn back.

Both daily surface heating/turbulence and low-pressure storm systems can move huge quantities of sand. In the Nafud, the wind moves an estimated 80 m^3 (262 ft^3) of sand across each 1-m width of the dune field each year. The dunes themselves move at a fairly high speed as the sand moves from the back of the dunes forward. In this area, dunes with an average height of 10 m (33 ft) are moving at a rate of 15–19 m per year (50–65 ft). The main highway across Saudi Arabia and the only railroad go through the Nafud, and both are frequently closed as the dunes encroach on the right of way.

On the south side of the Sahara Desert blows a dry northerly wind called the **harmattan**. Since it is a dry wind, it is cooling and welcome, although it may limit visibility and leave household furnishings covered with dust. This same type of wind blows out of the north side of the Sahara Desert. Occasionally, the winds travel across the Mediterranean Sea, bringing disaster to agricultural crops because they are so dry. They are frequent enough, and important enough, in the climate of the Mediterranean countries to warrant local names, such as *sirocco* in Italy and Yugoslavia and *leveche* in Spain. Selected local winds are described in Table 1.

TABLE 1
Characteristics of selected local winds.

NAME (ORIGIN)	LOCATION	CHARACTERISTICS
Bora (L, boreas, north)	Adriatic coast	Cold, gusty, northeasterly wind. Frequency at Trieste, 360 days in 10 yr. Mean winter speed 52 km per hr (32 mph), summer 38 km per hr (24 mph). May reach 100 km per hr (62 mph) in winter. Adiabatically warmed wind.
Chinook (from Chinook Indian territory)	Eastern slope of Rockies	A warm wind that may, at times, result in sudden and drastic rise in temperature. May attain 60°–70°F (15°–21°C) in spring with relative humidity of 10%. Predominate in spring. Adiabatically warmed wind.
Etesian (Gr, etesiai, annual)	Eastern Mediterranean	Cool, dry, northeasterly wind that recurs annually. Summer and early autumn. Associated with regional low-pressure systems.
Fohn (German, possibly from Latin *favoniun* = growth, i.e., favoring wind)	Alpine lands	Similar to Chinook. Characterized by warmth and dryness. Most frequent in early spring.
Haboob (Arabic)	Southern margins of Sahara (Sudan)	Hot, damp wind often containing sand. Of relatively short duration (3 hr). Average frequency of 24 per year. Early summer with the advance of the ITCZ.
Harmattan (Arabic)	West Africa	Hot, dry wind, characteristically dust-laden. All year, but most notable in the low-sun season. Associated with air flow from subtropical highs.
Khamsin (Arabic)	N. Africa and Arabia	Hot, dry, southeasterly wind. Regularly blows at a 50-day period (Khamsin = 50). Temperatures often 100°–120°F (38°–49°C). Same wind with adiabatic modifications includes Ghibli (Libya), Sirocco (Mediterranean), Leveche (Spain). Late winter, early spring. Regional low-pressure systems.
Levanter (from Levant, eastern Mediterranean)	Western Mediterranean	Strong easterly wind often felt in Straits of Gibraltar and Spain. Damp, moist, sometimes giving foggy weather for perhaps two days. Fall, early winter to late winter, spring. Regional low-pressure systems.

TABLE 1
(Continued)

NAME (ORIGIN)	LOCATION	CHARACTERISTICS
Mistral (maestrale of Italy = master wind)	Rhone Valley below Valence	Strong, cold wind channeled down Rhone Valley. May reach 100 km/hr in north. Can cause sudden chilling in coastal regions. (Note also the Bise, and equivalent cold north wind in other parts of France.) Most frequent in winter. Regional low-pressure systems.
Norther	Texas, Gulf of Mexico to W. Caribbean	Cold, strong, northerly wind whose rapid onset may suddenly drastically lower temperatures (also Tehauantepecer of C. America). Winter. Related to circulation pattern over United States.
Pampero	Pampas of S. America	Southern Hemisphere equivalent of the Norther. Winter. Related to large-scale pressure patterns.
Zonda	Argentina	A warm, dry wind on lee of the Andes. Can attain 120 km per hr (75 mph). Comparable to Chinook and Fohn.In dry weather carries much dust. Winter.

Chaotic Behavior in the Atmosphere

The notion of a standard atmosphere suggests a uniformity that is elusive. A fluid is any substance that has no rigidity. Gases are fluid, hence the atmosphere is a fluid. It is like a river flowing around Earth. The atmosphere acts according to all the principles of fluids as well as the laws of gases. Fluids have two classes of flow: laminar and turbulent. In the first, **laminar flow**, each particle in the fluid moves essentially in a smooth line without any rotation. The flow of air over the wing of an airplane is in part laminar. The flow of a thin sheet of water down the face of a spillway or dam is laminar. If small disturbances start to develop in laminar flow, the tendency is for the smooth flow to counteract the disturbance and maintain the smoothness.

Turbulent Flow

When fluids moving with a laminar flow encounter unequal pressure or heat, they begin to develop motion in three directions as convection currents or eddies. The flow is now turbulent, with disordered flow in small and large swirls and swirls within swirls. In a fast-moving river, eddies are almost always imbedded in the overall downstream movement of the river. Around a bridge pier, swirls of water or standing eddies may remain in nearly the same position downstream from the pier. In paddling a canoe, or rowing a boat, small eddies form around the end of the paddle or oar in response to the differential pressure of the blade. In **turbulent flow**, individual particles may move in any direction—sometimes upstream, the opposite direction of the stream's mean flow. Whenever the velocity in a moving fluid reaches a critical point, it will become turbulent.

The Butterfly Effect

When small-scale disturbances develop in turbulent flow, they often grow into larger and larger disturbances. Since the atmosphere is a turbulent fluid, many small disturbances grow into large ones. The *butterfly effect* is the principle that a butterfly stirring the air in one place can cause a change in the weather at some later time and at a different place. While this is an exaggeration, it illustrates the principle that very small eddies in the circulation can grow to be major weather systems.

In response to differences in temperature and pressure, the troposphere is especially turbulent. Very large standing eddies and counter currents make up the general circulation. Some of these vortices may be thousands of kilometers across. Tropical cyclones are eddies or vortices measuring several hundred kilometers in diameter. Tornadoes are intense vortices usually no more than 2.0 or 3.0 km (1.2–1.5 mi) across. Many of these storms grow from relatively small initial disturbances.

Sensitivity Dependence on Initial Conditions

The conditions of the atmosphere at any point are sensitive to previous conditions. Every child has taken a dry dandelion blossom and blown the seeds from the stem. All the seeds begin their journey at nearly the same place and diverge further and further as distance from the stem increases. The seeds do not all take the same path because each begins at a slightly different place and hence becomes subject to different air currents as it moves away. The position of the seeds can be forecast much more accurately at a distance of 1 m (3.3 ft) than at 10 m (33 ft).

Initial Conditions and Climate Forecasting

Global Computer Models (GCMs) were introduced in Chapter 4 when we discussed forecasting the effects of increased CO_2 on the energy balance. Chapter 20 contains a more detailed examination of these models. One of the problems with these models is that they all depend upon initial conditions established when the model is started. Unfortunately, the initial conditions used in numeric weather forecasting are only approximate. All computer models of the atmosphere sample points over Earth's surface and at different heights in the atmosphere. The distance between sample points on the surface ranges upward from 100 km (60 mi). A rectangle 10,000 km^2 (3080 mi^2) can contain thunderstorms, hailstorms, and tornadoes within its boundaries; sampling may completely miss these atmospheric storms. The thickness of clouds or amount of cloud cover may also be different. The ground surface may be all water or all land, or more likely part of each. The models assume that all of the atmospheric conditions within the area are the same, and a single number represents all of the sample area. To predict the weather accurately, we need to sample pressure, temperature, and moisture at every point in the atmosphere. Since the atmosphere contains an infinite number of points, the task becomes impossible.

Weather follows an overall pattern, but it is never quite the same two times in succession, nor does it ever repeat itself exactly. If the initial conditions are even only slightly different, the end product may be very different. Since initial conditions are never the same, the future state of the system can never be the same. Atmospheric forecasting is difficult because the initial conditions are never the same, and because we cannot know the initial conditions at every point in the atmosphere. Each set of even slightly different initial conditions produces different weather.

For the same reasons, a weather forecast for three hours from now is more accurate than one for three days from now. The more stable the atmospheric conditions are, the better the forecasts are. In a stable atmosphere, we can forecast several days in advance with some degree of accuracy. Stormy weather and an unstable atmosphere make it more difficult to forecast, since in stormy weather the system can change very rapidly. During an active storm a few hours ahead may exceed the range of credible forecasting. Forecasting the weather for the future is similar to—but much more difficult than—predicting the path that a dandelion seed will take.

Individual particles in a flow of water or air tend to take deviating paths. The more they deviate, the likelier they are to deviate still more. That chaotic nature makes predicting the future state difficult.

Persistence (the Joseph Effect)

Long-range weather forecasts employ a property called persistence. Persistence involves clustering periods of the same kind of weather. It is the tendency for a particular type of weather to continue or repeat itself. The simplest means of forecasting atmospheric conditions for an hour from now, or a day from now, is to forecast it to be the same as it is now. In fact, this method of forecasting is correct most of the time, since all weather systems last for some amount of time. In the midlatitudes, large weather systems exist for five to seven days. Heat waves, cold waves, floods, and droughts are examples of persistent weather systems. When an area has more rain than usual, that area is likely to have more rainy weather; the same is true of dry weather.

This persistence is know as the *Joseph effect*. The name comes from a chapter in the book of Genesis in the *Bible*.

> There came seven years of great plenty throughout the land of Egypt. And there shall arise after them seven years of famine. . . .

The seven wet years followed by seven dry years suggests a periodic phenomenon. We can forecast periodic phenomena with accuracy in amplitude and wavelength. Unfortunately, this is not the case with weather, because it is aperiodic or cyclical. For example, droughts repeat, but their intensity and the intervals between them vary. The reference to seven wet years followed by seven dry years indicates that persistence takes place over time intervals as short as minutes and hours to as long as years. Other evidence clearly shows that persistence operates over periods of decades and centuries. Geological evidence shows it works over periods of thousands and millions of years.

■ Applied Study

Energy from the Wind

Wind was a major source of energy for industrial societies until fossil fuels came into use. The famed windmills of Holland provide a good example of the use of the wind. Wind power has once again come into use as an alternative energy source, with researchers seeking efficient ways to harness wind power. The successful use of wind depends upon the wind characteristics at a given location.

The following equation provides the amount of power generated by a uniform flow of air.

$$P = \tfrac{1}{2}D \times V^3$$

where

P = available power in watts per square meter (W/m²)

D = density of air (1.29 kg/m³)

V = wind velocity (in m/sec)

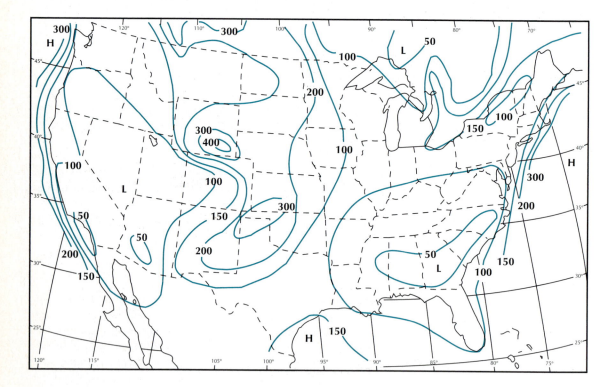

FIGURE 1

Average annual available wind power (watts/m²) in the United States. (Data from Reed [1974].)

FIGURE 2
Wind-driven electric power generators, Pacheco Pass, CA.
Source: John Elk III/Bruce Coleman, Inc.

Average wind speed is not a good indicator of how much power can be produced. Consider the following example. Site A has a wind speed of 7.6 m per sec for a unit of time. Site B has the same mean value, but the wind blows at a velocity of 5 m per sec half the time and 10.2 m per sec the other half. The total power produced at site B is more than a third again as much as at site A:

Example A
$$P = \frac{1}{2}(1.29) \times (7.6)^3$$
$$P = 283 \text{ W/m}^2$$

Example B
$$P_1 = \frac{1}{2}(1.29) \times (5)^3$$
$$P_1 = 81 \text{ W/m}^2$$
$$P_2 = \frac{1}{2}(1.29) \times (10.2)^3$$
$$P_2 = 685 \text{ W/m}^2$$
$$\frac{P_1 + P_2}{2} = 383 \text{ W/m}^2$$

Note that the equation uses the cube of the wind speed. Cubing shows that as wind speed increases, the amount of energy generated increases geometrically. High-velocity winds are thus the key to producing large quantities of electrical energy. For example, a uniform wind moving at 2 m per sec (4.5 mph) provides 5.16 W for each square meter. At a velocity of 4 m per sec (9 mph), the amount is 41.28 W per m². At double the wind's speed, the energy generated is eight times more.

Successful use of wind for commercial power generation depends upon the velocity and variability of wind at a given location. Figure 1 shows the geographic distribution of potential wind power in the United States. Several areas with high wind-power potential can be identified. The main areas of high potential are in an area extending north–south through the western high plains. The peak values are in southern Wyoming and the Oklahoma panhandle. Secondary areas of high potential include the New England coast and the northwestern states. Lowest values are in the southern states and the mountainous west.

The distribution suggests that the potential for wind power in populous areas such as Phoenix, Los Angeles, and Atlanta is poor. It is perhaps unfortunate that the regions with highest potential are far from heavily populated areas (see Figure 2). ■

Summary

The atmosphere is never still. It moves whenever and wherever pressure differences exist. Once air is moving, several processes affect its direction and velocity. The Coriolis force deflects the wind to the right or left, depending upon the hemisphere. Friction reduces the effects of the Coriolis force and determines how directly air will flow down the pressure slope.

Some wind systems blow regularly in a daily pattern. Two of these winds are the land and sea breeze and the mountain and valley breeze. Winds in tropical deserts also follow a daily pattern. At night, when the surface is cool, winds are calm. In the daytime, when the ground surface heats extensively, turbulence develops and blows dust and sand.

Various wind systems throughout the world have local importance. They influence many economic activities and have a major effect on human comfort. Some of these winds are peculiar to a single place, while others occur under similar circumstances at different places.

Winds are a major variable in weather. High winds can be either beneficial or destructive. Both the temperature and speed of the wind are important in determining the benefits or hazards.

GLOBAL CIRCULATION OF THE ATMOSPHERE

Chapter Overview

The Circulation Model

Tropical Circulation
- Trade Winds
- Intertropical Convergence Zone

Midlatitude Circulation

Polar Circulation

Seasonal Changes in the Global Pattern
- Monsoons
- Asian Monsoon
- North American Monsoon

The general circulation of the atmosphere is extremely important to Earth. The general circulation carries water from the ocean over the continents to provide precipitation. It also moves heat energy from the tropical regions toward the poles, warming the high latitudes.

The general circulation changes constantly. Next to the seasons, changes in the general circulation are the most significant weather-related changes. Chapter 1 showed how a standard atmosphere provides a model of an average or mean atmosphere. This chapter will do the same with a simple model of the how the general circulation works and changes. The final chapters of the book will examine how changes in the general circulation effect climatic change.

Between latitudes of about 35° N and 35° S, incoming solar radiation exceeds Earth radiation to space (Figure 8.1). Toward the poles from 35°, a net loss of energy to space occurs. For incoming solar radiation to exceed outgoing radiation in equatorial areas over a prolonged period, energy must be removed in some manner besides radiation. For outgoing radiant energy to exceed incoming radiation in the polar regions, energy must be

FIGURE 8.1

(a) Latitudinal distribution of incoming and outgoing radiation. Between latitudes of about 30° N and S, incoming solar radiation exceeds outgoing Earth radiation. Toward the poles from 35°, outgoing radiation exceeds incoming radiation. (b) For the distribution shown in a to exist, there is an equator-to-pole movement of energy by the ocean and atmosphere. The total amount of energy transported is shown by the solid line. The dashed line shows the amount of heat transported by the atmosphere. The shaded area is the amount transported by ocean currents.

Source: Modified from Gill, 1982.

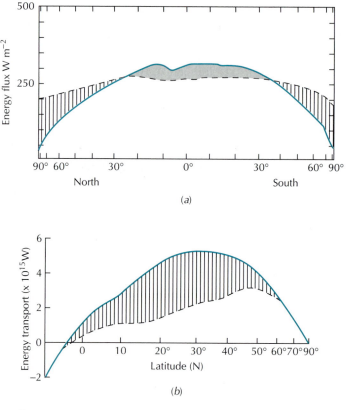

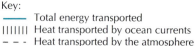

transported in. The mechanism for transporting this energy is the general circulation of the atmosphere and oceans. The general circulation tends to equalize the distribution of energy over Earth's surface.

The atmosphere transfers about 60% of the energy, and the ocean moves the rest. This complex circulation system consists of several semipermanent areas of convergence and divergence and the air flow in and between them. The energy moves primarily as sensible heat and latent heat of water vapor. Internal friction or friction with the ground surface converts the kinetic energy of the wind system to sensible heat. The energy lost from the system by friction balances the rate of kinetic energy generated within the atmosphere.

The Circulation Model

In 1735, George Hadley proposed a cellular model, now known as the **Hadley cell**, to explain the primary circulation of the atmosphere (Figure 8.2*a*). He based the model on the basic pressure differences brought about by uneven heating of Earth's surface. Hadley postulated that cold air descended at the poles and flowed along Earth's surface toward the warmer equator. This flow was countered by rising warm air at the equator and a flow aloft toward the poles. This simplified model corresponds only in part to the actual general circulation.

Earth's rotation prevents development of a single convective cell, and Hadley's model was gradually altered to include a three-cell structure between the poles and the equator. The revised model still shows subsiding cold air at the poles and rising warm air at the equator. While the three-cell structure shown in Figure 8.2*b* is still being refined, it crudely illustrates the essentials of the primary circulation.

The more detailed circulation pattern shown in Figure 8.2*c* actually represents a long-term average of the *meridional* (north–south) movement over the globe. In reality, the circulation is much more complex than this model suggests. While the low-latitude cell exists, the circulation of the midlatitudes consists essentially of a west-to-east flow of upper air that determines the position and location of moving high- and low-pressure systems at the surface. Embedded in these upper-air westerlies are corridors of rapidly moving air, or **jet streams**.

In the midlatitudes, an upper-level air flow making up the westerlies dominates the circulation. This flow is not constantly zonal, because large waves form in the flow pattern, with a jet stream associated with the leading part of the waves. These are **Rossby waves**, so named for the meteorologist Carl-Gustav Rossby. His theoretical work explained the mechanisms for their existence. Figure 8.2*c* combines the tropical and midlatitude regimes to provide a model of the global general circulation.

Tropical Circulation

The circulation closest to the equator closely resembles those proposed in the three-cell model and Hadley's model. The cell extends from around 25°–30° to near the equator (Figure 8.3). It features surface flow toward the equator and counterflow aloft, with ris-

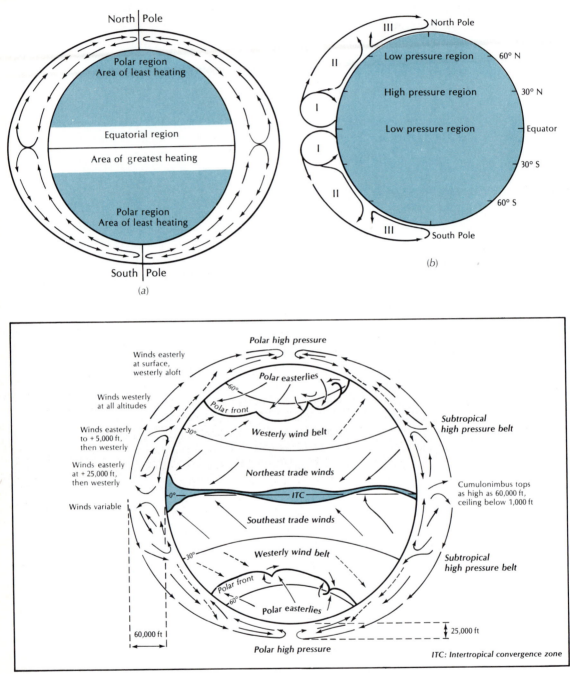

FIGURE 8.2
(a) The fundamental driving mechanisms of the general circulation. (b) A three-cell model that results from differential heating, Earth's rotation, and the fluid dynamics of the atmosphere. (c) The vertical and horizontal winds in the idealized general circulation.

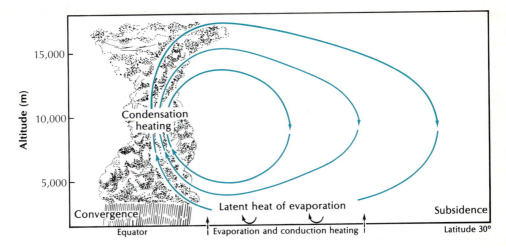

FIGURE 8.3

An idealized cross-section of the Hadley cell with the vertical scale greatly exaggerated.
Source: Herbert Riehl, *Introduction to the Atmosphere*. (New York: McGraw-Hill Book Company, 1972. Reprinted with permission of the publisher.)

ing air at the equator and subsiding air near the Tropics of Cancer and Capricorn. This section of the primary circulation is called the Hadley cell, since it operates effectively as Hadley outlined it in 1735. Associated with this tropical cell are two semipermanent belts of surface divergence and above-average sea-level pressure located near the two tropics. These semipermanent high-pressure zones on either side of the equator act as a source region for air flowing both north and south at the surface. As the air subsides, it heats adiabatically and grows quite warm. Since the temperature is increasing and no moisture is being added, the relative humidity remains low. These zones have high amounts of sunshine, clear skies, and infrequent precipitation. Major deserts occur in these areas: the Sahara and Sonoran deserts in the Northern Hemisphere and the Great Australian, Atacama, and Kalahari deserts in the Southern Hemisphere. If the air that flows out of these areas of subsidence has a trajectory over a land mass, it remains warm and dry and becomes a continental tropical air mass.

The high pressure in these tropical areas is due to the subsidence, which gives rise to divergence at the surface. The subsidence and divergence result more from the motion of the atmosphere and the Coriolis force than from thermal factors. Surface winds are light and variable over the oceans. In the days of the wind-driven sailing vessels, these zones were known as the *horse latitudes*. When ships were becalmed, sailors threw horses overboard to lighten the load.

In addition to moving heat, the primary circulation also moves moisture—from low to high latitudes and from the oceans to the land masses. The ocean yields much more water to the atmosphere through evaporation than does land. Evaporation from the ocean is highest where the differences between the vapor pressure of the air and of the water are greatest. Ocean temperatures are the highest at latitudes about 30° from either side of the equator. This is a result of the high-intensity radiation and clear atmosphere associated with divergence in this region. The high temperature of the water and the low moisture content of the air create a large difference between the vapor pressure of the air and the water. Thus, the subtropical oceans near the divergence zones are the major source of the precipitation for land areas of low and midlatitudes.

Trade Winds

Lying between the subtropical belts of subsidence and the equatorial convergence zone is a region of surface winds flowing toward the equator known as the **trade winds**. These wind systems are most frequent in belts 10° in width and centered on 15° to either side of the equator. They are the most regular wind systems found at Earth's surface. The winds, which average 4–7 km per hr (2–4 mph), are most persistent over the eastern half of the oceans and are more dependable there than elsewhere. These winds are less distinct at latitudes higher than 15°. Their origin is in the subsidence of the subtropical high-pressure zones, and they flow into the equatorial low-pressure belt. The Coriolis force gives the winds a westward turn, so they become northeasterly winds in the Northern Hemisphere and southeasterly in the Southern Hemisphere. The trade winds often have layers, with a surface layer that becomes increasingly moist and unstable as the air moves toward the equator. The layer above is dry and stable.

As soon as an air mass begins to move from the source region, its temperature and moisture content begin to change. Heating or cooling from the surface, either by conduction or radiation, alter the temperature of the air mass. Evaporation, condensation, and precipitation add water to or remove water from the air mass. At any given time, an air mass reflects the nature of the source region and the changes that the air mass has undergone while moving over the surface.

As the air from the subtropical high moves toward the equator, its moisture and instability increase, as does its latent heat. The following excerpts from *Mutiny on Board the HMS Bounty* illustrate the effects of the sea surface on the overlying air. The entries in the log were made as the *Bounty* sailed south in the North Atlantic toward the equator while in the trades.

> The thermometer was at 82°F in the shade and 81½° at the surface of the sea, so that the air and the water were within a half a degree of the same temperature. Monday the 4th. Had very heavy rain; during which we nearly filled our empty water casks. So much wet weather, with the closeness of the air, covering everything with mildew. The ship was aired below with fires, and frequently sprinkled with vinegar; and every little interval of dry weather was taken advantage of to open all the hatchways, and clean the ship, and to have all the peoples wet things washed and dried. Monday the 18th. The weather, after crossing the line, had been fine and clear, but the air so sultry as to occasion great faintness, the quicksilver in the thermometer, in the daytime, standing at between 81° and 83°F, and one time at 85°F. In our passage through the northern tropic, the air was temperate, the sun having then high south declination and the weather being generally fine till we lost the Northeast trade wind; but . . . such a thick haze surrounded the horizon, that no object could be seen, except at a very small distance. (pp. 30–31)

The trades are known for providing the route across the Atlantic for sailing ships during the age of exploration. These same winds play a major role in the Pacific Ocean as well: They carried Thor Heyerdahl and his crew across the Pacific Ocean on the first modern crossing by raft.

> The wind did not become absolute still—we never experienced that throughout the voyage—and when it was feeble we hoisted every rag we had to collect what little there was. There was not one day on which we moved backward toward America, and our smallest distance in

twenty-four hours was 9 sea miles, while our average run for the voyage as a whole was 42½ sea miles in twenty-four hours. (Heyerdahl 1950, 41)

Intertropical Convergence Zone

Near the equator, trade winds from both hemispheres converge to form a low-pressure trough with a gentle upward drift of air. This convergence area is also called the **intertropical convergence zone** (ITCZ). It is the area in which the trade winds from north and south of the equator merge. The area of convergence is quite broad along the equator because of the extensive area of warm surface. Although this low-pressure trough is not very deep, it is consistent over the oceans and is deep enough to produce convergence most of the time. Where the converging trades have a trajectory over the ocean, the air contains large amounts of moisture, cloud cover is extensive, and precipitation is frequent.

Less evaporation occurs from the ocean along the equator than near the tropics. The reason for this disparity is that the air is humid and the vapor pressure at the water surface is lower near the equator. Sea temperatures are lower near the equator than they are closer to the tropics. Precipitation and streams draining from adjacent land masses add fresh water that cools the ocean. Less solar radiation reaches the surface as a result of the more extensive cloud cover, further reducing water temperature.

A major source of water vapor in the stratosphere is the warm, moist air rising in the ITCZ. The tropopause marks a transition in the vertical temperature distribution. The temperature in the stratosphere is higher than below in the tropopause. This acts as a cap on rising air. The heated air in thunderstorms in the ITCZ routinely breaks through the tropopause into the stratosphere. Since the air rising through the tropopause cools at the adiabatic rate, it rapidly becomes cooler and denser than the surrounding air. It then drifts back downward. Some of the air is carried away from the equator and subsides in the subtropical high. The rising air in the ITCZ gains a lot of heat from condensation, then radiates away much of that heat to space as it flows toward the subtropical high. This is how Hadley cells take heat from the ocean surface near the equator and move it to the upper troposphere and toward the poles.

In the center of the ITCZ over much of the Atlantic and Pacific we find the *tropical easterlies*—stable, low-velocity winds moving from east to west. Do not confuse them with the trade winds. These easterly winds are quite regular in direction. Occasionally, a change in circulation will produce unstable weather in the form of squalls and general rainstorms. The converging winds on occasion will rise some distance away from the center of the convergence zone. This phenomenon may in turn result in stagnant conditions near the surface. Sailors of the sixteenth century called these stagnant conditions—very low-velocity winds with ill-defined direction—the **doldrums**. While traveling to the Western Hemisphere from Europe, sailing vessels became becalmed in this area, just as they did farther north in the horse latitudes.

Midlatitude Circulation

The three-cell model of the atmosphere has inadequacies, especially regarding the midlatitudes. The model calls for a mean flow of air between the semipermanent high-pressure cell at 30° and the semipermanent low-pressure trough near 60°. In reality, the mean north–south (or meridional) flow in these latitudes is very weak. Surface winds vary in both direction and velocity. The greatest frequency and highest average velocity occur west to east. Maximum westerly velocities occur at or about 35°N. Wind velocities at this latitude become more pronounced with height up to the tropopause. Thus, the primary flow in the midlatitudes is zonal (west to east).

The maximum transfer of energy takes place in the latitudes of 35°–40°N and S. With a relatively weak meridional flow, energy must be transported through the zone by some other means. This region contains the mixing zone between warm, tropical air and cold, polar air known as the *planetary frontal zone*. The midlatitude westerlies are strongest here. The westerly flow increases up to about 12 km (7.2 mi). In the upper part of the troposphere and the lower stratosphere, Rossby waves develop in the planetary frontal zone (Figure 8.4). These waves have an amplitude and wavelength that vary up to 6000 km. They are always in a state of flux, sometimes standing and sometimes moving with the westerlies.

The basic flow of the upper air westerlies is from west to east at a mean latitude of about 45°. The westerlies and the subpolar front divide the cold, dry arctic air from the warmer, moister tropical air. When the westerlies are blowing primarily west to east, it is called a **zonal flow**. Zonal flow over North America keeps arctic and tropical air separated; cold air remains north of the polar front and warm weather south of it. Since the dominant wind direction is west to east with zonal flow, air from the Pacific Ocean flows over the Rocky Mountains and crosses the continent. As it crosses the mountain ranges, the air warms and dries. This brings periods of warm, dry, stable weather to areas of the continent south of the jet stream.

When the westerlies begin to loop widely, they cause a greater meridional flow, even though the net direction of flow is from west to east. **Meridional flow** causes exchange of warm and cold air. Warm air, and hence heat, is carried northward, and cold air is carried southward. When meridional flow is high, temperatures may fluctuate rapidly and widely on a week-to-week basis. In the first week of February 1991, for example, temperatures went from record lows to record highs over much of the northern United States and southern Canada.

FIGURE 8.4

Model of the development of waves in the westerlies: (*a*) When the westerlies are zonal (flowing from west to east), there is minimal mixing of tropical and polar air. When waves form with airflow becoming more north–south, a great deal of cold and warm air is exchanged. (*b*) A wave has begun to form. (*c*) A wave exhibits large oscillations. The maximum exchange of warm and cold air occurs at this stage. (*d*) As the system begins to stabilize in a zonal pattern again, masses of cold air become isolated toward the equator.

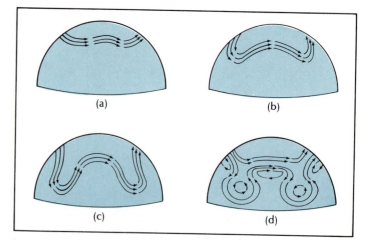

Figure 8.4*d* illustrates a situation in which the loops are large enough that it isolates cells of cold or warm air from the main airflow. Sometimes these large masses of air stagnate and control the weather of the area in which they are located for up to several weeks; they are called blocking systems. These masses of air can bring either very warm or very cold conditions. Blocking highs divert moisture-bringing storms away from the region, increasing the likelihood of drought.

The westerlies shift back and forth between a zonal pattern and a meridional pattern at irregular intervals, making long-range forecasting difficult.

Near the surface, large eddies or vortices develop that are 1000–2000 km across and have lifetimes of several days. These eddies are the cyclones and anticyclones of the midlatitudes. They rotate, carrying cold air from the poles toward the equator and warm air from the subtropics toward the poles. While the eddies carry energy toward the poles, the net flow of air in that direction is very small: It is basically an exchange of warm for cold air. The high frequency of these cyclonic cells also forms the subpolar low-pressure zone in latitudes 50°–60°. This convergence zone differs markedly from the equatorial convergence zone. The air flowing into it has very different temperature and moisture conditions.

In the midlatitudes, the primary circulation is very different from that of the tropics. The subtropical high over the oceans and the polar high act as major source regions for air moving toward the convergence zone. These two source areas are very different. One is very warm and the other is quite cold. While this difference alone is enough to produce different airstreams in the midlatitudes, moisture properties of the airstreams magnify the differences. Thus, while in the tropical convergence zone converging airstreams have similar temperatures and moisture content, in the midlatitudes they are very different.

The upper-air westerly flows dominate the circulation of the midlatitudes. The temperature gradient between the equator and poles determines the strength and location of the subtropical jet stream. As polar regions warm in summer, the gradient is least and the strength of the westerlies diminishes. The jet migrates toward the pole to the mean position shown in Figure 8.2*a*. When a steep gradient exists, the westerly circulation is strong and the jet migrates closer to the tropics. Figure 8.2*b* shows the mean jet stream for the Northern Hemisphere winter. The limit of the midlatitudes is along the axis of the semipermanent high-pressure cell located near 30°. From here to the equator, the Hadley circulation operates.

The surface circulation depends on the flow of upper-level westerlies. The subtropical jet migration and the westerly circulation weakening produce very different conditions from season to season. The movement and relative dominance of air masses will vary, as will the frequency and paths of traveling low-pressure systems.

The movement of the polar front toward the equator results in the movement of polar air masses over much of the midlatitudes. The invasion of extremely cold air produces cold waves. Cold waves are any sudden drop in temperature within a 24-hour period that may require measures to prevent extreme distress to humans and animals. In the United States, winter cold waves claim more lives than any other weather-related event. Statistics compiled by the National Center for Health show an average of 355 deaths per year for the period 1949–78. While deaths directly attributable to cold take a high toll, evidence shows that cold indirectly affects the death rate as well, particularly in subtropical areas where cold waves are infrequent. In Florida and Alabama, death rates nearly doubled during one severe cold wave in 1965. The Applied Study in Chapter 16 contains more specific information on cold weather and human physiology.

Equally harsh outbreaks of arctic weather occur in Europe and Asia. In 763, a cold wave froze both the Black Sea and the Straits of Dardanelles in Europe. In 1236, the Danube River reportedly froze solid. In 1468, the winter in Flanders was so severe that wine rations issued to soldiers were cut with hatchets and distributed in frozen chunks. In 1691, crazed from the lack of food in the frozen countryside, wolves invaded Vienna, Austria, killing humans for food.

Polar Circulation

Thermally induced high-pressure systems exist over the north and south polar regions, although knowledge of them is limited. The high-pressure cells exist some distance from the geographic pole toward the *continental cold pole* (the point of coldest temperature). In the Antarctic, this displacement is not large, but in the Northern Hemisphere it is significant. In the Northern Hemisphere winter, heat always flows through the pack ice and from the open-water areas. The cold pole is on the land mass. Since the Northern Hemisphere has two major land masses, it often has two cold poles, one over North America and one over Eurasia. Upper-air divergence and temperature inversions are characteristic of the cold poles. High-pressure systems and divergent air flow occur over polar areas most of the year. The size of the area affected changes, expanding during the winter months. Outward from the polar highs are belts of weak easterly winds that are more pronounced in the summer in the Northern Hemisphere.

Pressure gradients are weak in the tundra, so this is not a stormy region. In fact, it is one of the least stormy areas of Earth. High-pressure systems dominate with cold, still, dry air prevailing, particularly in the winter. The North American Arctic is even less stormy than other areas. Because of the mountainous character of Alaska, the cyclones of the Pacific Ocean do not move over the Arctic coastline. Some of the storms penetrate far into the Arctic, but few compared to other areas. Weather is most active in spring and fall, as in most high-latitude locations. Winds are low-velocity and variable in direction. The average velocity is less than 15 km per hr, less than 10 km per hr in the Canadian tundra.

In the region surrounding Antarctica is an unbroken sea extending completely around Earth. In the winter, storms are more frequent, bringing high winds from a variety of directions. Sailors of the sixteenth century named this part of the world ocean the "roaring 40s," the "furious 50s," and the "shrieking 60s." Sailors aboard vessels trying to sail around Cape Horn at the tip of South America named the winds. Discovery of the Strait of Magellan made the passage around South America considerably less trying, but by no means easy.

Winds are a predominant factor in the weather of the polar deserts. Westerly winds are most common at the surface to around 65°, where they give way to low-level easterly winds that extend to about 75°. Cyclonic storms develop over the oceans in the westerlies and move around the Antarctic from west to east. They normally do not penetrate far inland, and they account for much of the precipitation and weather along the coast. Winds of a high enough velocity to move snow and produce blizzard conditions occur at Byrd Station in the Antarctic about 65% of the time. They are of high enough velocity to produce zero visibility about 30% of the time. A slight increase in wind velocity brings a geometrically larger increase in blowing snow. Unfortunately for those working in the Antarctic, the most-accessible locations are also the windiest. Mawson's base at Commonwealth Bay experiences winds that are above gale force (44 kph or 28 mph) more than 340 days a year. At Cape Denison, the mean wind velocity is 19.3 m per sec (43 mph), and during July of 1913 it averaged 24.5 m per sec (55 mph).

Seasonal Changes in the Global Pattern

Some major elements of the general circulation shift location from season to season. The tropical Hadley cell shifts north and south with the seasons, a result of the shift in the thermal equator. As shown in Chapter 3, the latitude that receives the most sunshine during the year changes from about 30° N to about 30° S. This change in the heat equator causes the Hadley cell to move north and south as well. Most areas of Earth between the equator and 30° experience seasonal changes in wind direction and precipitation amounts. This is especially true over the continents, since the land surfaces heat and cool more rapidly than does the ocean.

Figure 8.5 shows the semipermanent highs and lows and the dominant winds that occur at Earth's surface. Note, however, that the actual system is not the same as the ideal shown in Figure 8.2c. Earth is not a smooth surface, and the distribution of land masses in the world ocean alters the winds.

The major semipermanent pressure zones tend to shift north and south with the seasons. Consider, for example, the location of the ITCZ. This convergence zone is mainly south of the equator in January (Figure 8.5a), but is north of the equator in July (Figure 8.5b). The subtropical highs and subpolar lows also shift location. This seasonal migration of the pressure belts and associated winds is a direct result of the changing temperature distribution that occurs with the seasons.

Along the equator, the ITCZ moves north and south during the year. The ITCZ results in the convergence of warm, moist air streams from both sides of the equator, producing frequent, abundant precipitation.

Monsoons

In addition to the north–south shift in the general circulation, the relative position of the subtropical highs and subpolar lows also shifts east–west. In the midlatitudes, insolation differs a great deal from winter to summer. Because of the relatively low specific heat of Earth materials compared to water, the continents heat and cool more rapidly than the oceans do with the seasons. Heating of the land masses in summer weakens the subtropical highs, in some cases breaking them down altogether and replacing them with a low. During the winter half-year, the subtropical high and polar high expand and often meet over the land masses that have undergone major cooling. In midwinter, high-pressure ridges or cells exist over the Northern Hemisphere land masses. During the summer months, the subpolar low expands over the continental areas. Sometimes tropical lows and subpolar lows merge to form a single trough of low pressure extending from the tropics to 60°–70° of latitude.

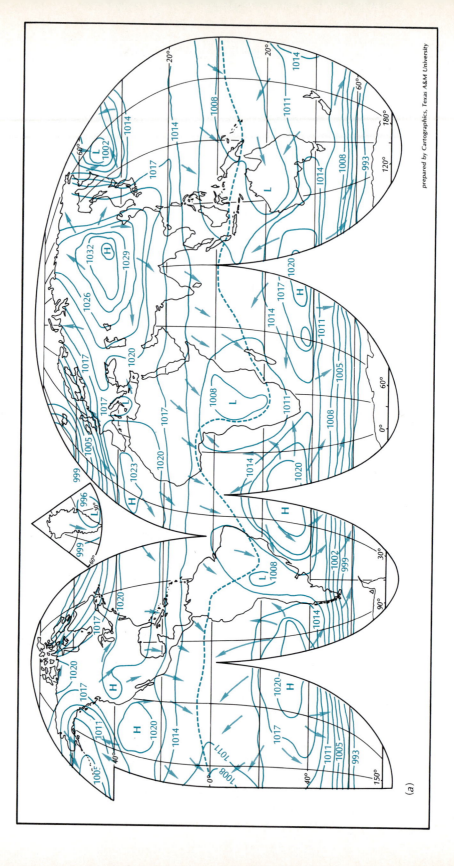

prepared by Cartographics, Texas A&M University

(a)

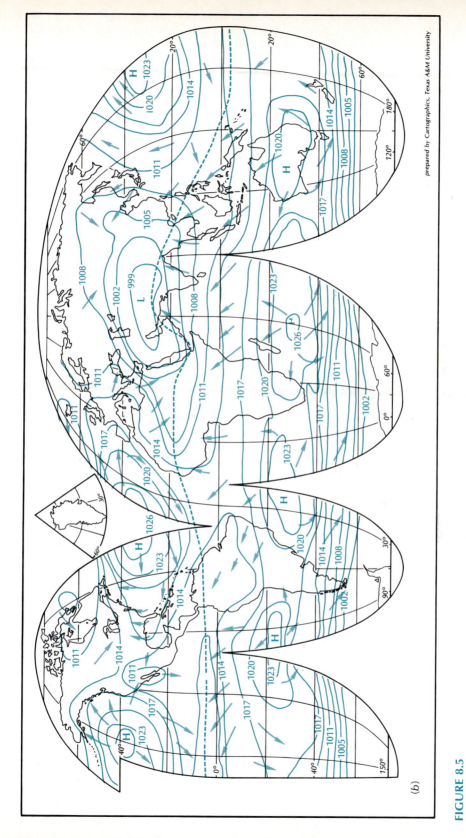

prepared by Cartographics, Texas A&M University

FIGURE 8.5

Mean sea-level atmospheric pressure: (a) in January; (b) in July. Pressure is in mb and arrows indicate prevailing airflow.

(b)

The change in pressure systems reverses wind and moisture conditions with the season. During the summer months, convergence predominates and high humidity and precipitation are common over the continents. In the winter months, when high-pressure systems predominate, divergence over the land is more frequent than convergence. As a result, humidity and precipitation are both lower in winter than in summer. This seasonal shift of pressure, with all its ramifications, is a monsoon. The largest land areas in the midlatitudes are in the Northern Hemisphere, and in these areas the monsoons are most pronounced. A large part of Asia and Africa is subject to this seasonal shift (Figure 8.6).

Asian Monsoon

Seasonal shifts in pressure, wind, and humidity are greatest over Asia. During the winter, subsiding dry air covers much of the continent. This results in a dry season that is comparatively much drier than the North American winters. The subsidence produces off-

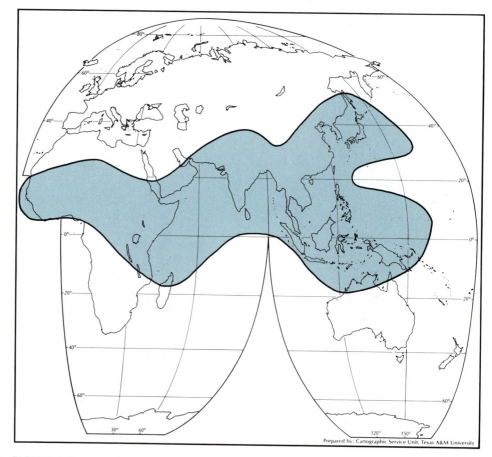

FIGURE 8.6

The monsoon region of Africa and Asia. Within this extensive area, seasonal reversals of wind direction are common and pronounced.

shore winds over much of coastal Asia. As these winds move out over the oceans, they evaporate large amounts of water from the sea surface and provide moisture for the off-shore islands. During the summer months, the low-pressure system and convergence are strong, causing a large onshore flow of moisture (Figure 8.7). Moisture-laden winds blow north from the Indian Ocean and the Arabian Sea. They converge over the northern plains of India and the slopes of the Himalayas. One of the rainiest regions of the world is found in the Assam hills. Cherrapunji, India, is an extreme example of seasonal rainfall: Annual precipitation there has reached 25 m (82.5 ft), and most of that falls in four months. Resulting precipitation on the coastal areas and oceanic sides of offshore islands is heavy.

In the spring and fall, periods between the rainy season and the dry season have unsteady (but not too unpleasant) weather. The highest temperatures of the year often occur late in the spring or in early summer. This is before lots of moisture begins to move onto the continent. The clear skies before development of the onshore winds allow more radiation than does the moist maritime air that follows. The rains and humidity, which reduce absolute temperatures, actually increase sensible temperatures.

The Asian monsoon is probably the product of a variety of factors working together. One is that the ITCZ migrates to between 25° and 30° N. A second is that the interior of the Asian land mass records a large change in pressure that reinforces the migration of the ITCZ. The Himalaya Mountains are probably also a very significant topographic barrier that splits the circulation over Asia. In the winter, the jet stream is over the mountains. The polar high expands and covers much of north-central Asia. Divergence from this strong system north of the Himalayas produces extremely strong offshore surface winds. South of the Himalayas and the jet stream, the anticyclonic flow is much weaker, though offshore flow is most common. In the summer, as the temperature differential over the continent diminishes, the jet stream breaks down. The subtropical high and ITCZ shift rapidly northward and bring the sudden shift of winds onshore over the Indian sub-continent. North of the Himalayas, a weaker convergent system develops, causing an onshore flow of air. The precipitation is primarily convectional rainfall, randomly distributed within the flowing moist air. Some traveling cyclones associated with the ITCZ add further rainfall. The peak season of rainfall moves northward over the subcontinent as the zone of convergence moves northward toward the slopes of the Himalayas.

North American Monsoon

The monsoon applies to North America as it does to Asia. In summer the subtropical high shifts north over the Pacific coast, causing virtual desert conditions over much of southern California. Summer is the dry season for all of the United States west of the front ranges of the Rockies. On the eastern side of the continent the low-pressure trough is more persistent, and convergence is frequent in this region. Onshore winds from the Atlantic high-pressure system centered over the Azores Islands carry considerable moisture. The convergence and lifting over the land areas produce considerable rainfall during the summer over much of the eastern part of the country. During winter, the subtropical high over the Pacific and the subpolar low move toward the equator as the solar equator moves southward. As the subpolar zone of convergence moves southward, precipitation increases along the West Coast. To the east of the mountains, winter high-pressure systems are more frequent, and the Great Plains experiences a season of low precipitation, although not absolute drought.

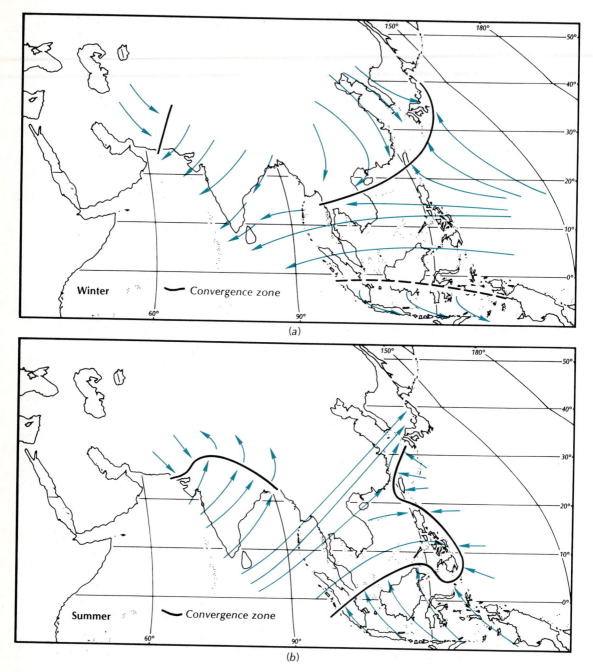

FIGURE 8.7

(a) Winter and (b) summer monsoons over Asia. The normal location of the convergence zone shifts substantially, so most of the area has distinct rainy and dry seasons.

Summary

The general circulation is important in transporting heat energy from the equator toward the poles. This heat transfer reduces the imbalance in energy between the tropics and the polar regions. The general circulation also transports water from ocean to land. The winds follow a pattern resulting from the temperature difference over Earth, the Coriolis force, and the differential heating of land and water.

Winds in equatorial regions are controlled by the subtropical highs and the intertropical convergence zone. In the midlatitudes, the airflow is between the subtropical high and subpolar low. The convergence of air into the subpolar low takes on sharply contrasting temperature and moisture. This results in the development of the lows that produce so much of the variable weather in the midlatitudes. Over the poles, subsidence predominates, producing a flow of cold, dry air toward the equator.

The seasonal imbalance in energy between the Northern and Southern Hemispheres causes the planetary circulation system to shift north and south. As the solar equator moves north toward the Tropic of Cancer, the general circulation moves northward. Because of the extensive amount of land in the Northern Hemisphere, a land-to-sea shift occurs in the relative position of the subtropical high and subpolar low. This produces a seasonal reversal of pressure and wind direction known as the monsoon.

9

INTERANNUAL VARIATION IN WEATHER

Chapter Overview

Variation in Solar Activity

Variation in Atmospheric Dust

Changes in the Ocean–Atmosphere System: The Southern Oscillation

Interannual Variation in the Monsoons

Chapter 8 discussed the general circulation of atmosphere as a semipermanent system that shifts location with the seasons. This is a simplification of the real-world system, representing the most-frequent circulation rather than the actual flow at any given moment.

The general circulation system never repeats itself exactly. It changes from day to day, week to week, and month to month. No two years of weather are ever alike. We need only examine the record for the past 50 years or so to find extreme short-term shifts in the weather. Exceptionally cold and warm or wet and dry years have occurred. Clusters of years similar in temperature or precipitation have also happened. These periods of weather that deviate from the normal create a hazard to human activity because so many of our activities are adapted to the normal pattern, including the basic seasonal changes.

Whenever the atmosphere changes dramatically from the usual, it causes all kinds of problems for living organisms. Changes benefit some areas, but have significant costs for others. Hazards such as floods, droughts, cold waves, and heat waves are examples of such deviations in the general circulation.

As yet, we have no satisfactory explanations for why the general circulation changes as it does. For this reason it is not possible to forecast future changes. Many processes and events can cause the planetary atmosphere to change. Some of these events and processes are external to the planet, such as changing solar activity. Other processes that affect weather result from change on the planet itself, such as changes in the dust content of the atmosphere or ocean currents. Some events cause changes in the weather to occur rather quickly. The resulting weather changes may last for months, or in some cases for several years. Some changes in the weather between one year and the next do not herald a change in climate.

Variation in Solar Activity

The solar energy received at Earth's surface can change due to the amount of energy given off by the sun, changes in the transparency of the atmosphere, or changes in the distance of Earth from the sun. There is little doubt that solar irradiance affects weather. The average temperature of the sun is near 5438°C, but varies slightly, as does its energy output. The actual variation in solar irradiance is small and difficult to measure. Most measurements are made from within earth's atmosphere, and the turbidity of the atmosphere affects the measurements. Until recently, measured differences in solar energy were less than the accuracy limits of the instruments used to record the variations. Beginning in January 1977, the temperature of the surface of the sun fell 11°C in a single year, as measured by Kitt Peak National Observatory near Tucson. This is the first time the observatory measured such a change, but measurements only began in 1975.

During the 1980s, the sun's brightness faded. The irradiance of the sun declined 0.07% from 1981 to 1984 as measured by satellites outside Earth's atmosphere. A change as small as 0.1% for a decade or more might visibly change Earth's climate. In the fall of 1986, the number of sunspots declined. As sunspots decline, so do other elements at the surface of the sun. The net effect is less irradiance and less energy reaching Earth.

Computer models show that a drop in solar irradiance of 1%–2% would bring about conditions similar to those of the Little Ice Age. Snow and ice would spread over high latitudes in the Northern Hemisphere. A decline of 2% for 50 years would be enough to revive glaciation. A drop of 5% would be adequate to bring about a major glaciation of Earth.

Scientists have suggested for a long time that sunspots are responsible for changes in weather patterns and for climatic cycles. Detailed analysis of the outer surface of sun, or photosphere, shows dark, circular areas, known as sunspots. These are areas in the surface where temperatures drop some 1400°C lower than surrounding areas. Intense magnetic fields are associated with sunspots. Sunspots were seen and recorded as early as 28 B.C. in eastern Asia. They have been studied intensively since invention of the telescope shortly after 1600. The number of sunspots occurring at any one time varies from as few as five or six to as many as 100.

Sunspot activity has both long- and short-term fluctuations. Two periods of major deviation from normal existed in the past 1000 years. An unusually high number of sunspots formed in a 200-year period around 1180. Between 1645 and 1715, no sunspots occurred for years at a time. The total reported for the entire 70 years was less than what usually occurs in a single year. This time saw minimal auroral activity, and the solar corona was less visible. This period, known as the Maunder Minimum, is the only such period in historic times. It was also a period of exceptional cold known as the Little Ice Age. The Maunder Minimum appears in both historical records and in studies of tree-ring growth using carbon-14 dating methods. Using a record of tree growth rates dating back 8000 years, we see that each period of weak sunspot activity correlates with periods of cold as shown by tree-ring growth and glacial advances. For more information on using tree rings for dating, see Chapter 18.

Some signs indicate that sunspots follow an 11-year cycle, and multiples of that cycle appear at 22 and 33 years. Still other cycles appear at periods as short as 5.5 years and as long as 90.4 years. The key to sunspot cycles is that magnetic fields on the surface of the sun reverse their magnetic polarity every 22 years, an interval twice the apparent sunspot cycle.

Repeated studies trying to correlate rainfall with the fluctuation in sunspot cycles have not yet produced statistically significant results. As Tannehill (1947) stated more than 40 years ago, "Some drought years have come close to the top of the sunspots (1893 and 1917), some near the bottom (1901 and 1933), some with increasing spots (1925 and 1936), and some with decreasing spots (1910 and 1930). No matter how we select the years or how we group them, we see no obvious relation to sunspots." Controversy surrounds the relationship between sunspots and weather because no clear physical connection exists between them.

Variation in Atmospheric Dust

Changes in transparency of the atmosphere result from changes in its dust content, in cloud cover, and in the ozone content of the upper atmosphere. Reducing radiation absorbed by Earth by as little as 1% can change surface temperatures by as much as 1.2°–1.5°C.

The amount of fine ash injected into the atmosphere from major volcanic eruptions is sometimes very large. This dust absorbs and scatters a significant portion of solar radiation. Although most of the scattered and absorbed energy eventually reaches the ground, a small part goes back into space without affecting temperatures in the lower atmosphere. Maass and Schneider (1977) examined temperature records for 42 weather stations scattered over Earth, each with a temperature record at least 85 years long, and compared these data with levels of atmospheric dust. They concluded that stratospheric temperatures have increased as a result of volcanic aerosol injection. They also found that annual temperatures at the Earth's surface drop following major volcanic events. Some individual cases support these findings, others do not. Mt. Aso-san in Japan erupted in 1783, and cold years followed from 1784 to 1786. The famous year without a summer in 1816 came the year after the massive eruption of Mt. Tambora in the Dutch East Indies. So much dust blew into the air that it produced almost total darkness for three days at distances up to 500 km (311 mi) from the mountain. The explosion of the volcano on the island of Krakatoa in the East Indies blasted some 53 km^3 (40 mi^3) of solid debris more than 30 km (18.6 mi) into the atmosphere. Winds in the upper atmosphere distributed the dust over the planet. For the next three years, Montpelier Observatory in France recorded a 10% drop in the intensity of solar radiation. Unusually cold years from 1884–86 accompanied the drop in insolation. Mt. Katmai in Alaska erupted in 1912, ejecting 21 km^3 of material. Observations at Mt. Wilson in California and at Bassour, Algeria showed a 20% drop in solar radiation during the following months. It is perhaps worth mentioning that unusually cool weather occurred for a month before Mt. Katmai erupted. Mt. Agung in Bali erupted in 1963 and observations at Mauna Loa, HI, show that the receipt of direct solar energy dropped sharply by nearly 2% for a period afterwards. However, since much of the scattered and absorbed radiation eventually reached the lower atmosphere, total radiation reaching the surface dropped by only 0.5%.

Other volcanic eruptions, including some more forceful than those already mentioned, produced no such cooling effects on the weather. For example, Mt. Cosequina in Nicaragua in 1835 blew 49 km^3 of ash into the atmosphere, but no atmospheric after-effects were reported. Other notable eruptions have also occurred without any undisputed effect on weather.

The eruption of El Chichon in Mexico in 1982 blew nearly 10 times as much ash and gas into the atmosphere as Mt. Saint Helens did in 1980; it was the largest volume of rock ejected since the eruption of Mt. Katmai in Alaska in 1912. The impact of these two eruptions on the atmosphere was very different. The blast from Mt. Saint Helens went laterally, and most of the gases and ash stayed in the troposphere, where they were rapidly precipitated out. Perhaps the most significant aspect of El Chichon is that the main eruption was vertical; as a result, most of the sulfuric gases and sulfuric acid aerosols went into the stratosphere, where they remain for much longer periods. Determining the effect this cloud's debris had on insolation is easier than determining its effect on actual surface temperatures. The reduction in solar radiation at the surface could have reduced Earth temperatures by about 0.25°C during 1983, but no actual temperature drop was measured. Volcanic eruptions probably affect weather only for short periods. There is not yet any solid evidence to show any long-term effects of volcanic activity.

Changes in the Ocean–Atmosphere System: The Southern Oscillation

Changes within the planetary circulation systems of the atmosphere and ocean cause some changes in weather patterns. Both the atmosphere and ocean are fluid, moving freely in response to internal and external influences. Neither remains in one state for very long; rather, they are each dynamic systems, and together they form a very complex, ever-shifting system. There is no normal condition. By far the most important process at work in the combined system is the heat exchange between the ocean and atmosphere. Seventy-one percent of Earth's surface is water, and that water controls atmospheric circulation. The ocean absorbs far more solar radiation than does the atmosphere, and it has a much higher capacity to store heat than does the atmosphere. The ocean is the major source for heat in the atmosphere, so small changes in the ocean temperature can cause major changes in atmospheric heating. This in turn alters the general circulation of the atmosphere.

El Niño

Off the west coast of South America, an event takes place at irregular intervals that the local people call El Niño. It appears as a warm current of water that replaces the normally cold rising water off the coast of South America. Fishermen of Spanish descent named this event after the Christ Child, since it appears most often in late December. El Niño has occurred frequently since first reported in 1541.

Every few times, El Niño is stronger than usual—it goes further south and is exceptionally warm. It produces rain over the coastal desert. The rains bring a period of profound growth, known as *años de abundancia* (years of abundance). The desert provides abundant grazing for herds of sheep and goats.

In 1904, Sir Gilbert Walker became director general of observatories in India. He studied the monsoon system of Asia partly as a result of the severe famines of 1877 and 1899. By the 1930s he was able to show a cyclical, interannual variation in the atmosphere over the southwest Pacific Ocean, which he called the southern oscillation. This oscillation brings major changes in pressure, winds, and precipitation over the southwest Pacific and the Indian Ocean. By the 1960s, enough data and information were available for scientists to conclude that the southern oscillation extends across the Pacific Ocean. It includes the event called El Niño and the years of abundance.

The Walker Circulation of the Equatorial Pacific Ocean

In the Pacific Ocean, the Hadley cells operate similarly to the general circulation model. The trade winds blow from the subtropical high toward the equator. They blow from the northeast in the Northern Hemisphere and from the southeast in the Southern Hemisphere, resulting in a westward drift along the equator. In the upper troposphere there is a counterflow of air from west to east. This system is called the Walker circulation.

Over the Pacific Ocean there is also a zonal, or east–west shift in the heart of the ITCZ (Figure 9.1). This seasonal shift is due to a shift in the area of ocean surface

FIGURE 9.1

Seasonal shift in the ITCZ over the Pacific Ocean. (*a*) From December through February, the warmest water near the equator is in the western Pacific Ocean, and this is where the convergence is strongest. (*b*) In the summer months of June, July, and August, the area of warm water spreads eastward, and the convergence becomes stronger over the eastern and central Pacific Ocean.

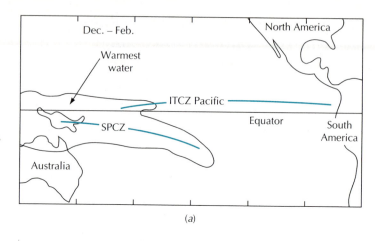

(a)

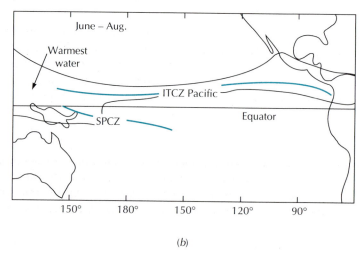

(b)

with a temperature of more than 27.5°C (81.5°F). There is also an upper limit to the temperature of the sea—a little above 30°C (86°F). The ocean temperature will not rise much above this because evaporation cools the surface, and the warmer the water gets, the more it evaporates. The southeast trade winds blowing westward across the Pacific Ocean move the surface water westward. This westward movement of surface water results in sea level being 1 m (3.3 ft) or more higher on the western side of the Pacific Ocean than on the eastern side. The water also warms as it moves westward.

Along the coast of South America, cold, nutrient-rich water from below the surface replaces the westward drift of warm water at the surface. The temperature of the cooler water from below may be as low as 20°C (68°F). As this cooler water moves westward, solar radiation warms it; by the time it reaches Micronesia it has warmed to 24°C (75°F) or more.

The layer of warm water at the top of the Pacific Ocean is fairly thin, usually less than 100 m (330 ft). At the bottom of this layer of warm water is a fairly sharp boundary called the thermocline. The water below the thermocline is a lot colder. The westward drift of warm water alters the depth of the thermocline south of the equator. The thermocline drops to a depth of 200 m (660 ft) in the western Pacific. At the eastern edge of the Pacific Ocean, the thermocline rises almost to the surface. To counterbalance the westward drift of water at the surface, water flows eastward along the thermocline. This is the Cromwell current, also known as the equatorial undercurrent, a strong current that reaches a velocity of 1.1 m per sec. in midocean. Since the winds along the west coast of South America are weak and the cold water offshore stabilizes the lower atmosphere, precipitation along the coast is fairly low. The western side of the Pacific Ocean has much more precipitation, since it has an onshore flow of air and the air is moister from traveling over the ocean.

El Niño–Southern Oscillation

In low latitudes, surface temperatures of the ocean average around 27°C (80°F). Marked changes in sea-surface temperatures in the Pacific Ocean accompany the southern oscillation. These surface-temperature changes happen mostly within 7.5° of the equator. Temperature increases of more than 3°C take place in the warm phase of the El Niño–southern oscillation (ENSO), and the warm pool shifts location. During El Niño, the area of warm water expands eastward and the area of convergence and precipitation drifts eastward (Figure 9.2). Rainfall increases over the eastern Pacific Ocean and along the west coast of South America. Pressure differences decrease and the trade winds die down. Pressure differences between Darwin, Australia and the island of Tahiti illustrate effects of the southern oscillation. During El Niño, pressure increases at Darwin and rainfall decreases. Pressure decreases at Tahiti and rainfall increases eastward toward the international dateline. Sea-surface temperatures also increase toward the dateline. In the western Pacific, rainfall decreases and pressure increases (Figure 9.3).

Temperature decreases occur during the cold phase of the southern oscillation. During El Niño, atmospheric pressures are higher than normal over southeast Asia.

The ENSO weakens when the supply of warm water moves away from the equator; the oscillation lasts anywhere from 14 to 22 months.

El Niño Episode of 1982–83

Atmospheric pressures began to rise over the Indian Ocean in the summer of 1981. In the spring and early summer of 1982, water temperatures stayed near normal. Across the ocean, the usually heavy rains of summer did not develop in Australia and Micronesia. Shortly after the summer solstice, the wind system reversed itself over the western Pacific, but water temperatures off South America remained normal.

When more uniform sea-surface temperatures occur across the Pacific, the trade winds weaken or even reverse. If the trade winds blow toward the east, they produce a surge of warm water known as the Kelvin wave. The Kelvin wave and the warm coastal current are not the same. They combine to push warm water further south than usual.

FIGURE 9.2
Circulation during an El Niño. The thermocline drops in the eastern Pacific Ocean. Cloud cover and convectional rainfall spread east across the Pacific. The warm pool of water shifts location. The surface winds slow or reverse direction, changing the surface temperature structure of the equatorial Pacific.

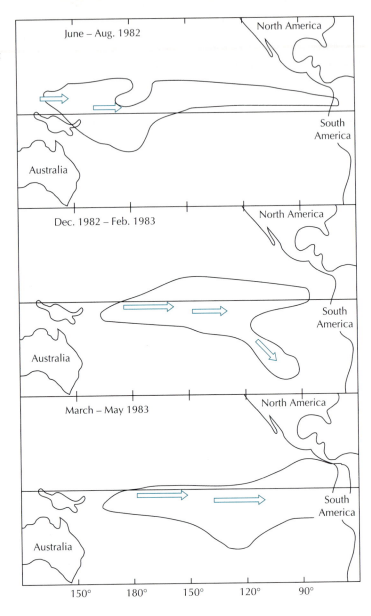

On September 25, 1982, sea-surface temperatures near the village of Paita, Peru rose 4°C (7.2°F) in one day. By the first of December they had risen 6.5°C (11.7°F). In some parts of the Pacific Ocean, surface temperatures rose as high as 31°C (88°F) and were as much as 8°C (14.4°F) above normal. Eventually, a swath of unusually warm water 13,000 km (8078 mi) long straddled the equator. By the end of Dec. 1982, the thermocline off the coast of South America pushed downward 150 m (492 ft). One result of this was that the eastward flowing Cromwell current came almost to a halt.

FIGURE 9.3

Shift in the surface current and warm water pool during the El Niño of 1982–83.

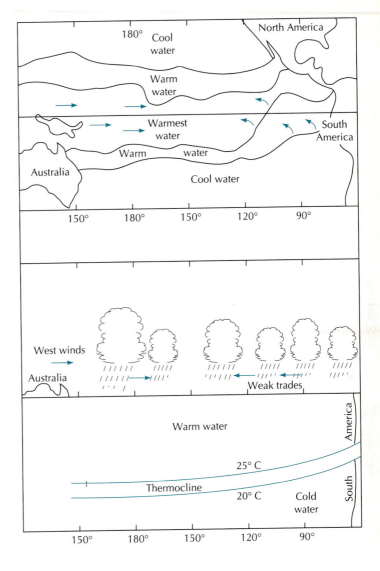

Counterpoint—La Niña

In the summer of 1987, the temperature in the Pacific Ocean along the equator dropped 4°C (7.2°F). This brought unusually cool weather to the eastern Pacific. They call this sudden cooling of the equatorial water La Niña, Spanish for "the girl." The name was applied to the phenomena for the first time in 1986. La Niña exaggerates the normal southern oscillation. During La Niña, the trade winds are stronger, the water off western South America is colder, and water in the western Pacific near the equator is warmer than normal. In the western Pacific, surface pressures are lower and heavy rainfall occurs (Figures 9.4 and 9.5).

The strengthening of the Walker circulation causes the coastal deserts of Peru and Chile to become even drier, if that is possible. On the western edge of the circulation, southeast Asia gets even more summer precipitation than usual. In Bangladesh, more than 1000 persons perished in floods.

Some scientists held La Niña responsible for the drought in the U.S. Midwest in summer 1988. The change in water temperature altered the jet stream from a zonal flow to a pattern containing a huge loop over North America. A high-pressure system developed in the loop that resulted in higher temperatures and blocked moisture from entering the Great Plains (Figure 9.6). La Niña was also blamed for the severe hurricane season, which included Gilbert, and for the abnormally cold winter in Alaska and western Canada. El Niño may also affect global warming because it will increase the absorption of CO_2 (i.e., plant growth will increase with more rainfall).

FIGURE 9.4

Ocean temperatures and wind directions during La Niña. The thermocline reaches the surface far west of South America. The warm pool of water is limited to the western Pacific Ocean. This restricts cloud cover and precipitation to the far western Pacific Ocean.

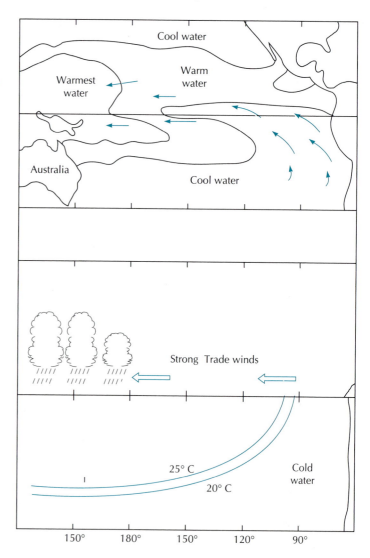

FIGURE 9.5

Temperature anomalies over North America and the Pacific Ocean in the summer of 1987. The shift in the circulation resulted in unusually warm temperatures over western Canada and the north central United States.
Source: Data from R.W. Reynolds, "A Real-Time Global Sea Surface Temperature Analysis," *Journal of Climate* (1988): 75–86.

The Impacts of El Niño

The El Niño phase of the southern oscillation does bring *anos de abundancia* to the Peruvian portion of the Atacama Desert, but the costs of El Niño far exceed the benefits. General-circulation changes cause widespread social, economic, and environmental disruption. Even in Peru, the benefits to some areas are outweighed by high costs in other parts of the nation.

FIGURE 9.6

During the drought of 1988, a standing wave developed in the jet stream with alternating high- and low-pressure systems along it. Over central North America, a large high-pressure system blocked out the normal flow of moist air from the Gulf of Mexico.

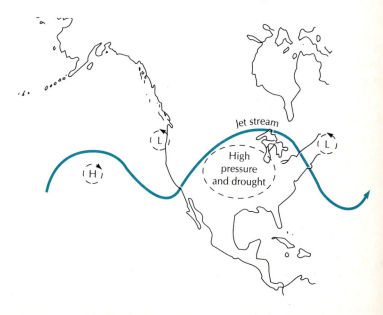

While El Niño brings abundance to the desert, it creates havoc in the marine industries. A major anchovy fishing ground lies off the coast of Peru and Ecuador. Anchovies flourish in the cold, upwelling water, which is rich in nutrients. The nutrients provide food for the plankton, which in turn provides food for the anchovies. During El Niño, upwelling continues, but it is restricted to the warm upper layer, some 125–150 m (410–492 ft) deep. The warm water contains fewer nutrients, the supply of plankton drops, and the anchovies either leave the area or die in the alien environment. When the anchovies disappear, the large flocks of birds that feed on them also disappear. That dries up the steady supply of guano that supports the fertilizer industry. Thus, both the fishing and fertilizer industries suffer from El Niño.

The same heavy rains that cause the coastal desert to bloom also cause massive landslides and floods in the Andes Mountains inland from the desert.

El Niño and Northern Hemisphere Weather

The southern oscillation directly affects atmospheric temperatures. In its immediate area, annual temperatures change by as much as 0.5°C (0.9°F). The ENSO affects atmospheric temperatures over the land masses within 20° of the equator, with temperature anomalies reaching a maximum late in the calendar year and extending into the spring. For this reason, calendar-year data tend to obliterate the effects of El Niño. The effect of the southern oscillation show up much better in winter (Oct.–March) data. Winter data, when the southern oscillation is strongest, show interannual variations of 0.3°C.

Some studies show that the southern oscillation changes temperatures in the Northern Hemisphere during the winter half-year by as much as 0.2°C. It affects northwestern North America and the southeast United States. Temperature changes over the Northern Hemisphere in 1987 were as expected from the model. Temperatures in large areas of the North Pacific Ocean increased by 1°–2°C (1.8°–3.6°F). Northwestern North America contained a large area of warmer weather, and the Gulf of Mexico had a small area of cooler-than-normal temperature (Figure 9.5).

The El Niño of 1982–83 was unusually severe, the most destructive in the last 100 years, and possibly the most extreme case yet documented. The event caused between 1000 and 2000 deaths. Ecuador and Peru sustained the highest loss of life and economic damage. In May 1983, Guayaquil, Peru, received almost 20 times the average rainfall for the month. The flow in some rivers increased to 1000 times the mean flow. The winds caused $400 million of damage in Ecuador alone.

In the western Pacific Ocean the effects of the shift in the ocean-atmosphere system were equally destructive. Six typhoons hit Tahiti during the season. This was after decades without a single occurrence. Eastern Australia suffered severe drought. The Southern Hemisphere subtropical high-pressure system dominates the weather in Australia. In the summer this high usually weakens to permit low-pressure systems to move onshore, bringing rain. In the 1982–83 season, few low-pressure systems reached the continent.

Many parts of the world experienced severe weather in 1982–83; much of that was attributed to El Niño. Table 9.1 lists some of the events. Their occurrence at the same time as El Niño is not enough reason to attribute them all to the phenomenon. For example, the unusually warm weather of the eastern United States during the winter of

TABLE 9.1
Worldwide events attributed to El Niño in 1982–83.

- Above-normal temperatures in Alaska and northwestern Canada.
- The warmest winter in 25 years in eastern United States.
- Drought in S.E. Africa.
- Drought in Australia and New Zealand.
- Drought in Southeast Asia, particularly Sri Lanka, India and Indonesia, Phillipines.
- Drought in Mexico.
- Floods in Louisiana and Cuba.
- Excessive beach erosion in California.
- A 200-mm (8-in.) rise in sea level off California.
- Death of coral reefs in the Pacific Ocean.
- Slowing of the rate of Earth's rotation in late January.
- Reduction in the salmon harvest off western North America.
- Encephalitis outbreaks in eastern United States.
- Increased incidence of bubonic plague in southwestern United States.
- High mortality of seabirds on the Farallon Islands.

1983–84 is attributed to El Niño. However, the last time El Niño occurred prior to 1983–84 was in 1976–77, and that winter was one of extreme cold in the eastern United States. It certainly is not scientific to attribute all of the environmental shifts that took place during the time to El Niño without further evidence to support the charge.

Forecasting El Niño

Like many cyclical phenomenon, El Niño occurs every few years. One usually occurs every four to five years, but they have varied from every other year to 10 years apart. It is not yet possible to forecast El Niño, although many have tried. In September 1982, a committee sponsored by the United Nations to discuss El Niño could not even forecast the event that began later that same month. Climatologists have been using mathematical models to try to predict the phenomenon, but the models have thus far failed. The most common modeling approach is statistical, drawing associations between past events and the current state of the atmosphere.

Part of the reason why forecasting the event is difficult is because we do not yet know what causes the system to form—we only have hypotheses. One suggests that El Niño results from huge amounts of heat released on the sea floor when magma pours out onto it. Another suggestion is that ENSO is a result of high snowfall over Asia the previous winter. When a lot of snow builds up on the Eurasian land mass in a given winter, the hypothesis says snow melt will be much greater; that, in turn, reduces the normal summer heating of the land mass. Unfortunately, it will probably take at least several more ENSO events to test any hypothesis adequately.

Forecasting changes in the southern oscillation is difficult for several reasons. First, it is not a periodic phenomenon; that is, it has neither a fixed interval nor a fixed amplitude. Each event begins a little differently and develops differently. Some events begin off South America, with the appearance of warm water, then develop westward. Other events, including that of 1982–83, appear first in the western part of the tropical Pacific Ocean. They begin with changes in the winds, leading to an eastward expansion of the pool of warm water and precipitation. This causes further breakdown in the trade winds. Then warm water progresses further eastward, and the system goes on to completion.

Interannual Variation in the Monsoons

Much of Asia, the Middle East, and equatorial Africa are subject to the effects of an atmospheric circulation system known as the monsoon. Figure 8.6 delineates the region of Africa and Asia affected by the monsoon. The monsoon is a very pronounced seasonal shift in atmospheric pressure, which in turn brings about a marked change in wind direction, atmospheric humidity, and precipitation.

Summer vs. Winter Monsoons

During the summer months, convergence predominates over Asia and high humidity and precipitation generally prevail over the continent. In summer, the low-pressure system and convergence of air are strong. The summer monsoon winds blow across the subcontinent of India from southwest to northeast from April to October. The winds reaching the west coast of India have traveled over the ocean and have a very high moisture content. As the winds move onshore, they produce considerable precipitation. At Bombay, the rains last from June until the end of September and total over 2 m (6.6 ft). The winds pass over the Western Ghats across the plains of Bengal toward the Himalaya Mountains. Annual rainfall varies tremendously as a result of the wind direction and the topography of the Indian subcontinent. The Western Ghats, which run nearly perpendicular to the direction of the summer winds, divide India into two distinct climatic zones: a narrow, very wet zone on the western side and a much wider and drier zone on the eastern side.

During the winter months, when high pressure dominates the Asian land mass, divergence is more frequent than convergence. Divergence from this strong system north of the Himalayas produces extremely strong offshore surface winds. South of the Himalayas and the jet stream, the anticyclonic flow is much weaker, though offshore flow dominates. When divergence is present, subsiding dry air covers much of the continent, resulting in lower humidity and less precipitation. The subsidence produces offshore winds along much of coastal Asia. As the air streams out over the ocean, it evaporates large amounts of water from the sea surface, thus becoming a source of moisture for the offshore islands and along some coastal areas. Parts of India have a secondary rainy season with the offshore flow of air as well as a second-season harvest in March and April.

Drought and Famine

Most rainfall in northern India occurs during the summer onshore monsoon. The time of year when the Asian monsoon begins and ends varies from year to year, as do the duration and intensity of the rainy season. Total rainfall varies as a consequence. The agricultural economies of northern India and Pakistan are cruelly subject to the whims of the system. This summer monsoon is important for most of India, since it is primary growing season for rain-fed crops.

The demand and supply of food have been in a delicate balance throughout history. When the food supply has increased, the population has grown, and when food has been scarce, some trauma has decreased the population. During the long period of the hunting and gathering societies, starvation was probably near at hand for individuals, family groups, and tribes. The development of agriculture allowed the world population to expand rapidly, but then the basis for the food supply became more directly dependent upon the weather. Famine did not become a part of human experience until after agriculture began.

Famine has many causes, but one of the major ones is drought, and most of the catastrophic famines have in fact resulted from drought. Drought affects the quality and quantity of crop yields and the food supply for domestic animals. Severe drought may mean a major loss of domestic animals, multiplying the effect of the drought. Major famines have hit Asia from the time agriculture spread over the continent. India, China, Russia, and the countries of the Middle East have all suffered from many drought-related famines. Temperature anomalies in 1987 led to a drought in North America in 1988 (Figures 9.5 and 9.6).

Summary

Weather changes from day to day, week to week, and year to year. Many factors cause these changes. The weather differs from year to year as a result of changes in radiation reaching the surface. This can be a result of changes either in output from the sun or in the atmosphere's aerosol content. Ocean currents change, which changes the atmospheric circulation. Changes in weather from year to year often bring disasters or economic stress. Perhaps the greatest impact of annual changes in weather is when precipitation over land areas falls below normal. In subsistence or near-subsistence economies, the result is sometimes famine. Famine has always plagued the human species. As the human population grows, ever-greater numbers of people are affected. Today, famine occurs almost every year in some part of the world.

10

AIR MASSES, FRONTAL SYSTEMS, AND SYNOPTIC CLIMATOLOGY

Chapter Overview

Air Masses

Fronts and Midlatitude Cyclones

Cyclogenesis and Storm Systems
- Measurement of Upper-Air Conditions
- Cyclone Development

Severe Weather and Frontal Systems
- Types of Thunderstorms
- Associated Hazards

Synoptic Climatology

Satellite Imagery and Synoptic Climate
- Satellite Orbits
- Type of Sensor
- Applications

The distribution of solar energy over Earth is the primary factor in the latitudinal variation in climate. This unequal energy distribution along with the general circulation of the atmosphere establish the basic characteristics of climate on Earth. In midlatitudes, the seasons dominate the weather through the year. Weather within the seasons also shows great variation from day to day and week to week. This is largely due to the weather produced by midlatitude cyclones. This chapter examines several elements associated with these storms: air masses, fronts, and waves in the polar front.

Air Masses

A significant step in understanding the workings of the atmosphere was the introduction of the concept of air masses. An **air mass** is a large, relatively homogeneous body of air that may cover thousands of square kilometers and extend upwards for thousands of meters. They are most homogeneous in terms of temperature and moisture. Air masses derive their properties from the surface over which they originate; they are identified and classified by that particular area.

Two categories of air masses are based on temperature and two on moisture properties. Air masses are classified as *tropical* (T) if the source is in low latitudes and *polar* (P) if the source region is in high latitudes. Air masses originating over land, and therefore relatively dry, are labeled *continental* (c); those originating over the oceans, and hence moist, are called *maritime* (m) air masses. This results in four individual kinds of air masses: maritime tropical (mT), maritime polar (mP), continental tropical (cT), and continental polar (cP). Two additional categories are sometimes used in reference to the extremes of continental polar air masses and maritime tropical air masses: Continental Arctic (cA) describes exceptionally cold and dry air; equatorial (E) indicates very warm, moist air. (See Table 10.1 and Figure 10.1.)

The atmosphere's general circulation requires that energy exchanges occur over the globe; as a result, air-mass movement from the source region acts as a mechanism for energy transfer in the atmosphere. The basic properties of the air mass change the further away it moves from the source region. To indicate how air masses are modified, their identification symbols sometimes include a third letter. If the air mass is warmer than the land over which it is moving, then the letter *w* is added. Thus an air mass designated mTw would indicate air that originated over the subtropical ocean and that is warmer than the surface over which it is moving. It may well be a warm, moist air mass moving onto land in the winter. If an air mass is colder than the surface it passes over, the letter *k* is added. Thus a cPk air mass might be passing from a Canadian source region and moving over warmer areas in the United States.

One of the best examples of air-mass modification occurs when cP air masses pass over the Great Lakes in winter. Although cold, the lake water is warm relative to the air, and evaporation supplies moisture to the air mass. Once the air leaves the lakes to pass onto the warmer land on the eastern or southern shores, it becomes unstable and produces snow flurries. As Figure 10.2*a* illustrates, greater snowfall may occur over the higher ground of the northern Appalachians. This lee-of-the-lake effect causes a remarkable gradient in snowfall amounts. Figure 10.2*b* provides snowfall distribution for the state of Michigan.

TABLE 10.1
Classification of air masses.

NAME OF MASS	PLACE OF ORIGIN	PROPERTIES	SYMBOL
Polar continental	Subpolar continental areas	Low temperatures (increasing with southward movement); low humidity remaining constant	cP
Polar maritime	Subpolar and arctic oceanic areas	Low temperatures, increasing with movement, higher humidity	mP
Tropical continental	Subtropical high-pressure land areas	High temperatures, low moisture content	cT
Tropical maritime	Southern borders of oceanic sub-tropical, high-pressure areas	Moderately high temperatures; high relative and specific humidity	mT
Equatorial	Equatorial and tropical seas	High temperature and humidity	E

In moving away from its source region, a mass of air does not pass into a vacuum; rather, it replaces an existing air mass. The leading edge of an air mass will thus be a site of conflict between air of different properties. This conflict site is where fronts form.

Fronts and Midlatitude Cyclones

During World War I, Norwegian researchers made remarkable advances in weather research. Cut off by the war from weather information from other countries, the Scandinavian countries established extensive networks of stations. This network produced data that meteorologists in Bergen, Norway, used to study the storm systems frequently encountered in northwest Europe. Their observations produced models detailing the structure of midlatitude cyclones.

The vocabulary the scientists used to describe their model reflects the war background of the time. The Norwegians developed the polar frontal model, with the term "front" used to identify a zone of transition between air of different properties; the analogy was to a front dividing the fighting armies. The Norwegians identified various types of fronts, as illustrated in Figure 10.3.

A warm front occurs when a warm air mass replaces colder air. Clearly, because warm air is less dense than cold, the warm air rises above the cold along a front which, in cross section, has a slope of about 1:100 (1 mi vertically for every 100 mi of horizontal distance), which means that high clouds forming along the front may occur some

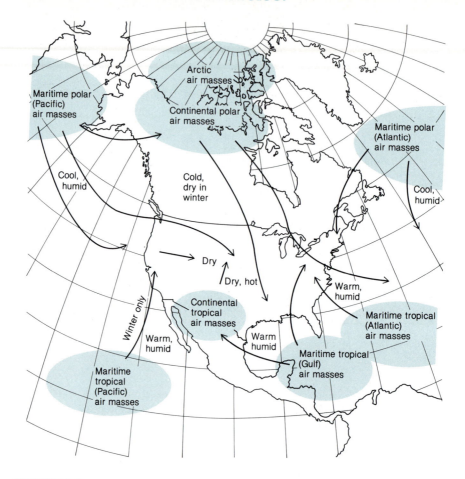

FIGURE 10.1

Air-mass source regions for North America.

Source: Reprinted with the permission of Macmillan Publishing Company from *Meteorology,* Third Edition by Moran and Morgan. Copyright © 1991 by Macmillan Publishing Company, Inc.

500–700 mi ahead of the location of the warm front at the surface. Generally, stable air ascends along the front, resulting in the formation of stratiform clouds. As the cross-section shows, clouds may range from cirrus (many miles ahead of the front) to nimbostratus (where the front intersects the surface). Widespread rainfall often occurs during the passage of warm fronts. The majority of warm fronts contain stable air, but if the air in the warm air mass is highly unstable, then cumuliform clouds will occur along the front. These may be embedded in stratiform clouds, creating problems for pilots of small aircraft.

Cold fronts occur when a cold air mass overtakes and replaces warmer air. Cold fronts are much better defined and generally move twice as fast as warm fronts. The front is much steeper, sometimes having a slope of 1:50. The rapid ascent of warm air along the front results in cumuliform clouds with the possibility of severe weather. In the Northern

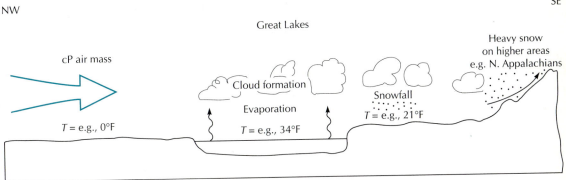

(a) T = surface temperature (over lakes = water temperature)

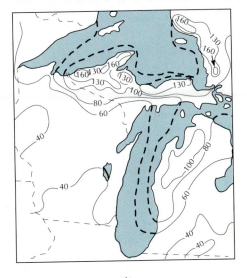

(b)

FIGURE 10.2

Snowfall in the lee of the Great Lakes: (a) Sample readings when winter cP air masses pass over the lakes. (b) Mean seasonal snowfall (in in.) in vicinity of Lakes Superior and Michigan.

After Michigan Department of Agriculture, Michigan Weather Service, and U.S. Department of Commerce, NOAA, National Weather Service.

Hemisphere, strong cold fronts are usually oriented in a northeast to southwest direction and move east or southeast.

At times, the forces are such that the frontal boundary neither advances nor retreats. This stationary front creates conditions similar to those along a warm front, and may persist for a number of days, causing low-ceiling weather over wide areas.

Analysis of these fronts, together with an understanding of the traveling high- and low-pressure systems, led to identification of the midlatitude cyclone. A **cyclone** is a low-pressure area with an organized wind-circulation pattern. In the midlatitudes, cyclones develop

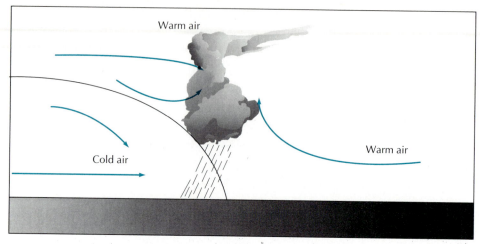

(a)

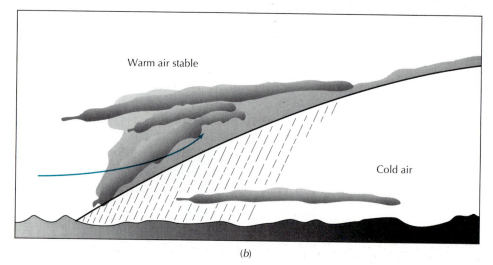

(b)

FIGURE 10.3

Schematic diagrams: (a) A fast-moving cold front with unstable warm air ahead of the front. (b) A warm front of warm stable air.

After *Aviation Weather*, Department of Commerce, Federal Aviation Agency (Washington, DC: Government Printing Office, 1965), 78 and 82.

at air-mass boundaries and are characterized at the surface by the formation of fronts. Figure 10.4 shows the surface history of the midlatitude cyclone developing. Subsequent to the front's formation *(a)*, a v-shaped pattern develops at the initial stage *(b)* together with a central low-pressure and two identified fronts. The wave then deepens, producing an open stage *(c)* in which a distinctive warm sector forms between the cold and warm fronts. The fronts move at different rates (in *d*, the cold front is advancing more rapidly than the warm); this leads to the occluded stage *(e)*, where one front overrides another. The

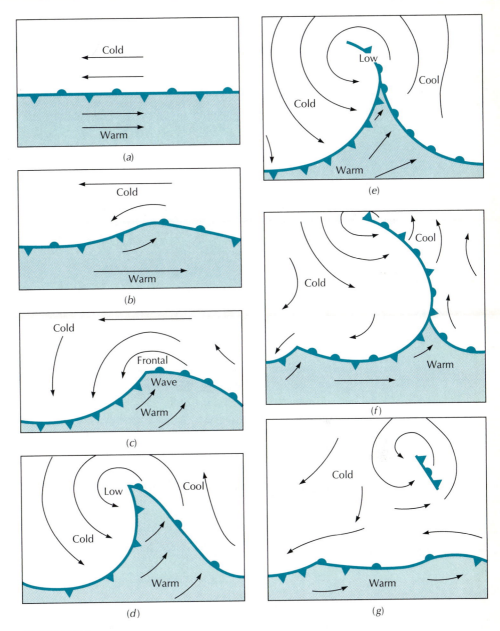

FIGURE 10.4

The life cycle of a frontal wave: (*a*) Formation of the front. (*b*) V-shaped pattern develops. (*c*) Open stage. (*d*) Cold front advances more rapidly than the warm. (*e*) The cold front overrides the warm front. (*f*) Warm air nearly cut off from the surface. (*g*) Dissolving stage. After *Aviation Weather*, Department of Commerce, Federal Aviation Agency (Washington, DC: Government Printing Office, 1965), 84.

overriding continues until the warm air sector has been almost completely cut off from the surface *(f)*. The dissolving stage *(g)* represents the end of the life cycle.

The open stage of the system best describes the types of weather associated with frontal passage. This sequence is illustrated in Figure 10.5, which shows typical temperatures, winds, and moisture characteristics.

While the surface characteristics of such systems were well-known, the cause for their formation required an understanding of upper-level circulation and modern ideas of *cyclogenesis*, the formation of midlatitude cyclones.

Cyclogenesis and Storm Systems

The discussion of the atmosphere's general circulation (Chapter 8) covered aspects of the upper-air circulation. Here, the discussion will emphasize the relationship between surface and upper-air circulation patterns.

Measurement of Upper-Air Conditions

In 1941, the U.S. Weather Service initiated a program to take soundings through the atmosphere; today, a network of stations routinely obtains information from which upper-air charts are constructed. Such was not always the case. A major breakthrough in monitoring the atmosphere came with development of recording instruments in the early 1900s. Carried upward by a balloon that eventually burst, the instruments were parachuted back to Earth. In the 1920s, the radiosonde was introduced. This instrument package carried a radio transmitter that sent back to Earth continuous measurements— called soundings—of temperature, humidity, and pressure. A radiosonde tracked by direction-finding ground equipment was introduced in World War II; named the rawinsonde, it measures wind direction and speed at various levels of the atmosphere. Radiosondes are launched daily at hundreds of ground stations throughout the world to provide the upper-air information required for synoptic analysis.

Upper-air conditions are shown on constant-pressure charts, which indicate the height contour at which a given pressure occurs. Three standard pressure charts commonly in use are the 850-mb, 700-mb, and 500-mb. The average heights at which these levels occur are 5000 ft, 10,000 ft, and 18,000 ft (1525 m, 3050 m, and 5490 m). A 500-mb chart is illustrated in Figure 10.6. The contours show pressure surfaces having the same geometric height above sea level.

Using what we have learned about factors that influence wind flow, we know that the balance between the pressure gradient force and the Coriolis force means that much upper-air flow is geostrophic; the wind will move approximately parallel to the contours. Observations of winds aloft (Figure 10.6) show this to be the case. The chart also shows that the circulation identified is wave-like and that troughs and ridges can be identified within the wave. These waves, with their ridges and troughs, migrate eastward across the United States in a complex of many interacting waves of variable lengths and amplitude.

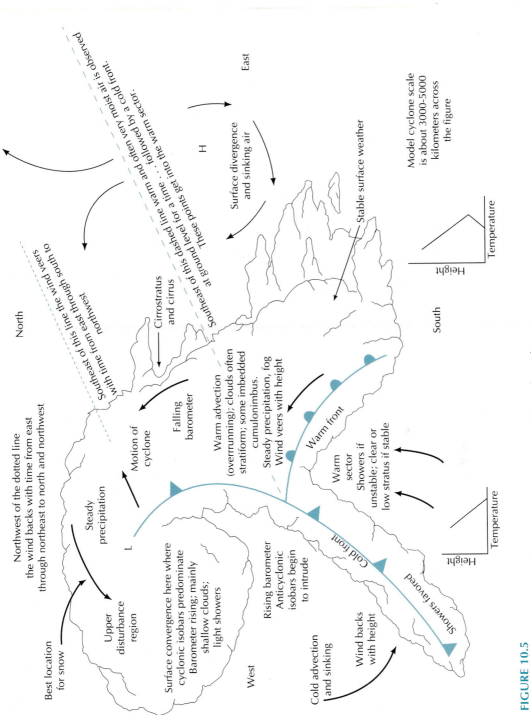

FIGURE 10.5

Distribution of clouds and weather around a mature midlatitude cyclone.

Source: Anthes et al., *The Atmosphere*, 3d ed. (New York: Macmillan, 1981), 119. Reprinted by permission of the author.

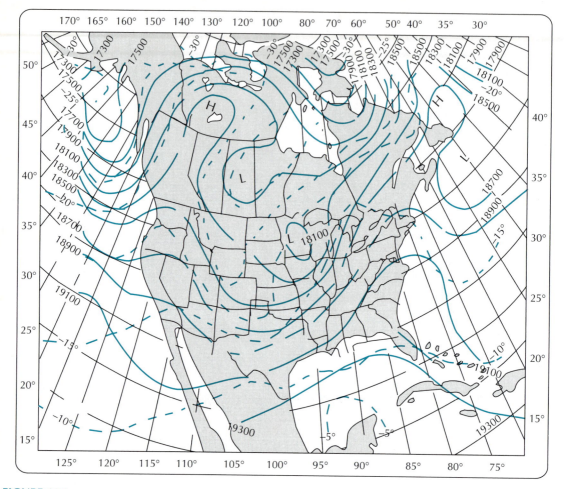

FIGURE 10.6

Upper-air (500-mb) map at 7:00 A.M. on May 2, 1973.

Source: Anthes et al., *The Atmosphere,* 3d ed. (New York: Macmillan, 1981), 225. Reprinted by permission of the author.

Cyclone Development

Winds in the upper air are strongest where the contours are closest together. The strongest bands within the westerly flow are jet streams, the cores of which lie above regions of strong horizontal temperature gradients. Such regions are located at air-mass boundaries; in other words, the location of frontal systems is closely related to the location of a jet stream (Figure 10.7).

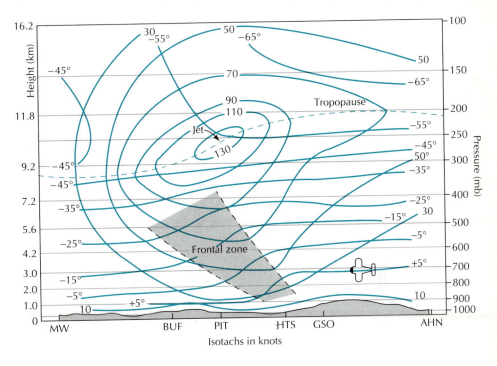

FIGURE 10.7

Vertical cross-section through a cold front at 7:00 A.M. October 16, 1973.

Source: Anthes et al., *The Atmosphere*, 3d ed. (New York: Macmillan, 1981), 181. Reprinted by permission of the author.

This relationship is significant in the formation of low-pressure areas along a frontal boundary and the eventual formation of a midlatitude cyclone. Figure 10.8 illustrates a mechanism for the origin of a midlatitude cyclone, the process of cyclogenesis. The figure shows that air moving within the ridge-trough pattern corresponds to the position of high and low pressures at the surface. As air converges aloft, it creates high pressure; conversely, divergence aloft corresponds to low pressure at the surface. Since the pattern of convergence and divergence rests with the position of the wave, the coincidence of a frontal boundary with the zone of upper-air divergence creates a low-pressure system (see Figure 10.9). The initial cyclone forms beneath the region of divergence aloft. A deepening upper-air trough amplifies the wave pattern of the midlatitude cyclone thereafter, as the trough deepens and a low forms aloft, the system becomes occluded. While the actual evolution of the system may vary from the example given here, it clearly illustrates the ideal environment for a low-pressure area to form along a frontal boundary.

Upper-air and surface conditions rely upon position of the upper-air waves and jet streams. This relationship is clear when severe storms form along frontal systems.

FIGURE 10.8

(a) Cross section of atmosphere showing how upper-level convergence and divergence are related to lower-level divergence and convergence. (b) Upper-air convergence and divergence are associated with ridges and troughs at the jet-stream level. (c) Surface pressure systems (heavy lines) are related to upper airflow (dashed lines). Surface high pressures (H) are associated with upper-air regions of convergence; surface low pressures (L) are associated with upper-air divergence.

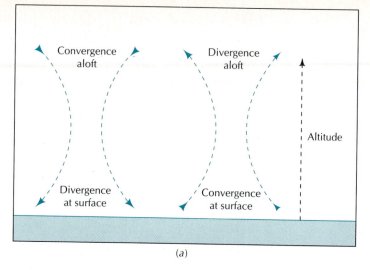

(a)

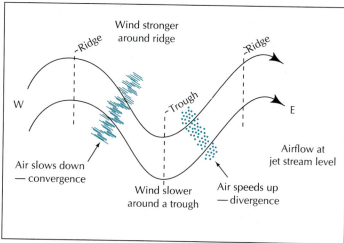

(b)

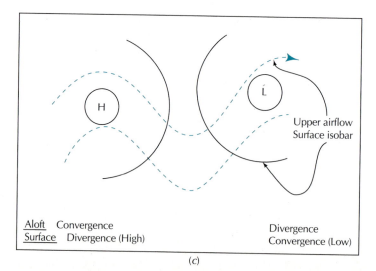

(c)

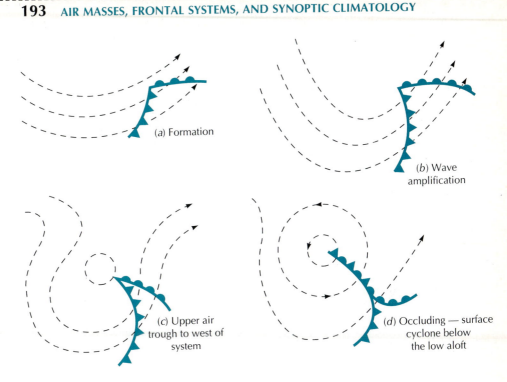

FIGURE 10.9
Development of a midlatitude cyclone along an upper-air trough.

Severe Weather and Frontal Systems

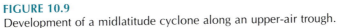

Observers for the National Weather Service consider a thunderstorm to begin when either thunder is heard or overhead lightning or hail is observed. The storm is considered ended 15 minutes after the last thunder is heard. Note that this definition makes no mention of rainfall; in fact, in dry climates, thunderstorms often occur without measurable precipitation.

Regardless of whether precipitation occurs, thunderstorms form from an initial uplift of moist, unstable air and the release of sufficient latent heat to cause continued uplift. The keys to storm formation are thus a source of moist air and a mechanism to produce the required uplift. The significance of the availability of warm, moist air in storm formation shows in the world distribution of thunderstorms (Figure 10.10): The greatest number occur in the moist, tropical realms, especially in Africa. The second requirement, a mechanism to initiate cumulonimbus cloud development and hence thunderstorms, is more varied. The three main types include air-mass thunderstorms, storms associated with fronts, and those that form ahead of fronts along squall lines.

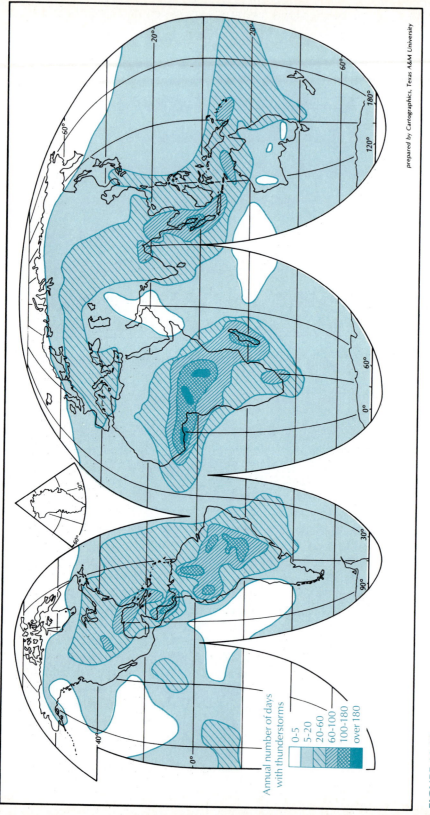

FIGURE 10.10

World distribution of thunderstorms.

Source: After S. D. Gedzelman, *The Science and Wonders of the Atmosphere*. (New York: John Wiley & Sons, 1980), 333.

Annual number of days with thunderstorms

0-5
5-20
20-60
60-100
100-180
over 180

prepared by Cartographics, Texas A&M University

Types of Thunderstorms

Isolated thunderstorms that occur in summer in the midlatitudes typify the air-mass thunderstorm. They occur in a disorganized manner, often consisting of a single cell or several distinct cells less than 10 km wide. While air-mass thunderstorms sometimes occur because of differences in surface heating, other trigger effects often cause their development. Alternate mechanisms responsible for growth of air-mass storms include converging winds and topography. An example of the former occurs in Florida, where convergence of the moist ocean air along both coasts produces frequent thunderstorms over the peninsula. The role of topography is apparent on the slopes of the Rockies. There, air near the slope is heated more intensely than air at similar levels over flat land, resulting in a distinct upslope movement and the potential for growth of cumulonimbus clouds.

Air-mass thunderstorms are generally much less violent than those associated with the forced upward motion of air that occurs along cold fronts and squall lines. Some of the most severe thunderstorms are associated with squall lines. These are lines of moving thunderstorms that occur mostly ahead of cold fronts and, unlike the isolated air-mass type, are an integral part of large-scale circulation patterns. The schematic sequence given in Figure 10.11 provides the essential details of the storm. Warm, moist air at surface levels lies ahead of an advancing cold front. At the 850-mb level (about 1500 m), this air flows from a warm, southerly source. In contrast, at 500 mb (about 5500 m), a westerly stream of cool, dry air—with divergence—flows across the surface systems. The combination of the unstable surface air and the divergence aloft leads to extensive

FIGURE 10.11
Schematic depiction of airflows, shown at three levels of the atmosphere, that give rise to severe weather.

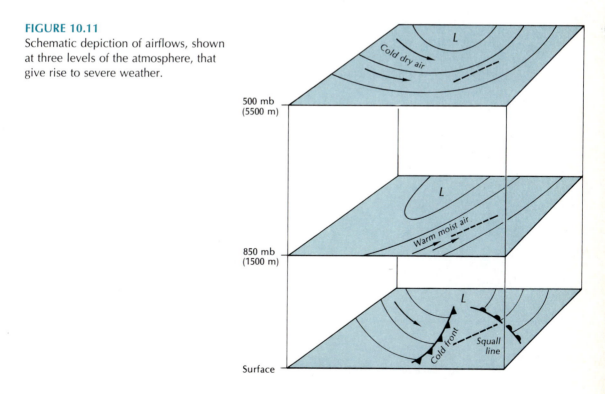

vertical development of clouds and a line of thunderstorms. The differential flow of air at varying altitudes adds to the severity of the storms. The westerly high-level flow tilts the top of the storm clouds so that falling precipitation does not slow the updrafts (as it does at the mature stage of air-mass thunderstorms), thus extending the storm's life and increasing the potential for hail to form. At times the size and severity of the storms allow them to be classed as supercells.

Note that the squall line may be located hundreds of kilometers ahead of the cold front. In fact, squall lines in tropical climates do not require the presence of a front for their formation. Clearly, however, the association of fronts and squall lines enhances the severity of thunderstorms.

Associated Hazards

The definition of a thunderstorm includes the presence of lightning, probably because thunder is the result of lightning, a sound resulting from violent expansion of air close to the lightning. Lightning is thus an integral part of severe storms and is itself a distinctive hazard (see Plate 7).

Lightning is an electrical discharge resulting from separation of positive and negative charges within clouds and between clouds and the ground. When the difference in charge is great enough to overcome the insulating effect of air, a lightning flash results. While the reason for the separation of charges is still imperfectly understood, a number of plausible theories try to explain it. One of these, illustrated in Figure 10.12, suggests that graupel and hail in the cloud become polarized, with negative charges at the top and positive charges at the bottom of the particles. At this stage, the graupel has no net charge, but, as Figure 10.12*a* shows, the larger falling pellets acquire a negative charge. The smaller, positively charged graupel is carried aloft while the larger particles accumulate in the lower part of the cloud (Figure 10.12*b*). The separation of charges leads to lightning inside the cloud and between the cloud and the positively charged Earth below (Figure 10.12*c*).

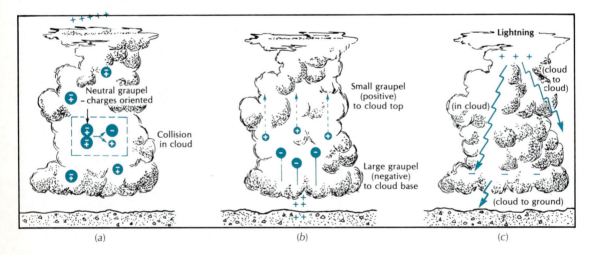

(a)   (b)   (c)

FIGURE 10.12

Lightning formation requires separation of electrical charges in the atmosphere. One potential mechanism is shown here.

The distribution of lightning is obviously associated with the distribution of thunderstorms. However, deaths resulting from lightning in the United States do not correlate with this distribution, since lightning-caused deaths are not concentrated in one area. This is probably due to the time of thunderstorm occurrence. A study of injuries and fatalities from lightning shows that 70% occur in the afternoon, and only 1% occur between midnight and 6:00 A.M. Thus, where the proportion of night storms is high, injuries are fewer because people aren't generally exposed to them.

Hail also results from thunderstorm activity. Its formation is illustrated in Figure 10.13. Essentially, hailstones form as a result of adding supercooled water to an initial nucleus; the eventual size of the stone depends upon the length of time spent on its passage through the cloud.

Hail exists in three forms. Graupel, or soft hail, is usually less than 5 mm (.2 in.) in diameter and has a crisp texture that causes it to be crushed easily when it strikes the ground. Graupel may serve as the nucleus for small hail, which is often mixed with rain. Because of the thin layer of exterior ice, a small hailstone remains intact on reaching the ground. Neither graupel nor small hail, which are about the same size, is large enough to

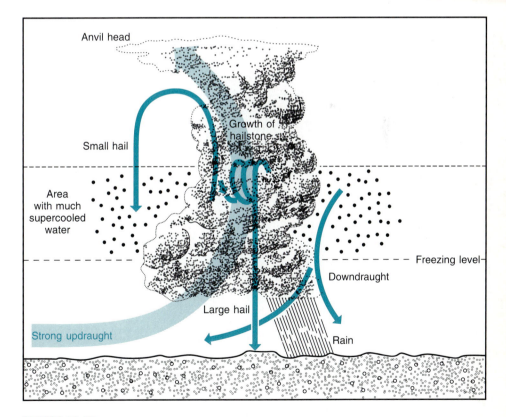

FIGURE 10.13
Simplified representation of the formation of hail in cumulonimbus clouds.

cause much destruction. Destructive hail, called true or severe hail, can attain large sizes: The remarkable Coffeyville, KS, hailstone weighed 3.7 kg (8.2 lb).

Neither the size nor the number of stones associated with severe hail have been systematically measured for long periods throughout the United States. The most general map of hail distribution indicates the average number of days with hail. Such a map is shown in Figure 10.14 and is based upon all occurrences of damaging hail, whether large or small. Texas, Oklahoma, Kansas, Nebraska, and Missouri (ranked in order from most severe) experienced over one-half of the total severe-hail occurrences in the 48 continental states over a period of 12 years.

Studies have shown that the diurnal distribution of hailstorms peaks between 3:00 P.M. and 6:00 P.M. local standard time. Of the severe hail that fell in the continental United States, 40% occurs within these three hours.

On the basis of both meteorological frequency and economic factors, we can divide the United States into 13 hail regions. These hail regions are delineated on the basis of (1) average frequency, (2) peak hail season, (3) primary cause of hail, and (4) regional hail intensity (see Figure 10.15).

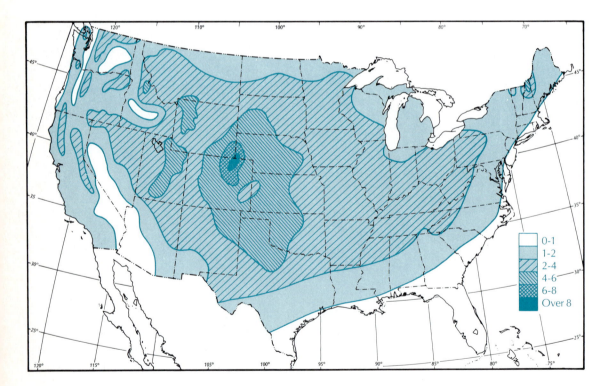

FIGURE 10.14
Average number of days with hail in the United States.

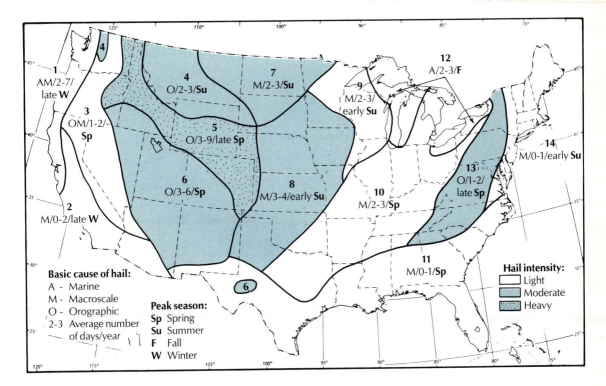

FIGURE 10.15

Hail regions of the United States as differentiated by four hail characteristics: frequency, cause, season, and intensity.

Source: S. A. Changnon, "The scales of hail," *Journal of Applied Meteorology, 16* (6) (1977): 630. Reprinted by permission of the American Meteorological Society.

Synoptic Climatology

Much of the analysis described in this text reflects the results of **synoptic climatology**, the study of local and regional climates expressed in terms of the properties and motion of the overlying atmosphere. The term synoptic climatology was first used in the 1940s, referring to analysis of past weather situations to assess the frequency of sets of conditions. The term *synoptic* is used by meteorologists to denote instantaneous weather conditions shown on a synoptic weather chart. By extension, synoptic charts are a major source for the study of synoptic climatology. Satellite imagery is another such source.

The basic procedures involved in synoptic climatic studies are (1) determining circulation types and (2) statistically assessing conditions in relation to the identified patterns. Clearly, an important step in an analysis is the classification of atmospheric properties and processes on the basis of synoptic patterns. Table 10.2 lists the classification methods. The initial division is according to scale, with subclassifications grouped by subjective and objective schemes. Typical subjective studies use numerous daily weather

TABLE 10.2
Classification methods used in synoptic climatology.

1. *Global scale*
   1.1 *Subjective schemes*
      Description of seasonal changes of pressure and circulation fields
      Characterization of typical circulation regimes (zonal, meridional blocking)
   1.2 *Objective schemes*
      Calculation of zonal and meridional circulation indices
2. *Continental scale*
   2.1 *Subjective schemes*
      Classification of pressure and circulation fields, based on the major centers of action
         (Grosswetterlage)
      Delimitation of zonal and meridional circulation types
      Assessment of weather conditions in relation to cyclone-anticyclone tracks
      Determination of the frequency of high and low centers
      Classification of air masses and fronts
   2.2 *Objective schemes*
      Classification based on derived parameters of the pressure and circulation fields
         (pressure gradient, relative vorticity, flow direction, etc.)
      Correlation of weather conditions with typical upper-level contour patterns
      Classification based on mathematical-statistical specification of pressure fields
         (orthogonal polynomial functions)
3. *Regional scale*
   3.1 *Subjective schemes*
      Grouping similar pressure fields or airflow patterns
      Air mass and frontal classifications
   3.2 *Objective schemes*
      Classification based on derived parameters of the pressure and circulation fields
         (pressure gradient, relative vorticity, flow direction, etc.)
      Classification based on upper airflow at selected stations
      Correlation of weather conditions with typical upper-level contour patterns
      Classification based on mathematical-statistical specification of pressure fields
         (orthogonal polynomial functions)
4. *Local scale*
   4.1 *Subjective schemes*
      Definition of weather types according to locally observed weather elements (complex climatology)
   4.2 *Objective schemes*
      Intercorrelation of locally observed weather elements and statistical condensation of
         these into local weather types

Source: After Wanner, from R. G. Barry "Synoptic Climatology" in J. E. Oliver and R. W. Fairbridge, *The Encyclopedia of Climatology*. (New York: Van Nostrand Reinhold, 1987), 825.

map sequences to identify a synoptic pattern associated with a distinctive climatic event and hence classify the types that occur. For example, to identify the type of circulation regime that gives rise to stagnant air masses, the researcher might need to classify conditions associated with blocking systems. Objective classifications take advantage of the availability of digital data banks, computers, and statistical packages to use correlation analysis, factor analysis, and various clustering techniques.

Many regional synoptic catalogs have been derived. Perhaps the best known is the *Grosswetterlage* (large-scale weather pattern), which involves the classified synoptic patterns over a large region. A daily catalog of Grosswetter is published in Germany to show conditions in Central Europe.

Because synoptic climatology is an integral part of climatic analysis, its methods appear in various parts of this text. However, to demonstrate the nature of the synoptic approach, it is useful to present appropriate examples.

An example of a subjective study was presented in the classic book *Compendium of Meteorology*. Figure 10.16 shows schematic diagrams of what are termed "weather types" but, in effect, demonstrate a synoptic classification identifying eight situations that typify winter circulation types for North America. The four diagrams on the left represent meridional circulation; those on the right represent zonal. Meridional flow types have upper-level troughs and ridges of large amplitude that tend to steer circulation systems toward the north and south (meridional). Zonal flow types have upper-air flows that are characteristically east–west and no large-amplitude waves. Each of the patterns is identified by a code and described according to the positions of the troughs and ridges (for meridional flows) and by cyclone tracks (for zonal patterns). Remarks about resulting surface conditions associated with each pattern would follow this classification. Clearly, a study like this has more detail than presented here, but the basic methodology used illustrates subjective classification.

Many of the recent objective classifications are used in an applied setting (see Figure 10.17). Here, eight predominant synoptic categories associated with a surface map were identified and a scheme prepared to delineate the predominant type. Then, a statistical technique for clustering used 24 weather variables for 141 stations in the United States to produce "cluster maps" for selected days (see Figure 10.18). The classification was then related to air-pollution conditions and correlations between the pollution levels and the synoptic types were made.

Midlatitude cyclones transfer heat and moisture through latitudes from 30° to 60°. The movement of the storms in the United States, especially those that influence the central and eastern United States, tends to follow four main tracks. Figure 10.19 shows the cyclogenesis area by name and common tracks of the storms.

Both the Alberta and the Colorado lows form in the lee of the Rockies. The Alberta system brings light snow and rain to the northern-tier states, a contrast to the Colorado low. This storm system is often regenerated in southern Oklahoma when moisture from the Gulf of Mexico is drawn into the circulation. In winter, these storms can produce blizzards and often produce severe thunderstorm activity in the Midwest. The gulf low originates in the Gulf of Mexico and moves northeast to bring copious rain to the east coast from Virginia to New England. The Cape Hatteras lows greatly influence weather in the Mid-Atlantic states, bringing abundant rain and snow; New Englanders often refer to these storms as "nor'easters."

Perhaps the most widely reproduced synoptic chart is that used to show how extreme conditions, such as polar cold or drought, are related to upper-air circulation patterns. Figure 10.20 provides an example of a bitter-cold winter that occurred in the United States. The normal circulation of upper airflows associated with the Rossby regime is

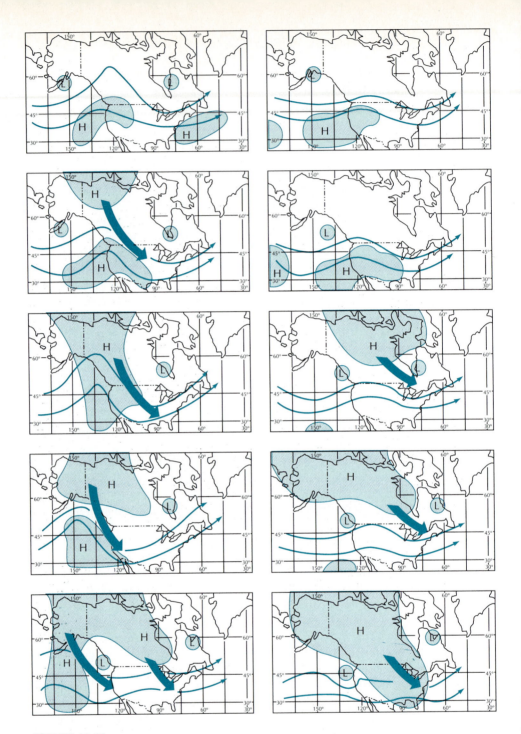

FIGURE 10.16

Principal circulation types for North America. Heavy lines indicate upper level mean flow. Shaded areas indicate quasi-stationary low pressure centers (L) and persistent surface anticyclones (H). Arrows show polar air outbreaks.

Source: From R. D. Eliot, "Extended range forecasting by weather types" in T. Malone, ed. *Compendium of Meteorology*. (Boston: American Meteorological Society), 1951. Reprinted by permission of the American Meteorological Society.

FIGURE 10.17

Categories identified in one synoptic analysis (upper) and their hypothetical locations on a generalized surface map (lower).

Source: R. E. Davis and L. S. Kalkstein, "Using a spatial synoptic climatological classification to assess changes in atmospheric pollution concentrations." *Physical Geography*, 11 (1990): 320–42. Reprinted by permission of V. H. Winston and Son, Inc.

I	Maritime tropical (cold front approaching)
II	Maritime (strong)
III	Modified maritime or maritime tropical
IV	Cold front passage (strong)
V	Polar influence (cyclone to the south)
VI	Weak cold front, trough, or thermal low passage
VII	Continental high
VIII	Transition to continental polar

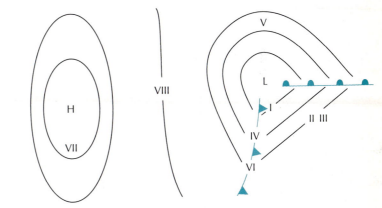

shown in Figure 10.20*a*. When we compare these normal conditions with those of the 1976–77 winter season, we can see remarkable differences, especially comparing the

amplitude of the wave pattern

location of the high-pressure ridge over the Pacific Ocean

southerly extent of the low-pressure ridge normally off the northern Canadian islands

intense high pressure in the polar area

deepening of the low-pressure system off the Aleutians

In all, the pattern of a frigid year differs dramatically from the normal, so winters in many places in the United States vary appreciably from what is normally expected. Comparing Figures 10.20*a* and 10.20*b* hints at how various locations felt the effects of different circulation patterns of 1976–77. Of immediate note are:

1. The high-pressure ridge that extends over the eastern Pacific Ocean shifts the jet stream path far north of its normal position. This means that the west coast of the United States does not fall in the path of frontal systems that normally bring rain from the Pacific. The area is dry.

2. The same ridging means that warm, moist air would extend to Alaska. Usually dominated by cold air at this time, the region would experience exceptionally mild conditions.

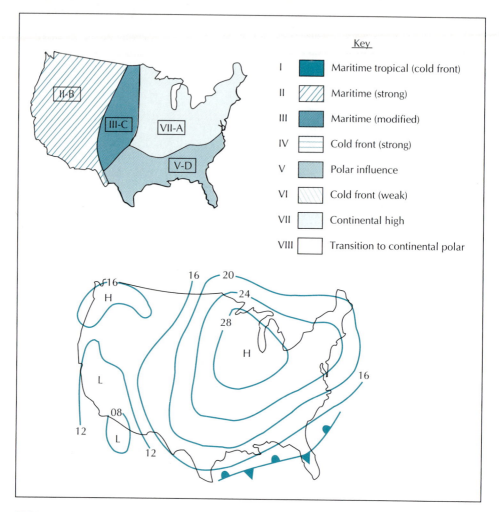

FIGURE 10.18

Application of the classification shown in Figure 10.17 to an actual situation. Upper is a cluster map derived from conditions on the existing surface weather map (lower).

Source: R. E. Davis and L. S. Kalkstein, "Using a spatial synoptic climatological classification to assess changes in atmospheric pollution concentrations," *Physical Geography,* 11 (1990): 320–42. Reprinted by permission of V. H. Winston & Son, Inc.

3. The exaggerated jet-stream axis would be maintained by the movement around the low pressure over Labrador, and cold Arctic air would stream across the United States.

These conditions, hypothesized from the observed circulation patterns, actually occurred. Figure 10.21*a* shows the temperature departure during January 1977. The departure from normal temperatures in the Midwest is especially noteworthy. Figure 10.21*b* shows the percentage of normal precipitation in January 1977. As anticipated, large parts of the west are unshaded, indicating rainfall less than 50% of normal.

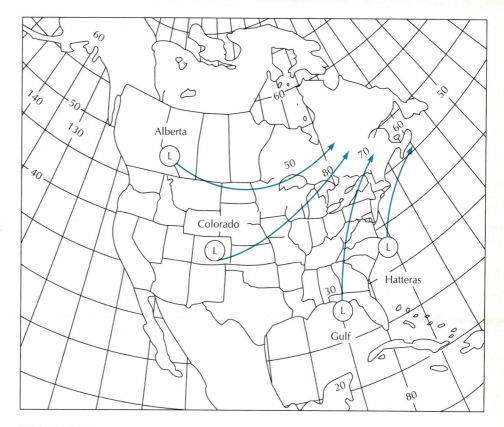

FIGURE 10.19
Typical storm tracks across the United States. The arrows depict the tracks of four major types of midlatitude cyclones that influence the central and eastern United States.

The synoptic conditions that produced the unusual January conditions are clear. Why did such conditions occur? In recent years, much research has been undertaken to understand why such patterns and anomalies occur; many of them are rooted in teleconnections, the influence on local climate conditions by events happening in other world areas. As already noted, the climates of the United States reflect ENSO events occurring in the Pacific Ocean far from the continent. Investigation of teleconnections, such as those dealt with in Chapter 9, is an integral part of synoptic climatology research.

Satellite Imagery and Synoptic Climate

World War II spurred the use of rocketry, and by 1947 a modified V-2 rocket took the first successful photographs of Earth's clouds from an altitude of about 100 mi (165 km). As rockets were used more, atmospheric scientists realized that viewing weather systems over wide areas from above could play a significant part in understanding the atmo-

FIGURE 10.20
Schematic diagrams illustrating upper air circulations in winter: (*a*) the "normal" pattern; (*b*) circulation that results in a severe winter over much of the United States.

(*a*)

(*b*)

sphere. Using rockets to place a satellite in orbit became a significant goal that was achieved when the Soviet Union launched Sputnik I in 1957.

Since the launching of the first specialized weather satellite, TIROS I, in 1960, images of Earth and its cloud cover have become a common feature of many TV weather pro-

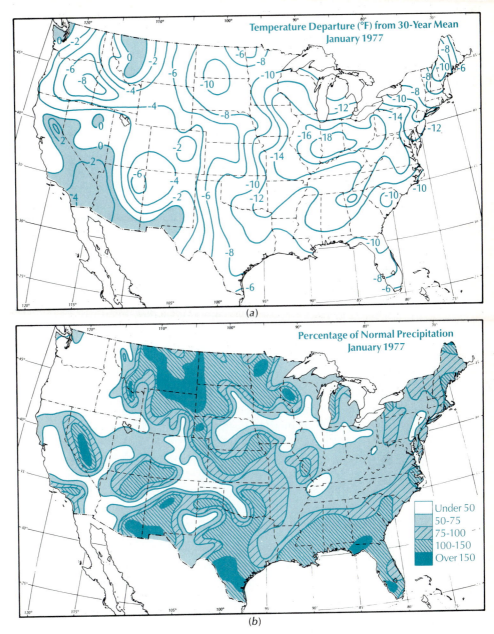

FIGURE 10.21
Unusual circulation patterns of winter 1976–77 resulted in significant departures from normal: (*a*) January temperature departures; (*b*) percentage of normal precipitation for the same month.

grams. Since that time, the orbits and sophistication of the sensors aboard the satellites have changed.

Satellite Orbits

The types of data obtained from weather satellites depend largely on their orbit. Two fundamental orbits are available—geostationary and low-level. Geostationary or equatorial geosynchronous orbits are when a satellite is in a fixed position at 35,400 km (21,983 mi) above selected points on Earth's surface. Such an orbit provides a full-disk image of the globe every half hour or so. This allows us to study sequential events the way we might with a film loop. The GOES satellite systems, first launched in 1975, were the first operational geostationary satellites. Figure 10.22*b* illustrates coverage by geostationary satellites. Low-level orbits are those from 800 to 1500 km (497–932 mi) above the surface. Several types of satellites use low orbits; probably the most useful are those classified as sun-synchronous. In these, the altitude of the sun in the sky is approximately the same within similar seasons at every crossing of a given latitude. The TIROS N satellites, introduced in 1978, are polar-orbiting, crossing the equator at approximately 90°. A polar orbiter may need 15 orbits to cover the entire globe, and may view a given location perhaps twice a day. Figure 10.22*a* illustrates typical paths of a polar-orbiting satellite. Both orbits have their own special uses, advantages, and disadvantages.

Type of Sensor

Satellites carry a wide variety of sensing instruments that are either *imaging systems* or *sounding systems*. Imaging systems are either (1) those that give a series of instantaneous images, such as the older vidicon systems, or (2) scanning sensors (i.e., scanning radiometers) that build up images line-by-line in tracks. Sounding systems measure emissions of radiation from different levels of the atmosphere to help us construct atmospheric profiles.

The NIMBUS series of satellites (1978 to the present) have been used as the basic research and development vehicle for sensors. The most recent satellites include

ERB (Earth Radiation Budget) for monitoring radiation budgets

SAMS (Stratospheric and Mesospheric Sounder) for temperature profiles

SBUV (Solar Backscatter, Ultraviolet Energy) for ozone profile retrieval and evaluation of ozone, terrestrial, and solar irradiance

TOMS (Total Ozone Mapping Spectrometer) for ozone monitoring

The workhorse of the weather satellites, in terms of data use, is the Geostationary Operational Environmental Satellites (GOES) series. These geosynchronous systems carry the Visible and Infrared Spin Radiometer (VISR) for visible and infrared images, an atmospheric sensor (VAS) to provide direct readouts of water vapor and temperature, and a Space Environment Monitor (SEM) to assess, for example, Earth's magnetic field and solar x-rays. In 1991, only one U.S. weather satellite (GOES 7) was operating, and NASA was in the process of building the newer modified geostationary sys-

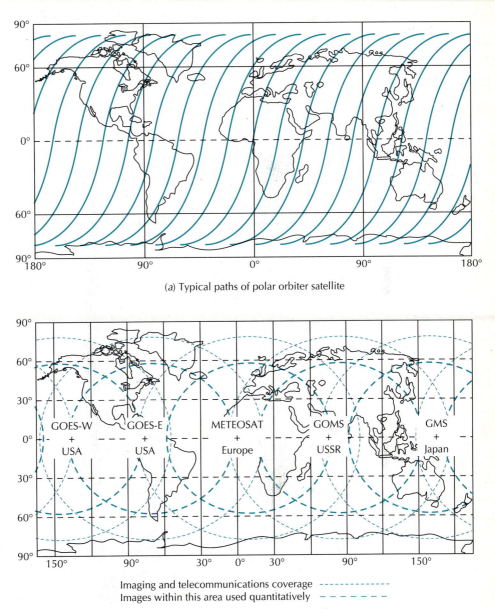

(a) Typical paths of polar orbiter satellite

Imaging and telecommunications coverage ─────────────
Images within this area used quantitatively ─ ─ ─ ─ ─ ─

(b) Coverage by geostationary satellites

FIGURE 10.22

(a) Typical paths of a polar orbiter satellite; (b) coverage by selected geostationary satellites.

Source: A. Henderson-Sellers and P. Robinson, *Contemporary Climatology*. (New York: John Wiley & Sons, 1986), 65. Reprinted by permission of Longman Group UK.

tems, the GOES-NEXT. This series will serve as remote sensors of weather into the twenty-first century.

Applications

While satellite climatology is accomplishing important work, its development is not as rapid as might be expected for a number of reasons. Climatology, by its very nature, demands long runs of homogeneous data, and weather satellite data still fall significantly short of the 30–35 years required for accurate evaluation of climatic norms. Frequent changes of satellite orbital patterns and sensors result in nonhomogenous data sets. Methods of data processing have changed greatly since 1960, and there have been few coordinated international programs, so archived data are extremely variable.

Despite this, satellite-based climatic studies have produced significant results concerning airflow and circulation, clouds and precipitation, and synoptic climatology. Satellite images presented in this text to illustrate such events as the distribution of stratospheric ozone represent a good example of what satellites can provide. Clearly, the availability of satellite-produced data over the oceans, the polar areas, and remote deserts is in itself of enormous value to climatologists. Such data are available from the Satellite Data Services Division of the National Climate Data Center, which stores a digital archive of satellite imagery.

Summary

Introducing the idea of air masses and developing the related concepts of fronts and midlatitude cyclones were major advances in the development of atmospheric science. Although the surface patterns of fronts and frontal systems were well-known, the initiation of midlatitude cyclones could not be fully explained until upper-air data accumulated. It is now known that convergence and divergence associated with the jet streams of the upper-air westerlies are important factors in cyclogenesis.

Air-mass and frontal analysis also helps explain the occurrence of thunderstorms. Studies of the climatology of thunderstorms provide important temporal and spatial information about their occurrence and related hazards. Such analysis is part of the study of synoptic climatology, an approach to climatic analysis concerned with the properties and motion of the atmosphere. Synoptic analysis can be both subjective or objective; the latter type of analysis is growing with the availability of rapid computing methods.

Satellite imagery results from both polar-orbiting and geosynchronous platforms. The type of sensors aboard the satellites provide data through images of the surface or as soundings through the atmosphere. Satellite climatology is becoming increasingly important as digital archives are growing.

TORNADOES AND HURRICANES

Chapter Overview

Severe storms are a series of disturbances in the atmosphere, including thunderstorms, hailstorms, tropical cyclones, and tornadoes. These storms represent an atmospheric response to an unequal distribution of energy. They are thus an integral part of dynamic energy exchange rather than simple isolated events. The energy involved in such storms is prodigious (Figure 11.1). The concentration of this energy in hurricanes and tornadoes causes loss of life and devastation. Storms are mechanisms for global energy exchange, making them an important climatological element as well as a meteorological event. Analysis of the storms provides equally important knowledge about their role in distributing energy.

Tornadoes

The **tornado** is the most intense vortex that occurs in the atmosphere. It is a converging spiral of air with wind speeds estimated at several hundred kilometers per hour. It is the most violent of atmospheric storms, but it is seldom larger than 1 km (0.6 mi) in dia-

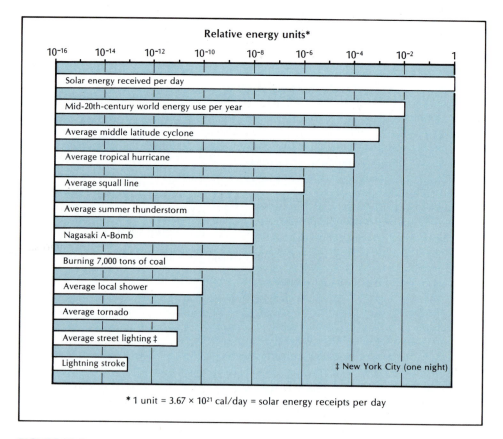

FIGURE 11.1
Energy equivalents of various phenomena relative to the amount of solar energy received by the earth in one day.
Source: W. D. Sellers, *Physical Climatology*. (Chicago: The University of Chicago Press, 1965), 106.

meter. The direction of rotation is counterclockwise, except in rare instances. The tornado depends upon moisture as an energy supply and occurs mostly in moist, tropical air. Funnel-shaped clouds form from other clouds and work down toward the ground. When a funnel cloud touches the ground, it becomes a tornado. The funnel cloud of condensed moisture hanging from the storm makes the tornado readily visible. The cloud may vary widely in thickness, and sometimes may be larger at the bottom than at the top. Most often condensed water vapor makes the cloud appear gray, but as the tip contacts the ground, the appearance changes as it picks up dirt and debris from the surface. In winter, a tornado may touch down on a field of snow and become brilliant white.

The storm develops extremely low atmospheric pressure in the center. The record known drop in pressure occurred in Minnesota in 1904, when the pressure dropped to 813 mb (against the mean sea-level pressure of 1013 mb). Wind velocities reach as high as 400 km per hr (250 mph). In most tornadoes, wind velocities are less than 145 km per hr (90 mph). Direct measurement is difficult because the sudden pressure changes and high wind velocities destroy the instruments.

Tornadoes move over the ground surface erratically. While they normally move parallel to cold fronts, they occasionally move in circles and figure eights and may even stay in one spot. Their speed over the ground ranges from nearly stationary to as much as 110 km per hr (65 mph). The average is 40–65 km per hr (25–40 mph). The surface path of most tornadoes is short and narrow, varying from a few meters to 2 km (1.2 mi) in width. The path averages less than 40 km (25 mi) in length. They stay on the ground an average of 15–20 min. In May 1977, a tornado traveled 570 km (340 mi) across Illinois and Indiana, existing for a total of 7 hr and 20 min. They stay on the ground longest and travel the straightest over flat, open land.

Formation of Tornadoes

Tornadoes occur most often with thunderstorm activity. Ideal conditions for tornado formation are those found ahead of a cold front. Tornadoes form when a mass of warm, very moist air collides with cooler, drier air from polar regions. Extreme turbulence develops along the air-mass boundary, and eddies occasionally develop into strong whirls through which warm air escapes upward. If the air is extremely unstable, the convergence intensifies and a storm forms.

Tornado formation requires several conditions: (1) A mass of very warm, moist air must be present at the surface; (2) the vertical temperature structure must be unstable; and (3) a mechanism to start rotation must be present.

The Great Plains is the foremost tornado region in the world; not surprisingly, these three elements are often present in the region's atmosphere. Low-pressure centers develop east of the Rocky Mountains. They typically have a cold front extending to the south and a warm front extending east from the low. The warm sector contains a south-to-north flow of warm, moist air from the Gulf of Mexico. Above this air is a stream of cold, dry air from the west. This air comes from the Pacific Ocean as cool, moist air. It loses its moisture as it crosses the western mountain ranges.

The boundary between the warm, moist air from the Gulf of Mexico and the dry air from the west is a very turbulent zone called the dry line. If the air pouring over the Rocky Mountains is warm enough, it will move out over the warm, moist air from the

Gulf, causing the dry line to extend more horizontally rather than vertically. This provides a stable—but potentially explosive—condition. If disturbance comes from the jet stream above or the surface below, a violent exchange of air may result.

One such disturbance is heat from the ground surface. As the ground heats during the day, it steadily radiates more and more heat to the air above. The warm, moist air moving northward heats during the day, and by late afternoon the surface air gets quite hot. The air near the surface becomes hot enough to break through the dry line above. This results in the explosive development of thunderstorms. The upward rush of air reaches velocities as high as 165 km per hr (100 mph). Under favorable conditions the storms will grow through the tropopause to heights of 18,000 m (60,000 ft). These *supercells* produce heavy rainfall and large hail. Violent updrafts develop in large thunderstorms and are one reason why commercial aircraft try to avoid flying through them.

For a tornado to develop, something must start the column rotating; often that something is **wind shear**. Wind shear is a change in wind speed and direction with height. It occurs when upper-level winds blow across the path of lower-level winds at higher speeds. That condition starts the rising column of air rotating counterclockwise. As more air flows in and the storm stretches in height, the rate of rotation increases. (See Chapter 7, on conservation of angular momentum.) Once the center of the system begins to spin, it becomes a mesocyclone and may be as much as 10 km (6 mi) across. The rotation starts in the middle level of the storm and works downward. If the process continues, the mesocyclone grows vertically through the thunderstorm and intensifies. In most thunderstorms, the supply of warm, moist air gradually shuts off, and the storm dies as the energy dissipates. If the supply of moisture continues, the rotating core stretches both upward and downward. The strengthening results in higher wind velocities, a smaller vortex, and lower pressure.

When turbulence and vertical motion become excessive in a severe thunderstorm, mamantus clouds appear at the cloud base. These clouds occur only with extreme turbulence and are a good sign of tornado potential. Rotating clouds at the base of a thunderstorm are an indicator that a mesocyclone exists. As vorticity increases, the rotating clouds drop below the rest of the thunderstorm to form a wall cloud. A funnel cloud may form within this spinning mass of air and drop below the cloud. If a tornado forms, it does so from this wall cloud (Figure 11.2). The presence of hail indicates the severity of the thunderstorm and the high potential for a tornado forming: Heavy hail accompanies most, but not all, tornadoes.

Distribution of Tornadoes

Every state in the United States has experienced tornadoes, but they occur only infrequently in regions north of 45° and west of the Rocky Mountains. As noted earlier, most of them occur in the Great Plains, a region often called Tornado Alley. Nine states—Kansas, Iowa, Texas, Arkansas, Oklahoma, Missouri, Alabama, Mississippi, and Nebraska—report an average of more than five tornadoes per year. Figure 11.3 shows tornado hazard potential. Table 11.1 provides data on the distribution of tornadoes and tornado fatalities.

The time that tornadoes most often occur in the United States is similar to that of thunderstorms: Most occur during a two-hour period between 4:00 and 6:00 P.M. About one-fourth of tornadoes occur during this time, and two-thirds of all tornadoes occur in the

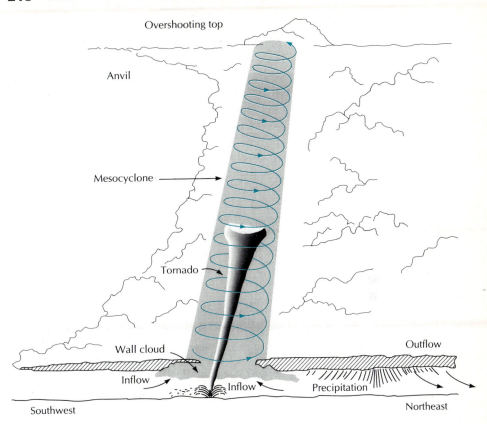

Overshooting top

Anvil

Mesocyclone

Tornado

Wall cloud

Inflow

Inflow

Outflow

Precipitation

Southwest

Northeast

FIGURE 11.2
Diagram of the development of a mesocyclone and tornado in a "supercell."
Source: C. Donald Ahrens, *Meteorology Today.* Minneapolis, MN: West Publishing Co., 1988), 447.

six-hour period from 2:00 to 8:00 P.M. This is the time when Earth radiation to the lower atmosphere is greatest. Tornadoes are least likely around dawn, when the air near the surface has cooled and is relatively stable.

Although tornadoes occur in all months, activity peaks from May to September (Table 11.2). The number of tornadoes varies from year to year. They also occur in different regions, depending on the season (see Figure 11.4).

Waterspouts

Waterspouts are intense vortices that form over the ocean or occasionally over very large lakes, such as Lake Michigan. They look like tornadoes, and in fact some are just that. Some tornadoes begin over land and move out to sea. The visible funnel cloud consists of condensed water vapor rather than water lifted from the sea surface.

Most waterspouts form over shallow water in the tropics. They are more common over shallow water because the water is warmer than in the deep oceans. They are common in the Florida Keys and occasionally in the West Indies. These water spouts are smaller and

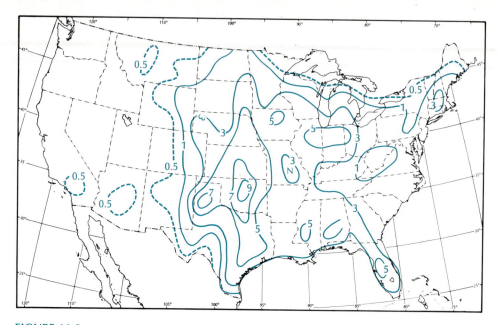

FIGURE 11.3
Number of tornadoes per year within a 90-km (56-mi) radius circle per year.
Source: *Weatherwise*, 3(2), April 1980: 54. A publication of the Helen Dwight Reid Educational Foundation.

weaker than true tornadoes at sea. They are usually less than 100 m (330 ft) in diameter and the winds are less than 80 km per hr (50 mph). Waterspouts develop in warm, unstable air when convectional clouds are building. A mesocyclone is not necessary to produce them. Their strength falls between the dust devils of land areas and true tornadoes.

Hazards of Tornadoes

Tornadoes are the most lethal atmospheric event in the United States. From 1916 to 1953, an average of 230 people died each year from tornadoes. The low was 36 in 1931 and the high was 842 in 1925.

Property damage from tornadoes is particularly serious: During 1916–53, tornado damage in the United States averaged $14 million per year. When a tornado is large and well-developed, it destroys most ground objects touched by the lower end of the funnel. If the funnel lifts off the ground, surface destruction decreases rapidly. Roofs often sustain considerable damage because most roofs are not fastened securely to the frame during construction.

Other factors that have increased the likelihood of property damage are the development of second (vacation) homes and the growth of mobile homes in North America. In addition, of the increasing number of mobile homes sold in the United States, more than half are outside standard metropolitan areas, where they do not need to meet building codes as stringent as they would need to meet within city limits. Mobile homes are more vulnerable to damage from wind than standard dwellings built on a foundation. This is easy to under-

TABLE 11.1
Number of tornadoes, fatalities, and fatalities per unit area, by state, 1953–80.

RANK	STATE	NUMBER OF TORNADOES
1	TX	3344
2	OK	1477
3	KS	1212
4	FL	1155
5	MO	758
		Number of Deaths
1	TX	371
2	MS	316
3	MI	231
4	IN	205
5	AL	202
		Deaths per 10,000 mi^2
1	MA	120
2	MS	66
3	IN	56
4	AL	39
5	OH	36

stand if one considers their shape and their construction: Often assembled with a stapling gun, in many cases the only steel used in their construction is the frame underneath.

In the late 1960s, T. Theodore Fujita of the University of Chicago introduced a method for assessing the relative severity of tornadoes. This scale is now known as the Fujita, or *F-scale* (Table 11.3). It determines a storm's severity only after the tornado is over, so it is not a forecasting device like the hurricane scale. Ratings reflect the worst

TABLE 11.2
Number of tornadoes and fatalities in the United States by month, 1953–80.

	TORNADOES		DEATHS	
	TOTAL	MEAN	TOTAL	MEAN
Jan.	416	15	90	3
Feb.	560	20	192	7
Mar.	1359	49	210	8
Apr.	2999	107	1063	38
May	4352	155	657	23
June	4046	145	446	16
July	2276	81	39	1
Aug.	1532	55	60	2
Sept.	1086	39	61	2
Oct.	659	24	62	2
Nov.	598	21	49	2
Dec.	476	17	96	3

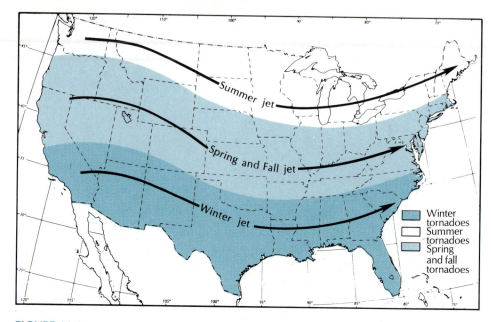

FIGURE 11.4

Migration of jet stream over the year causes areas of major tornado activity to follow a seasonal pattern.

Source: J. R. Eagleman, *Meteorology*. (Belmont, CA: Wadsworth Publishing Co., 1980), 246.

possible damage a tornado could produce based on wind speed. The most violent class describes the smallest group of tornadoes, while the weakest class covers the most storms. While only 2% are in the most violent class, that 2% is responsible for more than 60% of the deaths from tornadoes.

Deaths and damage from tornadoes undoubtedly will continue to increase due to the burgeoning world population, increased property values, and lack of adequate warning systems.

TABLE 11.3

Fujita scale for damaging wind.

CATEGORY		MI PER HR	KNOTS	EXPECTED DAMAGE
Weak	0	40–72	35–62	*Light*: tree branches broken, signs damaged
	1	73–112	63–67	*Moderate*: trees snapped, windows broken
Strong	2	113–157	98–136	*Considerable*: large trees uprooted, weak structures destroyed
	3	158–206	137–179	*Severe*: trees leveled, cars overturned, walls removed from buildings
Violent	4	207–260	180–226	*Devastating*: frame houses destroyed
	5	261–318	227–276	*Incredible*: structures the size of autos moved more than 100 m, steel-reinforced structures highly damaged

Tornado Outbreaks of 1925 and 1974

The most lethal tornado outbreak in North America occurred in the spring of 1925. Once thought to be a single tornado, it actually consisted of a series of perhaps seven tornadoes. They developed over Missouri and traveled northeast across Illinois and Indiana for a distance of 703 km (437 mi). More than 600 people in the path of the storms lost their lives.

On April 3, 1974 a strong cold front moved across the eastern half of the United States. The next day, an intense squall line formed ahead of the cold front (Figure 11.5*a*), accompanied by an extremely violent outbreak of severe thunderstorms and tornadoes. Over a period of 16 hr, 148 tornadoes and an unknown number of severe thunderstorms spread death and destruction. Tornadoes formed from Michigan south through Alabama and Georgia. Figure 11.5*b* shows the counties in which tornadoes occurred. The storms caused 307 deaths, 6000 injuries, and property damage of more than $600 million. The storms moved along a combined path of 4180 km (2598 mi).

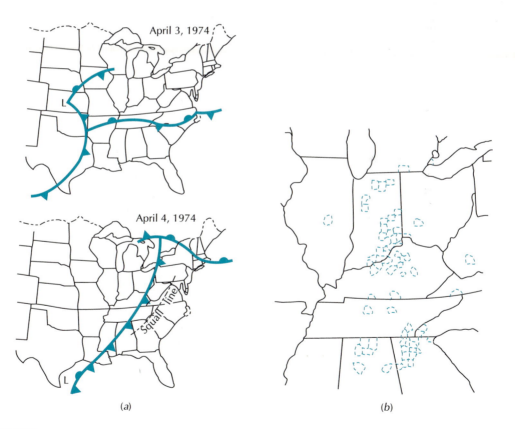

(a)   (b)

FIGURE 11.5

Severe weather systems of April 3–4, 1974: (*a*) Weather maps showing the location of the fronts on April 3 and 4. Most of the severe thunderstorms and tornadoes developed along the squall line ahead of the cold front. (*b*) Counties in which tornadoes touched down.
Source: J. J. Hidore, *A Workbook of Weather Maps*. (Dubuque, IA.: Wm. C. Brown Co., 1975), 140.

Tornado Forecasting

The main factor in the high death rate from tornadoes is our inability to predict and detect their occurrence. Tornadoes are so localized in extent and random in distribution that they are very difficult to forecast. Once a tornado is located, radar can determine its general direction and rate of movement. Because of their random nature, the National Severe Storms Forecast Center in Kansas City, MO, issues tornado watches. When conditions are ripe for a tornado, the center issues a tornado watch for the area concerned. When a tornado is actually sighted or detected by radar, a tornado warning is broadcast by the nearest National Weather Service office. This warning gives the storm's location and direction of travel (Figure 11.6).

Radar—A Tool for Analysis and Forecasting

Radar provides a means of examining what is happening inside storm clouds such as mesocyclones. A radar transmitter sends out radiation ranging from 1 to 20 cm in length, much longer than visible light. When the waves strike an object, part of the beam scatters back to the radar antenna. Cloud particles are very small and are detected with very short radar waves. Longer radar waves penetrate cloud particles and reflect from the larger raindrop-size particles. The harder it rains, the more of the beam reflects. The radar provides a map of where precipitation is occurring and how intense it is. When a tornado takes place

FIGURE 11.6
Distribution of wind velocity around a moving hurricane. The highest velocities are in the forward right-hand quadrant. This is due to the added motion of the storm to the winds within the storm.

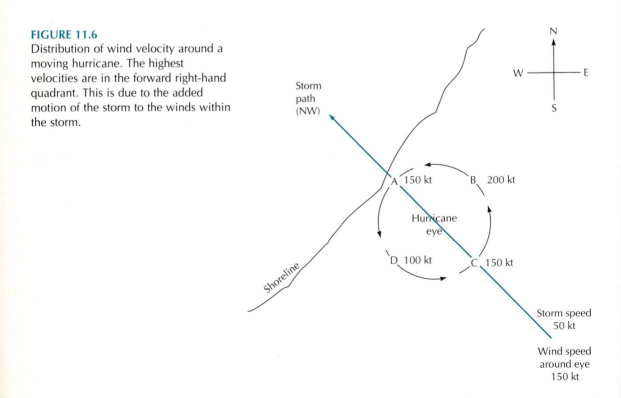

within a mesocyclone, a hook-shaped pattern of rainfall often appears. While the presence of this hook in a mesocyclone is a good indicator of a tornado's possible presence, it is not completely reliable. The hook-shaped pattern may appear without a tornado existing; conversely, a tornado may occur with no hook-shaped pattern of precipitation. Radar thus is not a very useful forecasting device because of these limitations. In addition, a tornado may be on the ground by the time the hook develops on a radar screen.

A more recent development in tornado research and forecasting is the use of Doppler radar. Doppler radar depends on a phenomenon known as the Doppler effect. According to the Doppler effect, the frequency of waves from an object coming toward a person is higher than that from an object going away from a person. A train whistle is a good example: As a train approaches, the pitch of the whistle is higher than after it passes and leaves an area. The frequency of radar waves reflected from rain varies with the direction in which the rain is moving. By converting the frequency of the radiation to a color on a computer screen, we can determine the direction of wind within a mesocyclone. If a storm develops into a mesocyclone with winds forming a vortex, the pattern of wind velocities shows this on the radar image. A mesocyclone develops before a tornado, sometimes as much as 20 min earlier. Therefore, when a mesocyclone forms with its rotating winds, Doppler radar can provide up to a 20-min warning for the affected area. Since radar can establish the speed and direction of the storm, we can make a fairly accurate forecast of the tornado's likely path. Once a tornado forms, a distinct pattern of very rapidly changing wind direction appears on the radar screen.

Because few Doppler radar systems are in use as yet, they are not a significant element in tornado forecasting. The National Weather Service plans to install a network of Doppler systems across the United States. The network, called the Next Generation Weather Radar, or NEXRAD, will be installed over the next decade.

Tropical Cyclones

The **tropical cyclone** is the typhoon of the Pacific Ocean and the hurricane of the Atlantic. It is a vortex, or circular storm, that rotates counterclockwise in the Northern Hemisphere and clockwise in the Southern Hemisphere. Its energy, which comes from the latent heat of condensation, may equal more than 10,000 atomic bombs the size of the Nagasaki bomb. These storms range in size from a few kilometers to several hundred kilometers in diameter. An eye in the middle can be as large as 65 km (39 mi) across, and the total area involved may be as much as 52,000 km^2 (20,000 mi^2).

The atmospheric pressure is nearly symmetrical about the center of a tropical cyclone. The pressure may drop as low as 900 mb, but this is rare. Such a drop in pressure represents a change of about 13% from normal. It is not an explosive drop, since it takes place over a distance of many kilometers. The wind system associated with hurricanes is one of contrasts. In the eye of the storm, the winds will be light and variable and of velocities not usually exceeding 25 km per hr (15 mph). The wind velocities increase rapidly away from the eye, reaching their highest velocities just outside the eye and at a height of about 0.8 km (2400 ft). To be classified as a hurricane, winds must exceed 125 km per hr (75 mph). The maximum winds in most well-developed hurricanes reach 200 km per hr (160 mph), but in extreme cases may reach 300 km per hr (240 mph). Hurricane-velocity winds may extend over an area 300–500 km (240–300 mi) in diameter, and gale-force winds 65 km per hr (40 mph) over an area 600–800 km (360–480 mi) in diameter. Anemometers reinforced against hurricanes have measured such winds at velocities greater than 250 km per hr. While the winds within the storm area are of very high velocity, the storm itself moves at a relatively slow speed, averaging only 15–30 km per hr (9–12 mph). In Hurricane Gilbert of 1988, the central pressure dropped to 885 mb (26.13 in.). This is the lowest pressure ever recorded in a Western Hemisphere hurricane. It produced 290 k/hr (175 mph) winds and pushed a hurricane surge of 4.5 m (15 ft) in front of it.

Some of the heaviest rains on record in low latitudes have come from tropical cyclones. Records of 500 mm (20 in.) of rain over a 48-hr period are relatively common. One typhoon in the Philippines produced more than 1600 mm (64 in.) of rain. Radar observations show that 500–800 km (300–480 mi) ahead of the storm there are often fairly well-defined lines of thunderstorms. These storms generate waves as high as 13 m (43 ft). Long sea swells as far as 1600 km (990 mi) from the center are evidence of the strength of the storms. Near the wall of the hurricane eye, wind often blows the tops off the waves, but in the eye itself the waves are often very high.

Surges of water 2–3 m (6–10 ft) high often form ahead of hurricanes, and flooding of coastal areas and islands is frequent.

Formation of Tropical Cyclones

There is no agreement yet as to why these storms form. The weather conditions required to produce them are known, and once formed, the storms can be tracked. The precise set of factors that triggers tropical storms requires further research. The storms form only over large bodies of warm water when both the air and water temperatures are higher than normal. Thus they form only in summer over tropical oceans. Temperatures only a few degrees above normal are enough. Hurricanes form in an atmosphere of essentially uniform pressure, and they are not associated with atmospheric fronts. They appear to build on a wave of low pressure or on a minor disturbance in wind circulation. These atmospheric ripples may result from local differences in water heating or instability in masses of tropical air.

The trade wind belt is typically a relatively shallow layer of warm, moist air above which lies a deep layer of warmer, dry subsiding air. This forms the trade wind inversion, a condition that limits the vertical development of clouds (Figure 11.7). The inversion is sometimes interrupted by a low-pressure trough, which allows thunderstorm development behind the wave. Increased convection and the normal pattern of high-altitude winds cause the trough to deepen, forming an isolated low-pressure system. If the pressure continues to fall, winds accelerate and a tropical storm is born. The change in status from tropical storm to hurricane requires a mechanism to stimulate vertical air motion and air convergence. Several possible trigger mechanisms exist, with an intruding high-altitude low-pressure system being the most-cited cause. Derived from the remains of an upper tropospheric cyclone wave, these abandoned waves act in two ways to promote instability: There is divergence on the eastern side of the abandoned system, and the low has a cold core that changes the lapse rate below. Both the divergence and the altered stability

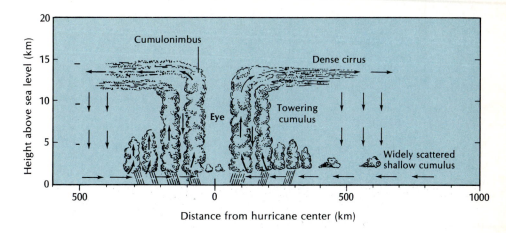

FIGURE 11.7
Vertical structure of a hurricane showing the cloud pattern and wind flow.

enhance surface-pressure differences enough to generate a tropical cyclone. Once a hurricane begins to form, moist air from all sides converges toward the storm center. Condensation supplies the energy needed to develop the storm, so a constant supply of water vapor is essential for the storm's continued existence.

The storm moves slowly at first, usually from east to west in low latitudes. As it gains strength, the speed increases and its path curves gradually toward the pole. As long as the storm remains over warm water, it can grow in intensity. A storm can travel far north along the east coast of the United States because it follows the warm Gulf stream. When a storm moves over the cold Labrador current, it dissipates rapidly. If it moves over land, increased friction with the land surface and loss of the energy supply causes the storm to dissolve.

Time and Place of Occurrence

Tropical cyclones occur only in certain regions. They start over the tropical oceans between latitudes of 5° and 20° and generally form over the western sides of the oceans. They are absent along the equator due to the weakness of the Coriolis force. They occur in six general regions: The Caribbean Sea and the Gulf of Mexico, the Pacific from the Philippines to the China Sea, the Pacific Ocean west of Mexico, the South Indian Ocean east of Madagascar, the North Indian Ocean in the Bay of Bengal, and the Arabian Sea. The central and western portions of the Pacific Ocean have an average of 20 tropical cyclones each year, mostly from June to October. In the eastern Pacific, southwest of Mexico, an average of three hurricanes form each season, but few reach land. Some hurricanes occur outside these regions, but these are the areas of highest frequency. It is of some interest that hurricanes do not occur south of the equator in the Atlantic Ocean. Apparently the equatorial convergence zone does not migrate far enough south to provide the necessary convergence (Figure 11.8).

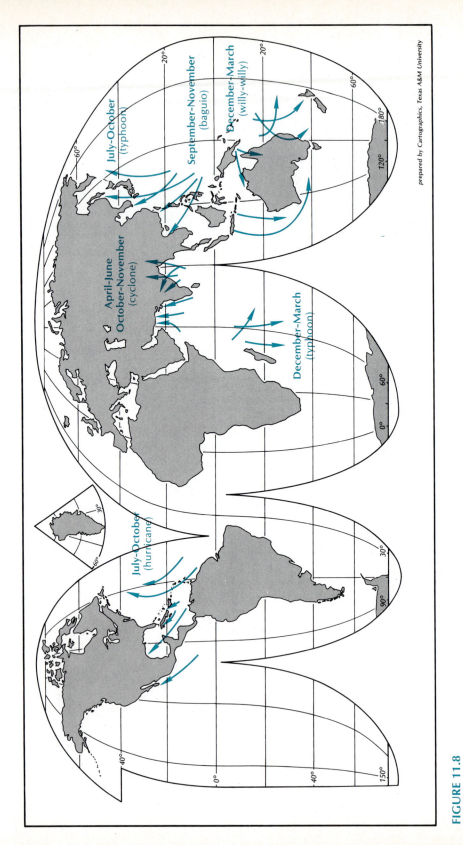

FIGURE 11.8

Areas of the world that experience the most frequent hurricanes.

July-October
(typhoon)

September-November
(baguio)

December-March
(willy-willy)

April-June
October-November
(cyclone)

December-March
(typhoon)

July-October
(hurricane)

prepared by Cartographics, Texas A&M University

Hurricanes also occur at particular times. Their peak frequency corresponds to the period of highest sea temperature during the year and the time of the maximum displacement of the convergence zone. Thus late summer and early fall are the seasons of maximum occurrence (see Plate 8).

Hurricane Surge and Coastal Flooding

Losses from hurricanes are due mainly to the high water and accompanying waves. Winds produce heavy seas with waves reported higher than 13 m (43 ft). Long sea swells as far as 1600 km (990 mi) from the storm center are evidence of both the strength of the storms and their far-reaching effects. Surges of water 2–3 m (6–10 ft) in height may form ahead of the storm, and the pounding of the waves on top of the surge do most of the damage. The problem is compounded if the storm arrives at high tide. Additional flooding results from the high amounts of rainfall that usually accompany the storm.

When a hurricane forms, its size, intensity, and path determine the potential for damage. The size and intensity of tropical cyclones provides a basis for classifying them. A widely used scale is the Saffir/Simpson scale (Table 11.4). This scale ranges from a value of one for the weakest storm to five for the most severe. The scale is based upon atmospheric pressure, wind speed, the size of the waves generated, and the relative damage the storm may cause.

North American Hurricanes

Major hurricanes are classified as three, four, and five on the Saffir/Simpson scale. Of the hurricanes influencing the United States for the period 1900–78, fewer than half were classified as major. The majority were class-one storms. For the same period, more than 50% of all hurricanes affecting the United States occurred in September.

To consider losses, we must distinguish between deadliest and costliest. Fatalities in the United States have been declining in recent years as a result of fewer severe hurricanes and better warnings. On the other hand, dollar damages have been increasing due to continued construction on coastal lands (Table 11.5).

Hurricane Hazard in the United States

The U.S. population has increasingly moved into the coastal areas along the Gulf of Mexico and the Atlantic Ocean, placing many people in jeopardy from hurricanes. The population along the coast has almost doubled in the years since 1930. In Florida alone, more than 5 million people live close enough to the sea to feel the effects of a hurricane surge. Nearly 90% of Florida's population is in urban areas, and a large proportion of these people are living in mobile homes. Coastal cities in Florida are among those with a high chance of being struck by hurricanes.

TABLE 11.4

Saffir-Simpson hurricane damage potential scale.

SCALE NUMBER	CENTRAL PRESSURE		WINDS		STORM SURGE		DAMAGE
	Mm	*In.*	*Mi/Hr*	*Knots*	*Ft*	*M*	
1	980	28.94	74–95	64–82	4–5	1.5	*Minimal:* damage mainly to trees, shrubbery, and unanchored mobile homes
2	965–979	28.50–28.91	96–110	83–95	6–8	2.0–2.5	*Moderate:* some trees blown down; major damage to exposed mobile homes; some damage to roofs of buildings
3	945–964	27.91–28.47	111–130	96–113	9–12	2.5–4.0	*Extensive:* foliage removed from trees; large trees blown down; mobile homes destroyed; some structural damage to small buildings
4	920–944	27.17–27.88	131–155	114–135	13–18	4.0–5.5	*Extreme:* all signs blown down; extensive damage to roofs, windows, and doors; complete destruction of mobile homes; flooding inland as far as 10 km (6 mi); major damage to lower floors of structures near shore
5	920	27.17	155	135	18	5.5	*Catastrophic:* severe damage to windows and doors; extensive damage to roofs of homes and industrial buildings; small buildings overturned and blown away; major damage to lower floors of all structures less than 4.5 m (15 ft) above sea level within 500 m offshore

TABLE 11.5
Hurricane data.

HURRICANES AFFECTING UNITED STATES (1900–78) BY CATEGORY (SAFFIR/SIMPSON SCALE NUMBER)

	1	2	3	4	5	All
United States	47	29	38	13	2	129
FL	18	11	15	5	1	50
TX	9	9	7	6	0	31
LA	4	6	6	3	1	20
NC	9	6	3	1	0	19

HURRICANES BY MONTH (MORE THAN 3 ON SAFFIR/SIMPSON SCALE)

	June	July	Aug.	Sept.	Oct.	All
United States	3	2	11	30	7	53
FL	0	1	1	14	5	21
TX	1	1	5	6	0	13
LA	2	0	3	4	1	10
NC	0	0	1	5	1	7

DEADLIEST AND COSTLIEST HURRICANES

Location (Name)	Year	Category	Deaths
TX (Galveston)	1900	4	6000
FL (Lake Okeechobee)	1928	4	1836
FL (Keys)	1919	4	900
New England	1938	3	600
FL (Keys)	1935	5	408

Location (Name)	Year	Category	Cost ($ millions)
Hugo	1989	5	10,500
Gilbert	1988	5	10,000
Agnes (FL, N.E. U.S.)	1972	1	2,100
Camille (MS, LA)	1969	5	1,420

■ Applied Study

Hurricane Hugo

Hurricane Hugo originated in the Atlantic Ocean off the coast of western Africa near the equator. It began as a tropical depression, then grew into a tropical storm. It soon reached hurricane status as the storm system fueled itself with warm, moist air from the ocean.

On September 17, 1989, Hugo hit the Leeward Islands of the Lesser Antilles with wind speeds more than 230 km per hr (140 mph). Guadeloupe and Dominica were the first two islands in Hugo's path (Figure 1).

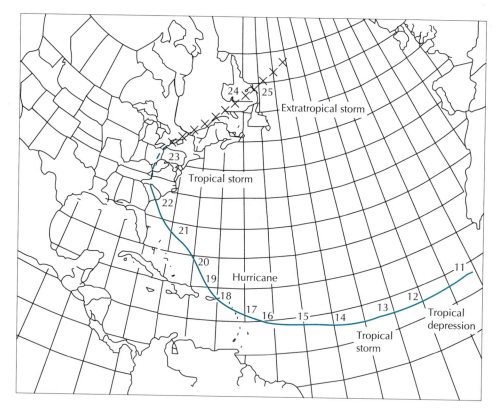

FIGURE 1
Hurricane Hugo began as a tropical depression on September 11, 1989. It grew into a tropical storm on the 12th and into a hurricane on the 14th. It reached land on the night of the 22nd. It was identifiable as a tropical storm and extratropical storm until September 25, when it passed out to sea off eastern Canada.

On September 18, the hurricane struck the U.S. Virgin Islands (Saint Croix, Saint John, Saint Thomas, and many small islets). Wind gusts of 360 km per hr (220 mph) battered Saint Croix. Following the storms, mobs began looting stores on Saint Thomas, and widespread violence and looting were reported on Saint Croix. Lawlessness remained unchecked until 1200 military police, U.S. Marshals, and FBI agents arrived on the islands.

Puerto Rico placed emergency plans in operation before the storm hit on September 19. Cruise ships steamed out of the path of Hugo. American Airlines shut down its San Juan hub and suspended all Caribbean services. In San Juan, the National Guard patrolled streets to prevent looting and helped in rescue efforts. Hugo hit the northeast portion of Puerto Rico with 200-km-per-hr (125 mph) winds.

After passing Puerto Rico, the hurricane turned north over the Atlantic Ocean and passed east of Florida. In the three days after Hugo passed over Puerto Rico, authorities and residents in the Carolinas were preparing to cope with the storm. One of the largest evacuation efforts of modern times in the United States took place. People evacuated coastal sites from the Georgia–South Carolina border as far north as Wrightsville Beach and Kure Beach in North Carolina. More than half a million people left their homes along the southeast coast.

When Hurricane Hugo hit Charleston, SC, about 12,000 people crowded into 83 evacuation shelters. Motels and hotels as far inland as Charlotte, NC, were full. Unfortunately, the evacuees who fled to Charlotte to escape found it wasn't safe even that far inland.

The storm slowly turned northwestward toward the coast on the morning of September 21. The storm center (now upgraded to a class-four hurricane) passed over Charleston at 10:00 P.M. on that day. At the time it was moving northwest at 36 km per hr (22 mph) with sustained winds of 225 km per hr (135 mph). The winds produced surges of water up to 4 m (12.9 ft) above normal. A 5.9-m (19.4-ft) tidal surge struck Awendaw, SC. Tornadoes imbedded in the storm touched down at several locations, including on James and Johns islands and in Berke and Connell Counties of North Carolina.

The storm crossed South Carolina in the early-morning hours of September 22. By the time Hugo reached North Carolina, it was downgraded to a tropical storm. The path went over Charlotte, NC, west of Winston-Salem, and then north across the Blue Ridge Mountains. Hugo reached Charlotte (200 miles from landfall) at about 8:00 A.M. with wind gusts of up to 150 km per hr (90 mph). Winds blew trees into houses, shattered glass in skyscrapers, snapped utility poles, and shredded awnings. Downed trees closed the Blue Ridge Parkway for 330 km (200 mi).

Remnants of Hurricane Hugo crossed Virginia and Pennsylvania on the afternoon of September 22. The storm was still powerful as it crossed the Virginias. Virginia's Washington and Taswell counties clocked winds of 130 km per hr (81 mph). Hugo took down trees and power lines. West Virginia's Mercer County had 75- to 80-km-per-hr (45–50 mph) winds.

Fatalities

The number of casualties from Hurricane Hugo was small, thanks to many people and agencies. Adequate warning preceded the storm, emergency plans were in effect at many places, and most people responded to the warnings and took appropriate action. Most fatalities occurred on the islands: 11 people died on Guadeloupe Island, 10 on Montserrat, 6 on the U.S. islands, and 12 on Puerto Rico. Only one death occurred in Charleston, SC, when a house collapsed on its occupant. One child was killed in North Carolina and two people were killed in Virginia.

Structural Damage

Property damage from the storm was the largest of any storm in U.S. history. Damage occurred to structures, boats, services, and vegetation. Virtually the entire population of 12,000 people in Montserrat was left homeless. Nearly every building suffered damage. In Antigua, 99% of the residents were homeless. St. Croix suffered the most physical damage with nearly all homes damaged or destroyed, leaving 90% of the population homeless. A resident on St. Thomas reported: "Whole buildings just picked up and left." Fifty airplanes were destroyed or damaged at the Isle Verde Airport on Puerto Rico, and about 80% of the population was left homeless.

In the Carolinas, damage occurred along the coast from Hilton Head Island, SC, north to Wrightsville Beach, NC. The most damage happened near Charleston. Johns and James islands just southwest of Charleston suffered some severely damaged homes and power outages. Folly Beach, an oceanfront town, lost 80% of its homes. Roads were stripped of pavement. Beachfront homes were flattened or carried away.

Eighty percent of the roofs in Charleston were damaged, and 128 buildings partially or totally collapsed. The Ben Sawyer Bridge, the only road leading from Sullivan's Island, was twisted and upended. Isle of Palms had a storm surge of 3.3 m (11 ft). Damage to a marina at Murrells Inlet created a 10,000-gal. diesel fuel spill. Hundreds of boats in marinas along the coast were damaged or destroyed. Beachfront property in Myrtle Beach sustained heavy damage. The Grand Strand lost all its piers and boardwalks to the 4-m (13.6-ft) tides. An estimated 50% of the housing was extensively damaged.

Damage to Vegetation and Wildlife

Thousands of acres of trees were destroyed in the path of Hurricane Hugo. The tropical forests of the Caribbean islands, including the Caribbean National Forest on Puerto Rico, were severely damaged, as were about 95% of the trees in Charleston. Charlotte streets and parks alone lost 20,000 trees. Seventy percent of the timber in the Francis Marion National Forest was destroyed. Orchards were blown down, and large areas of crops such as tobacco, corn, and soybeans were destroyed.

Wildlife also suffered from the storm. Many endangered species of birds suffered losses, including the Puerto Rican Parrot, the Plain Pigeon, Yellow-Shouldered Black Bird, Red Cockade Woodpecker, and the Bald Eagle. The Red Cockade Woodpecker population in Francis Marion National Forest was reduced by as much as 75%. ■

Summary

Two of the severe storms that develop in the atmosphere are the tornado and hurricane. Tornadoes are mainly a midlatitude storm, while hurricanes are primarily a tropical and subtropical storm. Both are low-pressure cyclonic storms.

Tornadoes are relatively small in diameter; hurricanes are quite large. Both storms form under specific sets of conditions, but when those conditions exist they do not always occur. Both occur more frequently at certain times than others, and in certain places. Because of their size, hurricanes are easier to spot, track, and forecast. Tornadoes, being much smaller and much shorter-lived, are less easy to spot, track, and forecast.

Both storms produce casualties and can cause a large amount of damage. Casualties and losses depend to a large extent on the time and place they occur. Because hurricanes are much larger, the potential from damage and loss of life is much greater than from tornadoes.

Part Two

REGIONAL CLIMATOLOGY

REGIONAL CLIMATES: SCALES OF STUDY

Chapter Overview

Climates can be identified and analyzed over a broad range of areal units, from a small vegetable garden up to a region the size of a continent. The climate over a plowed field is different from that over a field of clover or pasture, the climate of a city differs from that of a rural area, and the climate of a desert differs from that of a rain forest. Climatologists must deal with differences in climate at all levels. The analytic approach used often varies with the relative size of the study area. Before considering major world climates, we will briefly examine some other scales of study.

Definitions

Identifying and naming the range of scale in climatic studies has caused confusion, largely as a result of the difficulty of separating the atmospheric continuum into discrete units. The historic evolution of climatic studies has compounded the identification problem. Researchers in different countries have used different names for areas. Thus we find such terms as *gelendeklima*, *ecoclimate*, and *topoclimate* referring to areas of somewhat similar size. At the same time, the dimensions suggested as suitable boundaries for identified scales frequently differ from one source to another.

In a discussion of scale in climatology, M. M. Yoshino derived some generally accepted definitions; the description given here follows his grouping. Figure 12.1 provides examples of the scales that he suggests. The major subdivisions include the following:

Microclimate: Characterized by, for example, the climate that might occur in an individual field or around a single building, this describes the climates of an area that may extend horizontally from less than 1 m to 100 m (3.3–330 ft). Vertically the area may extend from the surface up to 100 m (330 ft.).

Local climate: This category comprises a number of microclimatic areas that make up a distinctive group. The climate in and above a forest or that of a city may fall into this division. Horizontal dimensions may extend from 100 m to 10,000 m (330 ft–6 mi), and the vertical extent is up to 1000 m (0.6 mi).

Mesoclimate: Such climates may range horizontally from 100 m to 20,000 m (330 ft–12 mi) and vertically from the surface to 6,000 m (20,000 ft). As Figure 12.1 illustrates, a great variety of individual landscapes belong in this category.

Macroclimate: The largest of the areas studied in climatology, extending horizontally for distances more than 20,000 m (12 mi) and vertically to heights in excess of 6,000 m (20,000 ft). Such areas can be continental in extent.

The dimensions provided in the listing are basic guides, with many studies overlapping the identified groups. Perhaps the best way to illustrate the role of scale in regional studies is to examine specific examples.

Microclimates

Studies of these small areal units inevitably begin with field work. The preliminary step is required to accumulate data and generally includes measuring variables. Such a proce-

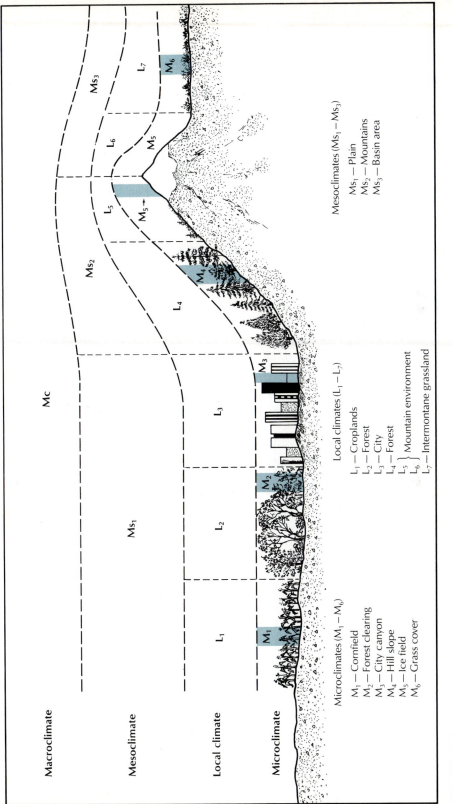

Macroclimate

Mesoclimate

Local climate

Microclimate

Mc

Ms₃

Ms₂

Ms₁

L₇

L₆

L₅

L₄

L₃

L₂

L₁

M₆

M₅

M₅ →

M₄

M₃

M₂

M₁

Mesoclimates (Ms₁ – Ms₃)

Ms₁ — Plain
Ms₂ — Mountains
Ms₃ — Basin area

Local climates (L₁ – L₇)

L₁ — Croplands
L₂ — Forest
L₃ — City
L₄ — Forest
L₅ }
L₆ } Mountain environment
L₇ — Intermontane grassland

Microclimates (M₁ – M₆)

M₁ — Cornfield
M₂ — Forest clearing
M₃ — City canyon
M₄ — Hill slope
M₅ — Ice field
M₆ — Grass cover

FIGURE 12.1

Area scales of climatic investigation.

dure is needed because published climatological data are almost entirely composed of readings taken at the standard height of the instrument shelter. The microclimatologist is often concerned with the state of the atmosphere below that level. Furthermore, the stations that report climatological data are widely spread, with one station perhaps representing many square miles. Obtaining data to help derive differences over small distances requires a fairly dense network of instruments within a small area.

Many studies of microclimatic environments have provided a generalized picture of climatic conditions near the ground. These findings indicate the types of conditions that will occur over a bare surface. Through analysis of surfaces covered by vegetation or synthetic materials it is possible to generate a more complete understanding of the variation of climatological processes that occurs at the microscale.

General Characteristics of Microclimates

Figure 12.2 shows the temperature characteristics at the interface of Earth and the atmosphere. A large diurnal temperature range occurs at the near-Earth levels; during the day, interface temperatures have been found to be as much as 10°C (18°F) warmer than the air only 1 m (3 ft) above the surface. At night, the situation is reversed; an inverted lapse rate occurs, with temperatures at the surface cooler than those immediately above. Such a response is to be expected. During the daylight hours, incoming solar radiation warms the surface and heat diffuses into the atmosphere, raising the temperature of the air by smaller amounts at increasing altitudes above the ground. After sunset, under clear-sky

FIGURE 12.2

The daily course of temperature on a summer day at four heights above the surface of the ground.
Source: Data from Geiger (1950).

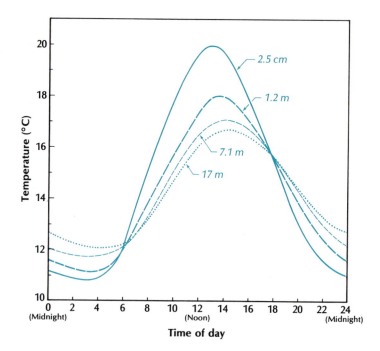

conditions, the surface cools rapidly by radiation and the atmosphere loses heat by diffusion to the cold surface.

Temperature changes in soils decrease with depth. As Figure 12.3 illustrates, the amplitude of diurnal temperature change is greater near the surface and decreases with depth until equilibrium is attained. A similar pattern is obtained for the annual cycle, although, of course, it follows the seasonal rather than the diurnal cycle.

Figure 12.4 is an idealized representation of wind near the surface. At the interface between the air and the ground, a thin layer of air adheres to the surface; in this layer, flow is *laminar*; streamlines are parallel to the surface and lack the cross-stream component of convective currents. The depth of this layer depends upon the surface roughness and wind speed, but it is seldom more than 1 mm (.04 in.) thick. Its significance lies in its role as an insulating barrier in which all nonradiative transfer is by molecular diffusion rather than the turbulent transfer typical of most of the lower atmosphere. The turbulent surface layer comprises a complex flow of swirling eddies extending up some 50 m (165 ft). In this zone, a general increase of wind speed occurs with height.

The exchange of moisture between the surface and the air above is reflected in humidity measurements at various levels. Figure 12.5 provides a generalized profile of water vapor concentration for day and night. During the day, the concentration decreases away from the surface similarly to the temperature profile. At night, if the dew point of the air is not attained, the humidity profile is somewhat similar to that in daytime. This occurs because evaporation will continue during the night hours, but at a lower rate than during the day. If the dew point is reached, an inverted moisture profile exists. The presence of dew causes a lowering of near-surface moisture content; if the moisture is replaced by downward movement of air, dew formation will continue. If downward movement ceases, the near-surface air is not replenished with moisture, and formation of dew stops.

Although it is possible to generalize about the nature of the microclimate above a bare soil surface, it is important to note that the actual characteristics will depend, at least in

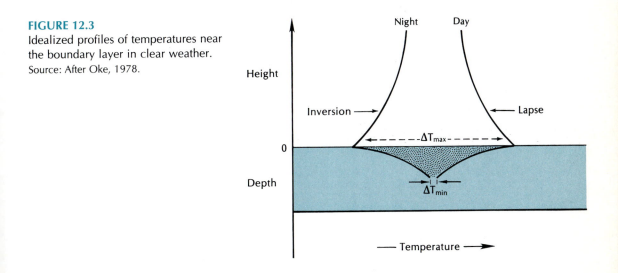

FIGURE 12.3
Idealized profiles of temperatures near the boundary layer in clear weather.
Source: After Oke, 1978.

FIGURE 12.4

Flow of air at the boundary layer from laminar to turbulent flow.

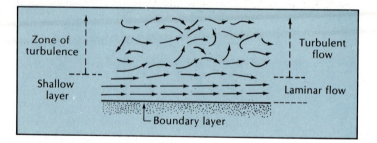

part, upon the type of soil surface exposed and the amount of water it contains. Sandy soils, for example, have a lower heat capacity and thermal conductivity than clays. This means that a sandy soil heats up rapidly in its top layers during the day but at night is cooler than less-sandy soils because it undergoes rapid radiational cooling. Organic matter in the soil reduces heat capacity and thermal conductivity, and the dark color increases the absorption of solar radiation. At night, such soils are relatively warm.

The presence of water in the soil or at the surface greatly modifies the exchange of energy that occurs. When water is present, incoming energy is used for evaporation, leaving less available for sensible heat. A comparison of the energy budgets for an irrigated field and one that is dry shows that the temperatures over the moist soil are usually lower than those over the dry.

These general principles apply to all microclimates; however, modification of the bare surface through vegetative growth or human interference alters the intensity and rates at which ongoing processes occur. Such changes are described in the following section.

Role of Surface Cover

The role of surface covering in creating microclimates is a response to how incoming energy is disposed at the surface and how the surface modifies airflow. The differences that exist can best be shown through illustrative data. Figure 12.6a shows, in schematic

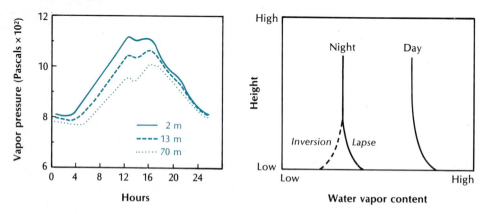

FIGURE 12.5

Generalized profile of water vapor concentration near the boundary layer.
Source: After Oke, 1978.

form, some modifications that occur when plants cover the surface. Of particular significance is the creation of a canopy layer in which the microenvironment reflects the nature and extent of the canopy. Clearly, the relative continuity of the canopy plays a major role, so plants with large leaves horizontal to the surface form a more effective canopy than those whose leaves are small and are aligned vertically (Figure 12.6b). The canopy becomes most effective when a plant stand has grown to the extent the ground is shaded.

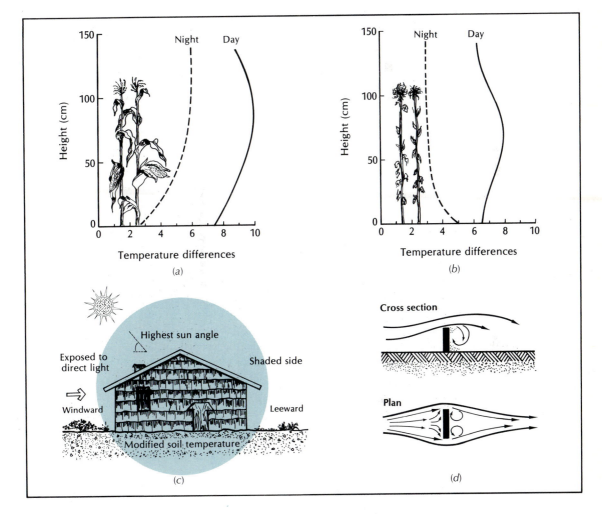

FIGURE 12.6
Schematic representation of microclimatic modifications caused by surface cover:
(a) Temperature profiles in a crop whose growth shades the ground. (b) Temperature profiles for crops with essentially vertical growth. (c) Climatic sheath around a building. Many microclimatic variations occur in the sheath. Examples show creation of windward-leeward sides, sunlit and shaded sides, and soil modification. (d) Airflow across a barrier at right angles to wind direction. Cross-section shows how such an obstruction is used as a snow fence.

This growth causes the highest temperature zone to move away from the surface to the canopy area. What effectively happens is that the top of the canopy, rather than the soil surface, becomes the energy-exchange layer.

A building or a fence, for example, has a marked effect upon the microclimate. A building creates a climatic "sheath" (Figure 12.6c) in which temperature variations will occur as a result of shading, humidity anomalies are found, and even a rain shadow effect may occur. When the changes in the microenvironment of a single building are assessed, it is not surprising that cities consisting of buildings can create their own climate.

Modification of wind in the microenvironment is well demonstrated by effects of a vertical structure on airflow. As the diagram in Figure 12.6d shows, the patterns change with the direction from which the wind is blowing. Such modification has been put to good use in the construction of snow fences to keep roads and other used areas clear of drifting snow.

Role of Topography

Chapter 3 noted that topography is important in determining the temperature distribution at a given location. The height of the snow line on the slopes facing the equator compared to those facing the poles was stressed. These differences can be translated to the micro-level. Most people have seen snow on one side of a small hill that remains in place long after that on the opposite slope has melted.

Topographic variations also modify thermal patterns in small areas. One of the best-known effects of this is the creation of an inversion that, at times, can create a distinctive airflow. On cool, still nights, air close to the surface becomes cooler than that above. If this cooling occurs in areas of uneven topography, the cold, dense surface air will tend to flow downslope and accumulate in the bottom of valleys and depressions. Such airflow is called a **katabatic wind**.

If the temperature of the ground is at or just below freezing, the collection of cold air in the valley bottom causes a localized frost. Such an effect in citrus-growing areas can cause appreciable crop damage. Topographically induced air drainage may also create potential for air-pollution. If the pool of cold air becomes deep enough, any effluent carried into the inversion layer will be trapped in it. Should the inversion remain in place for more than a few days, the stagnant air can become highly polluted.

Local Climates

Local climates encompass several microclimatic environments because they comprise a larger area. The organizational framework does have some overlap, but a further differentiation can be made because local climatic studies stress horizontal rather than vertical differences in climate. There are many examples of local climatic studies, and here the climate of a forest is used to illustrate this level of analysis.

Forest Climates

Forest climates differ from those of surrounding nonforested areas. The boundary layer of the forest is its canopy, and this is the level where energy exchanges occur. Some insolation is returned directly to space, with the amount depending upon the albedo, or reflectivity, of the canopy layer. In some forests the quantity will vary enormously from season to season. Some energy will be trapped within the canopy layer, and some will penetrate to the floor of the forest. As illustrated by the data in Table 12.1, the amount of penetration is characteristically low. The evaluation of the data is further complicated in that the amounts involved in the disposition of radiant energy vary both with the state of the sky and the amount of foliage that exists.

Compared to the flow of air over open areas, wind inside the forest is reduced to some degree, depending on the forest's type and structure. For example, in a deciduous forest the wind velocity is reduced by as much as 60%–80% at a distance of 30 m (100 ft) inside the forest. Similarly, in a Brazilian forest, the wind speed has been found to decrease from about 8 km per hour (5 mph) to 1.6 km per hr (1 mph) with the same distance. The flow of air is, of course, highly complex and varies in the vertical as well as the horizontal dimensions. Studies have shown that winds of 8–24 km per hr (5–15 mph) above the canopy are often less than 3.2 km per hr (2 mph) at the surface.

The forest environment also modifies local moisture conditions. Evaporation from the forest floor is relatively low because of reduced insolation and wind velocity. This effect is counterbalanced by the high transpiration that occurs with profuse vegetation. As such, the humidity within a forest depends upon the density of the forest and the rates of transpiration. Generally, the relative humidity in the forest is 2%–10% higher than that of nonforested areas, with the highest humidities occurring during the high-sun season. Note that comparisons of humidity using relative humidity values are not always meaningful, because the modified thermal environment directly influences the water-holding capability.

TABLE 12.1
Variations in radiation receipts within forests.

DAILY TOTALS OF NET RADIATION IN AND ABOVE A YOUNG PINE FOREST (LY/DAY)

Height (m)	10.0	5.0	4.1	3.3	2.1	0.2
July 7, 1952						
total	566	555	223	36	—	35
percentage	100	98	39	6	—	6
Nov. 9, 1951						
total	291	—	104	—	14	—
percentage	100	—	35	—	5	—

SOLAR RADIATION RECEIVED ON A HORIZONTAL SURFACE AT THE 1-M LEVEL IN AN OAK FOREST AS A PERCENTAGE OF INCIDENT RADIATION ABOVE THE CANOPY

	Clear Sky	Overcast Sky
Foliaged	9%	11%
Defoliaged	27%	56%

Source: After R.E. Munn, *Descriptive Meteorology* (New York: Academic Press, 1966).

The thermal differences in a forest result from a combination of the factors already outlined: shelter from direct rays of the sun, heat modification through water transfer, and the "blanket" effect of the canopy. Interaction of these factors results in moderated temperatures, lower maximums, and the higher minimums than in nonforested areas experiencing a similar macroclimate regime. The amount of variation is seasonal: The main difference in summer can be as much as 2.8°C (5°F) within a low-altitude, midlatitude forest. Exceptions to such moderating influences do occur; the Forteto oak forests of the Mediterranean, for example, experience higher temperatures than neighboring, non-forested areas. Such trees transpire slowly, negating the "usual" humidity and thermal conditions of the forest.

Forest temperatures vary vertically as well as horizontally. For the most part, temperature increases with height during the day and lapse rate conditions prevail at night, reflecting the role of the canopy in the energy exchange.

Urban Climates

The constructed environment of a city creates a totally different climatic realm from its natural predecessor. Consider the following:

Concrete, asphalt, and glass replace natural vegetation.

Structures of vertical extent replace a largely horizontal interface.

Large amounts of energy are imported and combusted. Combustion of fossil fuels creates pollution.

These related factors modify the climatic process in the urban environment.

The Modified Processes. Figure 12.7 provides a graphic summary of the modified flows of energy in a city. Of prime importance is the energy characteristics of the asphalt/concrete/glass environment as compared to that of the vegetation-covered surfaces of rural areas. The city surface generally has a lower albedo than nonurban areas, greater heat conduction, and more heat storage. During the day, the city surfaces absorb heat more readily; after sunset, they become a radiating source that raises night temperatures.

The energy flows are further modified by the geometry of the city buildings (Figure 12.8a). Walls, roofs, and streets present a much more varied surface to solar radiation than do undeveloped areas. Even when the sun is low in the sky—a time when little energy absorption occurs on flat land—vertical city buildings feel the full impact of the sun's rays. In the early morning and late evening, the city is absorbing more energy than surrounding rural areas.

The different surfaces that make up the city modify energy availability by changing the water balance (Figure 12.8b). Rain falling on urban and nonurban areas is disposed of quite differently. Some of the water on building-free rural surfaces is retained as soil moisture. Plants draw upon this source for their needs and eventually return the moisture to that air through transpiration. At the same time, standing water and soil moisture evaporate. Both transpiration and evaporation require solar energy. In the city, pavements and buildings prohibit entry of water into the soil; with limited green areas, transpiration is minimal. Most of the rainwater drains off quickly and passes to storm sewers, greatly reducing the water available for evaporation. Since both evaporation and transpiration

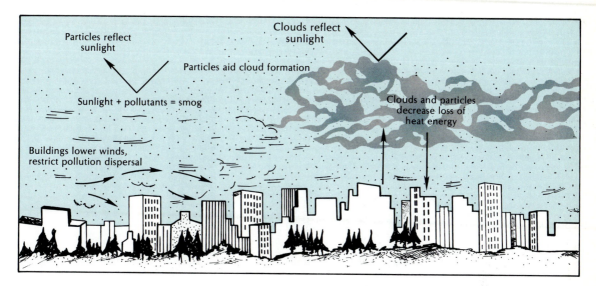

FIGURE 12.7
Schematic diagram showing the modified flows of energy in an urban environment.

amounts are decreased, the solar energy available for this process provides additional surface heating.

The ratio of energy used for sensible *heat flow* (which heats the atmosphere) to that of the latent *heat flux* (the latent energy in water vapor) is given by the Bowen ratio. In areas such as deserts, the lack of water yields a high Bowen ratio, on the order of 20. In most vegetated areas the value is usually less than 1, indicating more transfer by latent than sensible heat. In cities, the ratio is about 4.0, indicating a large decrease in latent-energy transfer and an increase in sensible heat. Again, this results in more available energy to heat the city atmosphere.

As areas of concentrated activity, cities are high consumers of energy, and enormous amounts of energy are imported to maintain their functions. In New York City, for example, the heat generated through the burning of fossil fuels in winter is two and one-half times the amount of heat energy derived from solar radiation for the same period. This high energy use means that much "waste" heat from factories, buildings, and transportation systems passes to the atmosphere, increasing the city's warmth.

Combustion of fossil fuels causes cities to have higher air pollution levels than their rural counterparts. Visible evidence of air pollution over cities is the dust dome that is often produced. This pall of smoke and smog modifies the urban climate in a number of ways. Some incoming solar radiation strikes particles in the air and is reflected back to space. Other particles act as a nucleus onto which water vapor condenses to form clouds. That cities tend to be cloudier places than rural areas is illustrated by data from London: One study showed that the city received 270 hours less of bright sunshine than did the surrounding countryside over a year.

The lower input of solar energy because of increased reflection from particles and clouds results in lower temperatures in cities; this lack of energy, however, is more than

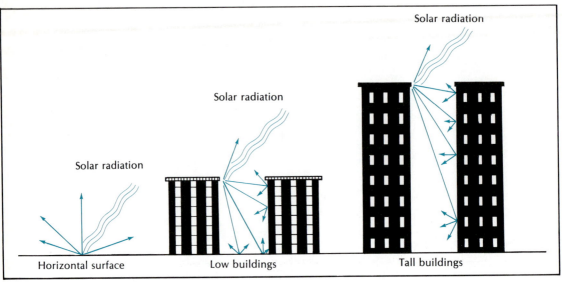

(a)

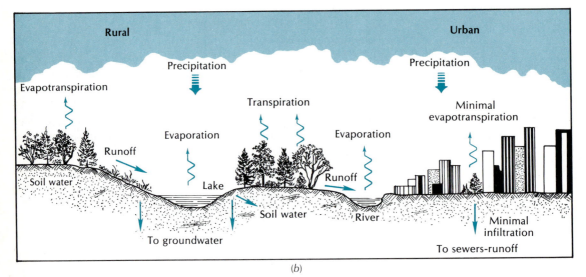

(b)

FIGURE 12.8

Modified climatic processes associated with urbanization: (a) the effects of horizontal and vertical surfaces on incoming solar radiation; (b) the disposal of precipitation on rural and urban surfaces.

Source: John E. Oliver, *Climate and Man's Environment*. (New York: John Wiley & Sons, 1973), 239.

counterbalanced by other effects. Some of the pollutants absorb rather than reflect energy, and the increased cloudiness reduces loss of long-wave energy to space.

The Observed Results.   Of the many changes that occur in the climate of cities, the modified temperature regime is the best known and most closely studied. Cities tend to be warmer than the surrounding nonurban environment. The higher temperatures are best developed at night, when, under stable conditions, a heat island forms. In passing from the center of the city toward the surrounding countryside, temperatures decrease slowly until, in rural areas, they remain about the same. The city is an island of warmth surrounded by cooler air. Figure 12.9a provides a schematic summary of this feature. It is possible to visualize the formation of an urban dome in which the normal lapse rate

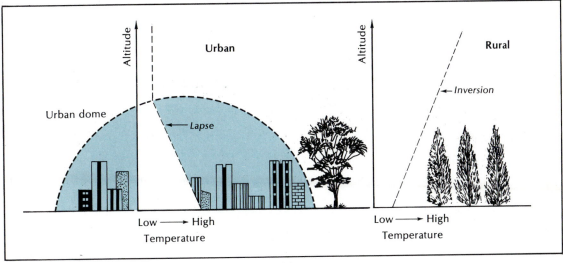

(a)

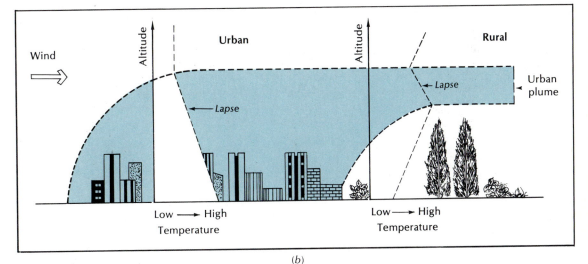

(b)

FIGURE 12.9

Comparison of lapse rates over rural and urban environments: (a) on a still night, normal lapse rate conditions occur in the city, and an inversion in the surrounding rural areas; (b) with a slight regional wind, the rural conditions are modified by the creation of urban plume.

Source: John E. Oliver, *Climate and Man's Environment.* (New York: John Wiley & Sons, 1973), 245.

occurs because of the warmth of the city. This effect is in contrast to rural areas, where, during the night, a low-level inversion often forms. If a slight regional wind is blowing, the warmer air of the city is swept downwind as an urban plume, modifying the rural lapse-rate conditions (Figure 12.9b).

The heat-island effect has been measured in many cities. Places of diverse size—from large cities like Tokyo, London, and St. Louis to smaller ones like Corvallis, OR, and San Jose, CA—show similar patterns. Generally, when the heat island is best developed, temperature differences between city and country are as much as 5.5°C (10°F).

Wind speeds in a city are, on average, lower than those in open surrounding areas because of the increased roughness of the urban fabric. A comparison of highest expected wind speeds in a city and its airport (usually located outside the city in flat, unobstructed land) shows this to be true. In Boston the highest airport wind speed is 46 m per sec (102 mph); the value for the city is 32 m per sec (71 mph). The same effect is found in Chicago (city 25 m/s [56 mph]; airport 31 m/s [70 mph]) and Spokane, WA (city 23 m/s [52 mph]; airport 35 m/s [78 mph]).

The wind in cities may also be modified by the urban heat island. As noted, the heat island is best formed under calm conditions; in such instances, the city may create its own wind pattern. A small pressure gradient results when the warm city air rises and air is replaced by that from surrounding areas. The uplift of air is not high, and a convective cycle results. This process means that the city air is recycled. If pollutants are being added to the air under these conditions, pollution levels can rise dramatically and produce a dust dome over the city.

Cities' influence on rainfall is still not completely understood. Generally, however, cities are thought to receive about 10% more rainfall than surrounding rural areas. Intensive studies of this aspect of urban climates were part of the research project called METROMEX, an in-depth study of the conditions in St. Louis. Many significant findings have resulted from this project (see the example in Figure 12.10), but the results cannot always be applied to other urban centers in different climatic regimes.

Table 12.2 lists some of the basic climatic modifications that occur within cities. Examples of changes associated with the elements outlined are given, together with other parameters. It is an imposing list, indicating the amount of change that can inadvertently occur when people modify the environment. As the urban population of the world continues to expand, even more widespread results might be anticipated.

Meso- and Macroclimates

Mesoscale climates are frequently identified with a distinctive geographic region in which the physical controls of climate are similar and are not modified by major differences within the region. A mesoscale climatic study might concern the climates of areas such as the Central Valley of California, the lands in the vicinity of the Great Lakes, or the Mississippi Delta lowlands.

Climatologists have completed extensive research on mesoscale climates. The type of research has evolved over time as newer methodologies have become available. Early studies often were merely an inventory of a region's climate. Classic works on such loca-

FIGURE 12.10

Examples of results of the METROMEX study of St. Louis: (*a*) Pattern of average summer thunder days, 1971–1975. St. Louis metropolitan area is shaded, and an outline of a circular rain gauge network is shown. (*b*) The 1972–75 summer pattern of lightning-caused power shortages, based on township frequencies, and the average thunder day pattern.

Source: Stanley A. Changnon, Floyd A. Huff, Paul T. Schickendanz, and John Voger, *Summary of METROMEX*, Vol. 1. Weather Anomalies and Impacts, 1977, Illinois State Water Survey.

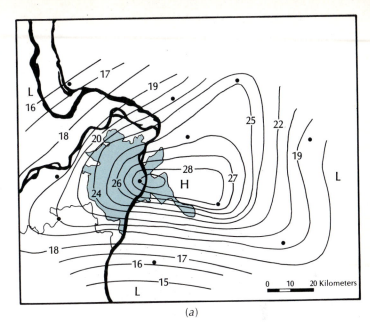

(*a*)

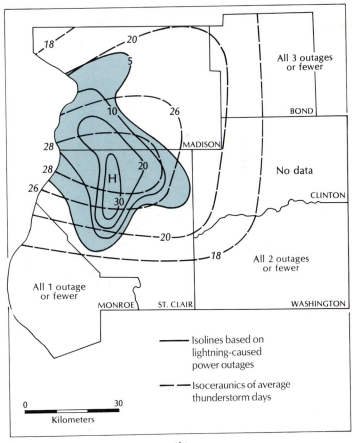

(*b*)

tions as the Paris and Thames basins provided the basis for the understanding of much regional climatology.

More recently, mesoscale climatic studies have concerned scales of motion rather than area. This development corresponds to the meteorologic concept of mesoscale phenomena, which includes analysis of such features as severe storms, mountain–valley winds, and the like. The climatological equivalents are seen in studies ranging from the cause of the Sahel drought to precipitation variations resulting from circulation patterns in the U.S. Midwest.

While the distinction between various scales of climate is not always clear, the identification of *macroclimates* is aided by the concept of *filtering*, the averaging of circulation features over successively longer time periods. As the time scale grows, the smaller scale patterns are filtered out and only the general characteristics remain. Thus a mesoscale feature such as a hurricane or a land-sea breeze would not appear on a macroscale representation.

The climatic analysis of macroclimates can follow two methods of approach. First, the analysis can consider the major surface climatic characteristics, ranging from temperature and precipitation patterns to statistical analysis of other elements. Such an approach provides the descriptive climatology of large climatic regions or even continents. Second, the analysis can deal with the dominant circulation patterns that influence the climate and provide the key to understanding its cause. This process is accomplished through analysis of the state of the atmosphere as inferred by pressure maps showing patterns and winds at various levels of the atmosphere. Alternatively, the patterns can be depicted as cross-sections through the troposphere and stratosphere. This approach obtains a synopsis, or

TABLE 12.2
Average changes in climatic elements caused by urbanization

ELEMENT	PARAMETER	URBAN VS. RURAL AREAS
Radiation	On horizontal surface	−15%
	Ultraviolet	−30% (winter); −5% (summer)
Temperature	Annual mean	+0.7°C
	Winter maximum	+1.5°C
	Length of freeze-free season	+2 to 3 weeks (possible)
Wind speed	Annual mean	−20 to −30%
	Extreme gusts	−10 to −20%
	Frequency of calms	+5 to +20%
Humidity	Relative—Annual mean	−6%
	Seasonal mean	−2% (winter); −8% (summer)
Cloudiness	Cloud frequency and amount	+5 to 10%
	Fogs	+100% (winter); +30% (summer)
Precipitation	Total annual	+5 to 10%
	Days (with less than 0.2 in.)	+10%
	Snow days	−14%

Source: H. E. Landsberg, "Climates and Urban Planning." In *Urban Climates*, Technical Note 10, p. 372. (Geneva: World Meteorological Organization, 1970).

condensed view, of the atmosphere at a given time and is referred to as the synoptic climatology. Elements of synoptic climatology were covered in Chapter 10.

As noted, one approach to describing the climates of the world involves descriptive climatology of large regions. Historically, the boundaries of such regions have been defined in terms of the surface conditions, and a number of classification schemes have been implemented. Such classification procedures are considered in the next chapter. The remaining chapters of Part Two will describe and analyze the climates of the region. The treatment is in terms of both the traditional descriptive regional climatology, which uses the major surface climatic characteristics, and dynamic analysis, which uses the principles of synoptic climatology. Of necessity, the scale used to provide an overview of the entire climate of the world is at the macrolevel.

Summary

To comprehend the incredible variety of climates that exist on Earth, climatologists work on a number of areal scales. Microclimatic studies deal with the energy and matter exchanges in a small area, concentrating upon the exchanges that occur at the Earth-atmosphere interface. Local climates represent the next size scale; their study incorporates the characteristics and processes that occur over a variety of surfaces and in a more extensive vertical space. The climate of forests represents an example of a natural local climate, that of the urban area, a local climate produced through human activities. Both are the result of a wide variety of microclimates within each setting.

Mesoscale climatic features can be explained in terms of circulation patterns. The largest scale of investigation concerns macroclimatic regions. Such regions are identified in a variety of ways, and their grouping permits the climates of the world to be studied in their entirety. Although many significant data are lost and while extensive generalizations must be made, climatic study at the macroscale allows a basic understanding of the variety of world climates.

REGIONALIZATION OF THE CLIMATIC ENVIRONMENT

Chapter Overview

The Rationale

Approaches to Classification
- Empiric Systems
- Genetic Systems

Climatic Influence on Other Environmental Variables
- Climate and the Distribution of Vegetation
- Climate and the Distribution of Soils

Chapter 12 examined the concept of areal scale in regional systems. Climates were formally classified beginning with macroscale climatic regions, which cover large geographical areas. This chapter concerns schemes for classifying the atmospheric environment of Earth into areas of similar climatic characteristics. The following chapters will examine each of these major climatic regions in detail.

The Rationale

To produce a useful classification of any data, we must first group those items that have the greatest number of common characteristics, then subdivide those groups on a uniform basis until we have reached a satisfactory degree of subdivision. In most areas of science, attempts have been made to produce classifications based upon the most fundamental characteristics possible, rather than on elements that might be more easily observed but of less intrinsic importance. Some items don't lend themselves easily to grouping or classifying. Attempting to group similar regional environments on Earth is much like trying to group students on the basis of height or weight. There are no clear-cut divisions.

The climatic elements of a region distinguish that region not by their presence or absence but by the difference in their character. Change from place to place is a basic assumption in regional study. This is not to imply that the differences from place to place are not without order. Climatic elements vary systematically from place to place, sometimes rather abruptly and in other cases over considerable distance. Since change with time and place is an integral part of Earth's environment, it is appropriate that a classification of climates should be based upon the patterns of variation.

Throughout this book, we have made repeated references to the annual and diurnal periodicities that exist. The motions of Earth in space give rise to periodic climatic, hydrologic, biological, and geologic events. If we consider the basic pattern of energy fluctuation, we find that it is responsible for the largest share of the periodicity and spatial variation in Earth's environment. The changing intensity and duration of solar radiation bring about changes in the atmosphere and seas, influence migration and hibernation habits in animals, and regulate the life cycle of much of the biota, including humans. The changing intensity of solar radiation between the Northern and Southern Hemispheres produces a shifting of the general circulation of the atmosphere, which in turn is responsible for periodic changes in Earth's hydrologic cycle. Thus, an energy flow that varies systematically through time and space results in an environment that also varies systemically through time and space. This chapter considers the major regions that result from periodic patterns of energy and moisture and shows how other environmental variables are related to these seasonal patterns.

Attempts to classify regional differences date back thousands of years. During the period of Greek civilization, Earth was divided into three broad temperature zones—torrid, temperate, and frigid. The use of the word temperate to describe midlatitude weather may not have been a particularly good choice, but the classification nevertheless persisted through the centuries. Since the recording of that very early division of Earth, classification schemes for organizing regional differences have appeared from necessity with ever-greater frequency.

Dividing Earth into three temperature zones is appropriate. The *tropics* are essentially winterless areas, as they are not directly affected by the outreach of cold polar air; they are dominated by air currents that originate in the warm areas between 30° N and S. The *polar* regions lack a warm summer with the incursion of tropical air currents. The *midlatitudes* are characterized primarily by the very marked summer and winter seasons, since these areas are dominated seasonally by tropical and polar air currents.

A second element that varies markedly over Earth is the seasonal moisture pattern. Some areas have a nearly equal probability of rain every day of the year. Most of Earth is subjected to a seasonal probability of rainfall as the primary circulation shifts back and forth with the solar energy supply. Some desert areas have nearly equal probability of having no rain on any day of the year. This breakdown includes perennial precipitation regimes, seasonal precipitation regimes, and dry regions. Such basic divisions of global climate into regions with varying temperature and precipitation form the basis of a number of well-known classification systems.

Approaches to Classification

Of basic importance in the classification procedure is the selection of the variables used in delineation. Temperature and precipitation have historically been the major variables. Reliable data have been available only over the last 100 years (much less for most climatic stations), and many early records consist only of temperature and precipitation. Much of the early classification work was limited to the use of these two variables. Furthermore, much of the early work was carried out by plant physiologists and plant geographers, who found a correlation between vegetation and temperature and moisture. Mid-nineteenth century researchers were influential in the development of climatic classification. As a result of plant geographers' influence on climatic classification, many climatic regions are identified by plant-associated names: It is not unusual to find a climate described as savanna, taiga, or tundra. Note that both plant and animal names were given to climatic regions. It would be quite unusual, however, to find a climate described as "yak" or "penguin" climate in modern literature. The correlation between climate and vegetation is still prevalent, and climates of the world are still described in terms of natural vegetation distribution.

It is evident that, in relating the distribution of natural vegetation to the distribution of climate, the *effect* of climate is being measured instead of the climate itself. It is assumed that a given climate gives rise to a distinctive vegetation association. To identify the climatic type, it is first necessary to determine the vegetation and then infer the climate. If climate can be so identified—that is, by expressing it as the result of the distribution of one selected component of the environment—then equally useful climatic regions may be identified through other measures of the effects of climate. It becomes possible to devise climatic schemes using factors ranging from the human response to climate to the effect of climate on rock weathering. Such systems would be based upon the observed effects of climate, and the criteria used to outline their boundaries established by best-fit properties. Systems derived through such methodology may be collectively termed *empiric classifications*. These classifications are the result of the observed effects of climate.

The empiric systems essentially identify similar climate types. An equally valid approach to climatic distribution asks why climatic types occur in distinctive locations. Systems that attempt to deal with the question "why" must concern the cause of spatial climate variation; we can collectively call them *genetic classifications*. As with the empiric systems, we can use a number of methods to examine the causes of climates, so the basis of genetic systems can vary appreciably.

Much controversy surrounds the relative merits of the genetic and empiric approaches to classification. If we accept that two approaches exist and that both are valid, then the genetic versus empiric argument is not of concern. What is important is which classification applies to a given distributional problem. A discussion of the climatological implications of the polar front migration, for example, may not use an empiric system; by the same token, a genetic approach may not prove of great value in discussing specific temperature requirements for plant types. In effect, the approach used and the classification selected depend on the purpose for which they are designed.

Because we can observe the effects of climate on a whole range of environmental phenomena, we can use many criteria in forming an empiric system, including

the human response to climate

climatic requirements for crop growth

water needs and precipitation effectiveness related to vegetation

study and identification of climatic analogs (e.g., agricultural analogs)

vegetation distribution related to climatic controls

geomorphic processes acting under different climatic conditions

climate and soil-forming processes

synoptic conditions and satellite imagery

Many other relationships are possible, and each probably has a climatic classification.

In recent years, the availability of extensive databases and electronic computing devices has resulted in *numerical classification*. This method uses many variables and numeric procedures, such as correlation and cluster analysis, to classify climates. With numerical classification the data and analytic method used are entirely objective. It is becoming an increasingly important method, especially in special-purpose classifications, and most future systems are expected to be of this type.

Empiric Systems

Of the many classifications that have been devised, it is inevitable that one or two develop into what might be termed "standard systems" because they are most widely used. It follows that these systems facilitate an orderly description of world climates. To achieve any prominence, such a system must, of course, be acceptable conceptually.

Wladimir Köppen developed one system along these lines. In its various forms, it is probably the most widely used of all climatic systems. A second scheme that is also widely used, largely as a result of its innovative methodology, is that devised by C. W. Thornthwaite.

The Köppen System.   Köppen made one of the most lasting and important contributions in the field of climatic classification. Trained as a botanist, in the early stages of his work he was strongly influenced by the writings of botanists. The systems he formulated range from a highly descriptive vegetation zonal scheme to a classification in which boundaries are defined in relatively precise mathematical terms.

Beginning with his doctoral dissertation in 1870 and continuing up to his death in 1940, Köppen proposed, modified, and remodified his system. The system's evolution shows that Köppen was not as concerned with precise boundaries as he was with attempting to use simple observations of selected climatic elements to provide a first-order world pattern of climates.

Köppen completed his early work at a time when plant geographers were first compiling vegetation maps of the world. His early publications discussed temperature distribution in relation to plant growth, and it was not until 1900 that any of his publications really addressed world climatic classification. The 1900 system, which did not get much notice, is a highly descriptive scheme that uses plant and animal names to characterize climates. In 1918, Köppen produced a system that is substantially the one in use today. Boundary values have changed and new symbols have been introduced, but the framework of the present system was clearly evident. The scheme demonstrates Köppen's major contribution to the systematic treatment of the world's climates. He recognized a pattern underlying world climatic regions and introduced a quantitative method that allows any set of data to be categorized within the system. A unique set of letter symbols obviates the need for long, descriptive terms.

The classification is based essentially on the distribution of vegetation. Köppen's assumption was that the type of vegetation found in an area is very closely related to the region's temperature and moisture characteristics. These general relationships were already known at the time Köppen developed his classification, but he attempted to translate the boundaries of selected plant types into climatic equivalents. The Köppen system is based on monthly mean temperatures, monthly mean precipitation, and mean annual temperature.

Köppen recognized four major temperature regimes: one tropical, two midlatitude, and one polar (Table 13.1). After identifying the four regimes, he assigned numerical values to the boundaries (Table 13.2). To be a tropical climate, the temperature must average 18°C (64.4° F) or above each month of the year. This temperature was selected because it approximates the poleward limit of certain tropical plants. The two midlatitude climates are distinguished on the basis of the mean temperature of the coolest month: If that mean temperature is below −3°C (26.6° F), the climate is *microthermal*; If it is above −3° C, the climate is *mesothermal*. The fourth major temperature category is the polar climate. The boundary between the microthermal and polar climates is set at 10°C (50° F) for the average of the warmest month, which roughly corresponds to the northern limit of tree growth. A fifth major regime, the dry climates, was based not upon temperature criteria but upon lack of moisture. Dry-climate boundaries are obtained using derived formulas. Figure 13.1 shows the distribution of the identified climatic types.

The system has drawn criticism in two areas. First, the distribution of natural vegetation and climate do not completely coincide. Factors other than average climatic conditions (soils, for instance) affect the distribution of vegetation. The system is also criticized because of its rigid boundaries. Temperatures and rainfall at any site differ from year to year, and the boundary based on a given value of temperature will change location

TABLE 13.1
Köppen's major climates.

A	Tropical rainy climates
B	Dry climates
C	Midlatitude rainy climates, mild winter
D	Midlatitude rainy climates, cold winter
E	Polar climates

PRINCIPAL CLIMATIC TYPES ACCORDING TO KÖPPEN'S CLASSIFICATION

Af	Tropical rainy	Cw	Midlatitude wet-and-dry, mild winter
Aw	Tropical wet-and-dry	Cf	Midlatitude rainy, mild winter
Am	Tropical monsoon	Dw	Midlatitude wet-and-dry, cold winter
BS	Steppe	Df	Midlatitude rainy, cold winter
BW	Desert	ET	Tundra
Cs	Mediterranean	EF	Ice Cap

from year to year. In spite of the criticisms and the empiric basis of the classification, it has proved useful as a general system.

The Thornthwaite Classification. In a paper published in 1931, C. W. Thornthwaite proposed a climatic classification much different from previous systems. Unlike most classifications available at the time, Thornthwaite based his system on the concepts of moisture and thermal efficiency. Many authors prior to Thornthwaite suggested that the relationship between precipitation and evaporation provided a useful measure of precipitation effectiveness, but few utilized the concept because of scarce evaporation data. Faced by the same problem, Thornthwaite produced a precipitation-evaporation index that could be determined empirically from available data. Using this index, Thornthwaite devised humidity provinces that formed the first-order division of his classification scheme. It differs from the Köppen system in that boundaries between provinces are not related to any practical vegetation or soil criteria; instead, they are based upon regular arithmetic intervals of values. Despite some adverse criticism, Thornthwaite's 1931 system is considered a major contribution to the process of climatic classification.

Of more significance today (because it has superseded the earlier systems) is Thornthwaite's 1948 classification, which made a radical departure from the 1931 system in using the important concept of evapotranspiration. The earlier system had been concerned with the loss of moisture through evaporation, whereas the new approach considers loss through the combined process of evaporation and transpiration. Plants are considered physical mechanisms that return moisture to the air. The combined loss is termed *evapotranspiration*, and when the amount of moisture available is nonlimiting, the term *potential evapotranspiration* is used.

The two major elements in the system are the use of precipitation effectiveness and temperature efficiency. Precipitation effectiveness was designed as an indicator of net moisture supply, taking into account both the actual amount of precipitation and the estimated consumption of moisture by evaporation. The precipitation effectiveness is determined by calculating the ratio of the *precipitation to evaporation* (P/E) for each month of

TABLE 13.2
The Köppen classification.

A   *Temperature of the coolest month above 18°C (64.4°F)*

    Subcategories:
        f: rainfall in driest month at least 6 cm (2.4 in.)
       m: rainfall in driest month greater than $10 - r/25$, but less than 6 cm when
               r = annual rainfall in cm
               OR
               rainfall in driest month greater than $3.94 - r/25$, but less than 2.4 in., when
               r = annual rainfall in inches
       w: rainfall in driest month less than 6 cm (2.4 in.), but insufficient for
               m and dry season in low sun period
        s: rainfall in driest month less than 6 cm (2.4 in.), but insufficient
               for *m* and dry season in high sun period
      w':   maximum rainfall in autumn
     w":   two rainfall maxima, with intervening dry periods
       i: annual temperature range less than 5°C (9°F)
       g: warmest month precedes summer solstice

B   *Evaporation exceeds precipitation for the year*

    Subcategories:
       BS (Steppe): Derived by the following, when r = annual rainfall in cm and
                                              t = annual average temperature
                                            in °C

          70% of rainfall in summer six months: $r = 2(t + 14)$
          70% of rainfall in winter six months: $r = 2t$
          Even rainfall distribution or
             neither of above: $r = 2(t + 7)$
                             OR, when r = annual rainfall in inches
                                 and t = annual average temp. in °F
          70% of rainfall in summer six months: $r = .44t - 3.5$
          70% of rainfall in winter six months: $r = .44t - 14$
          Even rainfall distribution or
             neither of above: $r = .44t - 8.5$

The value r is the BS/humid boundary. When the derived r is greater than the value on the right of the equation, the climate is humid; when less it is *B*. If *B*, then determine if BS by dividing the answer by 2. If, after dividing, r is greater than value on right, climate is BS; if less, climate is BW (desert).

       BW (Desert): Derived as indicated above
          Subcategories of BW:
               h: average annual temperature above 18°C (64.4°F)
               k: average annual temperature below 18°C (64.4°F)
              k': average of warmest month below 18°C (64.4°F)
               n: high frequency of fog
               s: 70% of rainfall in winter six months (summer dry season)
              w: 70% of rainfall in summer six months (winter dry season)

TABLE 13.2

(continued)

C   *Coolest month temperature averages below 18°C (64.4°F) and above −3°C (26.6°F); warmest month is above 10°C (50°F)*

Subcategories:

    f: at least 3 cm (1.2 in.) of precipitation in each month; or, neither *w* or *s*

   w: minimum of 10 times as much precipitation in a summer month as in driest winter month

    s: minimum of 3 times as much precipitation in a winter month as in driest summer month, and one month with less than 3 cm (1.2 in.) of precipitation

    x: rainfall maximum in late spring or early summer; dry in late summer

    n: high frequency of fog

    a: warmest month over 22°C (71.6°F)

    b: warmest month under 22°C, but at least four months over 10°C (50°F)

    c: only one to three months above 10°C

    i: mean annual temperature range less than 5°C (9°F)

    g: warmest month precedes summer solstice

   t': hottest month delayed until autumn

   s': maximum rainfall in autumn

D   *Coolest month temperature averages below −3°C (26.6°F) and warmest month over 10°C*

Subcategories:

    d: coldest month below −38°C (−36.4°F)

   Other subcategories same as for C

E   *Warmest month temperature averages less than 10°C (50°F)*

Subcategories:

  ET: average temperature of warmest month between 0°C (32°F) and 10°C (50°F)

  EF: average temperature of warmest month below 0°C (32°F)

the year and summing them to form the *precipitation effectiveness* (P-E) index. Temperature efficiency (T-E) in this classification indicates the energy or heat supply relative to evaporation rates. The T-E index is calculated in the same fashion as the P-E index, using temperature and evaporation data.

Nine moisture provinces were established on the basis of the precipitation effectiveness index (Table 13.3). The boundary of each more-humid region represents a doubling of the P-E index. Similar to the temperature efficiency index, there are nine major temperature provinces. As in the case of the P-E index, each progressively warmer province has an index double that of the preceding province. Thornthwaite used many of the same alphabetic symbols that Köppen used in his classification. To the two indexes of moisture and temperature are added a letter designation for rainfall distribution through the year. The initial classification yields 32 different climatic types.

Many other classifications of climate have been proposed and numerous modifications of these schemes suggested (Table 13.4). Any classification of continuous variables is

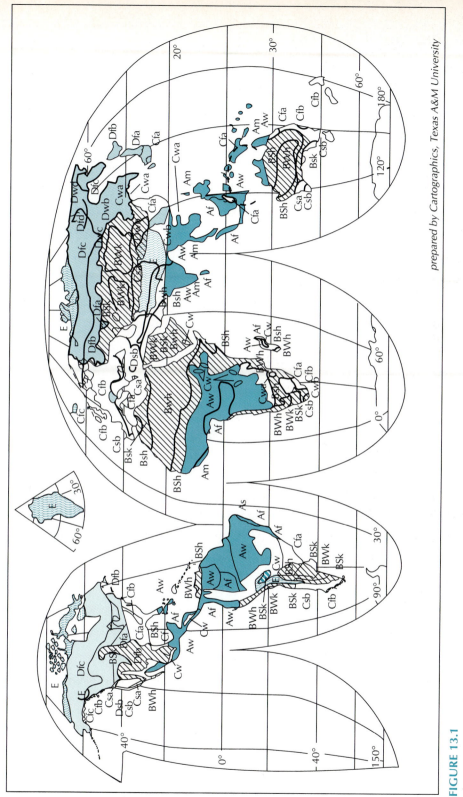

prepared by Cartographics, Texas A&M University

FIGURE 13.1

The Köppen classification.

Source: J.F. Griffiths and D.M. Driscoll, *Survey of Climatology*. (Columbus: Merrill Publishing Co., 1982), 198. Reprinted by permission of the authors.

TABLE 13.3
Thornthwaite's 1948 classification of climate.

NINE DIVISIONS BASED UPON MOISTURE EFFICIENCY					NINE DIVISIONS BASED UPON TEMPERATURE EFFICIENCY	
			T/E Index			
	Climatic type	Moisture index	(cm)	(in.)		Climatic type
A	Perhumid	100 and above	14.2	5.61	E'	Frost
B_4	Humid	80 to 100	28.5	11.22	D'	Tundra
B_3	Humid	60 to 80	42.7	16.83	C_1'	Microthermal
B_2	Humid	40 to 60	57.0	22.44	C_2'	Microthermal
B_1	Humid	20 to 40	71.2	28.05	B_1'	
C_2	Moist subhumid	0 to 20	85.5	33.66	B_2'	
C_1	Dry subhumid	−20 to 0	99.7	39.27	B_3'	Mesothermal
D	Semiarid	−40 to −20	114.0	44.88	B_4'	
E	Arid	−60 to −40			A'	Megathermal

SUBDIVISIONS BASED UPON SEASONALITY OF PRECIPITATION

	Moist Climates (A, B, C_2)	Aridity index
r	Little or no water deficiency	0–16.7
s	Moderate summer water deficiency	16.7–33.3
w	Moderate winter deficiency	16.7–33.3
s_2	Large summer water deficiency	33.3+
w_2	Large winter water deficiency	33.3+

	Dry Climates (C_1, D, E)	Humidity index
d	Little or no water surplus	0–10
s	Moderate winter water surplus	10–20
w	Moderate summer water surplus	10–20
s_2	Large winter water surplus	20+
w_2	Large summer water surplus	20+

Aridity index = water deficit/water need.
Humidity index = water surplus/water need.

arbitrary, and the variety of possible classes or climatic regions is infinite. No two square miles of Earth's surface are likely to have the same atmospheric conditions.

Genetic Systems

Compared to the empiric classifications, genetic systems are generally less formally defined and often less well-developed. As a result, they are used less widely in general descriptive climatology. The types that have been proposed are based either upon identi-

TABLE 13.4
Chronological development of some world climatic classifications.

AUTHOR(S)	YEAR	BASE OF SYSTEM
de Martonne	1909	First-order divisions based upon temperature and precipitation criteria; numerous subdivisions named for local areas in Europe; considerable attention given to desert limits, but most boundaries derived nonquantitatively
Penck	1910	Physiography and world climate; 3 main types of climates significant in determining weathering and erosion: humid, arid, and nival; each subdivided into 2 parts
Köppen	1918	Based upon vegetation regions, with specific climatic values for boundaries between regions
Vahl	1919	5 zones appraised by temperature limits, a function of the data for warmest and coldest months; subdivision by precipitation expressed as % of number of wet days in a given humid month
Passarge	1924	Recognition of 5 climatic zones subdivided into 10 regions, emphasizing vegetation distribution
Miller	1931	5 temperature years based upon vegetation; each temperature zone divided into 3 possible moisture zones: areas with rain in all seasons, marked seasonal drought, or drought all year
Thornthwaite	1931	Use of precipitation effectiveness and temperature efficiency to construct 5 humidity provinces and 6 temperature provinces; 30 major climatic types
Philipson	1933	Based upon temperature of warmest and coldest months and precipitation characteristics; 5 climate zones with 21 climatic types and 63 climatic provinces
Blair	1942	5 main zonal climates: tropical (T), subtropical (ST), intermediate (I), subpolar (SP), and polar (P); 14 types and 6 subtypes distinguished using letter notation; based on precipitation and temperature data and related to vegetation types
Gorsczynski	1945	10 "decimal" types associated with 5 main zonal climates; emphasis on continental vs. marine climates and definition of aridity
von Wissman	1948	Related to the Köppen approach; 5 temperature zones subdivided by precipitation distribution and temperature regimes
Thornthwaite	1948	9 primary climates based on a moisture index, 9 divisions based on temperature index; evapotranspiration introduced as an index

TABLE 13.4
(continued)

AUTHOR(S)	YEAR	BASE OF SYSTEM
Creutzberg	1950	Annual rhythm of climate based on identification of *Isohygromen* (lines of equal duration of humid months) and *Tag-Isochione* (lines of equal daily snow cover duration); 4 major zones differentiated, subdivided by monthly moisture values
Geiger-Pohl	1953	Modification of the Köppen system
Brazol	1954	Human comfort zones; use of wet and dry bulb temperatures to establish comfort months; 12 ranked classes ranked from No. 12 (lethal heat) to No. 1 (glacial cold)
Trewartha	1954	Major modification of the Köppen system
Budyko	1958	Distribution of energy in relation to water budgets
Peguy	1961	Modification of de Martonne's system
Troll	1963	Climates differentiated by thermal and hygric seasons
Carter and Mather	1966	Modification of the 1948 Thornthwaite system
Papadakis	1966	Agricultural potential of climatic regions; crop-ecologic characteristics of a climate based upon empirically derived threshold values; 10 main climate groups recognized, each divided into subgroups that are subdivided
Hidore	1966	Based on dominance of air masses; 10 primary climates identified by annual distribution of air mass frequency
Terjung	1968	Thermal classification of the earth's climates
Oliver	1970	Genetic classification based on air-mass frequency

fied physical determinants or upon air-mass dominance. The scheme described here, and used as a basis for classifying regional climates in the following chapters, is an air-mass approach, first formulated by John Hidore (1966).

On the basis of seasonal patterns of radiant energy and precipitation, Earth's environments fall into nine basic types (see Figure 13.2). In Figure 13.2 each row and column of three has the same seasonal characteristics of either temperature or moisture. Thus each of the three types of tropical systems has a corresponding midlatitude and polar system with a similar seasonal moisture pattern.

Figure 13.3 illustrates several characteristics of four of the regional systems. Each of these four areas tends to lack any strong seasonal variation in energy or moisture. Some differences between seasons do exist, of course. In the polar ice caps, radiation increases in the summer season; even so, it remains cold throughout. The increase in radiant energy simply is not sufficient to bring about real warming, and tropical air currents don't penetrate that far poleward. Each of the four systems is influenced primarily by one kind of air mass, and each is associated with one of the four semipermanent pressure zones in the atmosphere.

FIGURE 13.2

Schematic representation of the nine basic types of climate found on Earth showing the seasonal patterns of temperature and moisture, the types of air masses most common to the area, and the location in the general circulation.

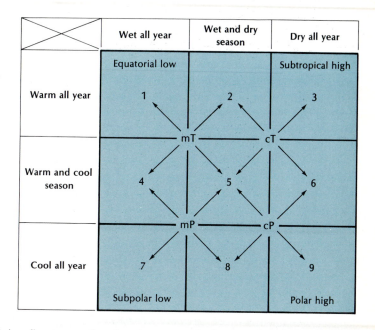

Four of the remaining five types of regions have a marked seasonal variation in either temperature or moisture, but not both. Two have pronounced seasonal moisture regimes and two have pronounced seasonal temperature regimes. Situated between the major pressure zones, these four environmental groups are affected by the seasonal migration of the primary circulation. They are subjected to air currents that are considerably different from one season to the next (Figure 13.4). One of the nine systems, the midlatitude seasonal rainfall regime, represents the maximum seasonal variation of weather systems. Periodic invasions of air currents developing in each of the major types of source regions bring a variety of weather.

FIGURE 13.3

Four of the climates are dominated primarily by single types of air masses. Each of these is found in the core area of one of the semipermanent pressure zones.

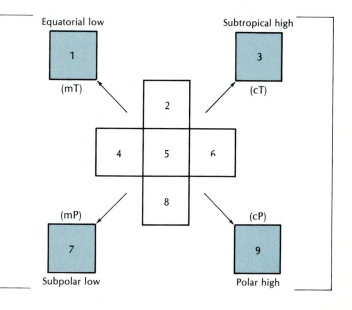

FIGURE 13.4

Four of the climates have marked seasonal change in temperature or moisture associated with a seasonal change in air mass control. The other climate, shown in the center of the diagram, is subject to weather associated with all four basic kinds of air masses. It has both wet and dry seasons and summer and winter seasons.

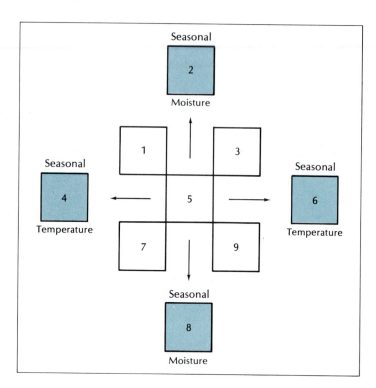

Using the model, we can identify the climatic types (Figure 13.5) and list their characteristics (Table 13.5). When the identified types are located on a world map, a highly generalized distribution results, with large areas grouped as a single climatic type. Given that the distribution is based upon only nine major climatic types, such might be expected. The resulting map does, however, provide a useful guide to the classes identified within the model (Figure 13.6). The climates in each region can be subdivided to whatever scale is needed using selected criteria. In Figure 13.6, the midlatitude wet climate has been divided with summer conditions as a criterion.

Each type of climate is detailed in the chapters that follow. Mountain regions are not identified because (1) the seasonal patterns of energy and precipitation will be

FIGURE 13.5

Climates differentiated in the air-mass model.

	Wet ⟵—————⟶ Dry			
Hot	Tropical wet (equatorial)	Tropical wet and dry	Tropical desert	**Group I**
	Midlatitude wet	Midlatitude summer or winter dry	Midlatitude desert	**Group II**
Cold	Polar wet	Polar wet and dry	Polar desert	**Group III**

TABLE 13.5
Classification of climates by air mass control.

TYPE		AIR MASSES
Group I	Tropical air masses dominate	
	1. Tropical wet	mT or mE air masses dominate
	2. Tropical wet and dry	mT/cT seasonally
	3. Tropical dry	cT dominates
Group II	Tropical/polar air masses seasonally	
	4. Midlatitude wet	mT or mP dominates
	5. Midlatitude wet and dry	
	5S summer dry	cT/mP
	5W winter dry	mT/cP
	6. Midlatitude dry	cT/cP seasonally
Group III	Polar air masses dominate	
	7. Polar wet	mP dominate
	8. Polar wet and dry	mP/cP seasonally
	9. Polar dry	cP dominate

the same as those of the surrounding lowlands; and (2) mountain regions may be, and probably are, found in each of the regions differentiated in the previous paragraphs. The altitudinal variations that occur in mountains are similar wherever they are found.

Climatic Influence on Other Environmental Variables

Climate is a concept derived from an orderly examination of atmospheric conditions over a long period. As a result, one cannot "see" climate. Its impact, however, is clear in other environmental systems, particularly in the vegetation and soils of a given locale.

Climate and the Distribution of Vegetation

Among factors that determine the distribution of flora are moisture availability, radiant energy, soils, parent material, slope, and other biota. Climatic factors are a major determinant in the distribution of individual species and communities. The implications of climate are most readily visible at the biome level. Where climatic conditions are similar from year to year, a distinct plant formation (ecosystem) evolves. Four such regions exist: the tropical rainy, the tropical desert, the polar rainy, and polar desert or ice caps. A biome associated with the tropical rainy climate is the tropical rain forest. The relationship between the rain forest and climate is so close that until recently the distribution of rain forest has been used to map the areas with a tropical rainy climate. We now know that the association between the two is not that perfect. The tropical deserts contain the

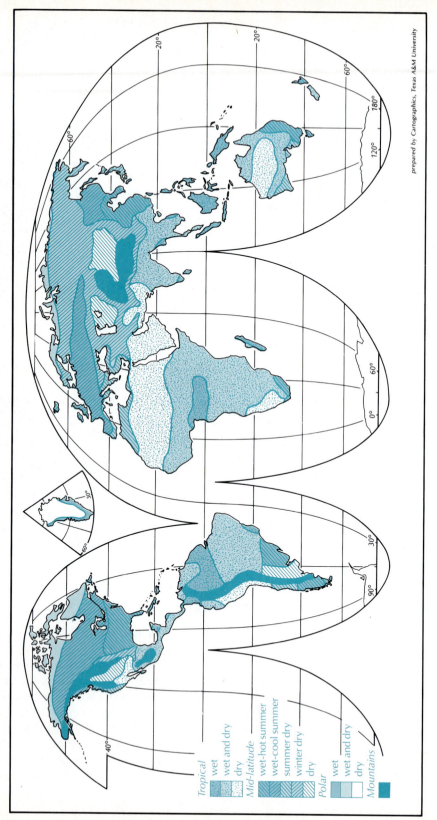

FIGURE 13.6
World climatic regions differentiated using the air-mass model.

prepared by Cartographics, Texas A&M University

Tropical
  wet
  wet and dry
  dry
Mid-latitude
  wet-hot summer
  wet-cool summer
  summer dry
  winter dry
  dry
Polar
  wet
  wet and dry
  dry
Mountains

desert biome, which covers large areas of Earth's surface. Associated with the polar rainy is the coniferous forest (taiga). Little or no vegetation is found near the ice caps as a result of very harsh conditions.

In the intervening areas between these four very different biomes, classification is more difficult and often confused by the intermingling of communities and species. Here the biomes present transitions from tropical to polar and humid to dry. Some of the different biomes found in the tropical seasonal rainfall regime include

seasonal forest

woodland

savanna

grassland

steppe

Since the communities form a continuum over the landscape, locating the boundaries is difficult. A brief description of some of the more widely recognized biomes follows.

Rain Forest.   Forests include formations in which trees are a prevalent plant form. This classification obviously covers a multitude of different ecosystems; here it is possible to mention but a few.

In rain forests, broadleaved evergreen trees dominate and the canopy is more or less continuous. Rain forests exist where moisture is adundant, if not year-round, at least during the greater part of the year. The foliage of the trees in rain forests is concentrated in the crowns, as is the foliage of the lianas, epiphytes, and parasites that live on the trees. The understory consists largely of young trees of the dominant species and a sparse ground cover of shade-tolerant or shade-demanding shrubs. These forests are very limited in extent.

Seasonal Forests.   The majority of the world's forests occur where the dry season is long enough to affect a seasonal change in the forest community. The seasonal forest may include evergreen, semideciduous, or deciduous trees, or some combination of these. Where a mixture exists, it is not usually random individuals of each species but mixed stands of one type or the other. Local differences in soil characteristics or other site characteristics often determine which community will persist. Since seasonal forests exist where seasonal precipitation occurs, the character of the forest is closely associated with the length of the rainy season. As the length of the rainy season decreases, the density of the canopy decreases. The most dense forests, or true jungles, are found in the seasonal forests where the dry season is long enough to spread the canopy and allow sunlight to reach the ground, but where there is not such a long dry season that edaphic drought occurs with any regularity.

Deserts.   A desert is characterized by discontinuous and usually sparse plant cover. Some desert plants, such as the creosote bush, have their own population-control devices. They produce hormones through their roots or leaves that inhibit the sprouting of other individuals within the proximity of the parent plant. This controls spacing of the plants and nutrient resources to increase the probability that some individuals of the species will survive.

Many of the animals found in the deserts are the same as those found in the grasslands but in greatly reduced numbers. Some animals, including the kangaroo rats, have adapted specifically to the deserts; such species are relatively few in number.

Savanna. Savannas are tropical grasslands with scattered trees or clumps of trees. Isolated trees are often found right at the desert edge. This scattering of drought-resistant trees gives the tropical grasslands a distinctly different appearance from the midlatitude grasslands. Fires are a recurrent phenomenon in the savannas; consequently, both the grasses and trees are fire-tolerant. The variety of species of trees and grasses is small compared to the tropical forests, but because they must adapt to drought and fire, the existing species are very hardy and respond rapidly after a disturbance. Acacia and baobab trees are among the species that thrive in the savannas.

Grasslands. The world's grasslands consist of communities in which herbs and shrubs dominate. Grasses and legumes, some of which reach considerable size, dominate these communities. Grasses ranging up to more than 3 m (10 ft) in height are not uncommon in the more humid grasslands. Perhaps the tallest well-known member of the grass family is bamboo, a phenomenally fast-growing plant that has been observed growing as much as 200 mm (8 in.) per day. Although they lack the upper story of trees, the grasslands are still multistoried, with usually more than one story of herbs or shrubs and some of the lesser structural forms closer to the ground.

Grazing animals grow to their greatest numbers in the grasslands. They tend to live in groups and depend upon speed for defense against predators. Some creatures have developed the abilities to leap high above the grasses for easier traveling and to see over the grass to watch for predators. The jackrabbit of the American and Australian plains is a good example. Many small animals burrow for shelter and concealment. Among this group are the prairie dogs of the Great Plains of North America; they live in groups and construct extensive underground towns.

Very few areas of natural grassland remain on Earth. The grasslands have proved to be the most useful of the biomes when measured in terms of agricultural purposes. The natural communities of the grasslands have been either burned off or plowed up and replaced by the simpler communities of the domestic cereals such as corn, rice, wheat, and barley.

The grasslands occur in places with a seasonal moisture regime. They occupy all of the continents except Antarctica, and convergent adaptation has led to similar species in each area and strange forms in some. Grasslands have been subdivided by secondary structural characteristics; grasses are either steppe and prairie, tall and short, or sod and bunch grass. The grasslands may border forests, seasonal forests, woodlands, or deserts.

Taiga. The dominant species of this forest are the needle-leaved evergreen trees. Members of the spruce, pine, and fir families are most common. A mature coniferous forest has very little understory as a result of the dense shade. The ground often is covered with a thick layer of undecomposed and partially decomposed needles. This forest is associated with cool, moist conditions poleward to the limits of tree growth. The forest extends equatorward considerable distances along mountain ranges, where favorable temperature and moisture conditions prevail.

Tundra.   Vegetation of the tundra consists largely of grasses, sedges, lichens, and dwarf woody plants. The tundra is associated with the seasonal rainfall areas around the Arctic Ocean and at high altitudes in mountains. It exists where the summer season is so short that it is too cool for trees to thrive. Tundra vegetation has distinct characteristics that allow it to survive the cold temperatures, wind, and the very long physiological drought of winter. Almost all tundra plants grow low and compact to escape the wind, reduce evaporation, and conserve heat. Perennials predominate, and many reproduce asexually by runners, bulbs, or rhizomes. Being perennial, the plants store food during the summer, and the buds are sheltered either underground or close to the surface.

Disturbed Formations.   Some scientists maintain that the grassland, savanna, and brush formations are disturbed formations that have persisted because of repeated razing by fire. The sharp boundaries that exist, the variety of formation boundaries, and the lack of woody plants have led to this hypothesis. The debate still seems a long way from being resolved. The suggestion that grasslands are a product of disturbance does not in any way change the fact that disturbance is a very real factor in formation structure. Disturbance affects formations in several ways, sharpening the boundary between formations, at least in some areas, and often favoring the intrusion of one formation into another. Local disturbances within a formation may allow establishment of an alternative formation. The effects of a disturbance vary, depending upon the type of formation. Some communities in midlatitude grasslands may completely reestablish a climax community in less than 50 years. The tundra, where ecological processes are very slow, may not recover from a disturbance for centuries.

Climate and the Distribution of Soils

Many different classifications of soils have appeared over the years, with each new system generally an improvement over the preceding ones. Until recent years those most widely used were variations of a scheme proposed by V. V. Dockuchaiev around 1870. The original proposal underwent additions and changes until around 1950, when soil scientists decided that the old system was not adequate to include all of the soils that were being classified during the rapid expansion of soil science following World War II. A new soil classification came into use in the United States in the 1970s (U.S.D.A. 1975). Based upon characteristics of the soil profile, it has six levels of categorization, given in descending order of generalization in Table 13.6. The system allows for the inclusion of soils worldwide. At the order level, which is the most general category, all of the soils

TABLE 13.6
Soil category levels.

LEVEL	CATEGORY
1	Order
2	Suborder
3	Great group
4	Subgroup
5	Family
6	Series

included must have some basic similarities—essentially in color and organic content. These two elements often, but not always, reflect climate.

In classifying vegetation over Earth, plant formations are most closely associated with climatic regimes where the weather conditions tend to be persistent through the year. This is also the case with soils. Associated with the tropical rainy regimes and rain forests are the *oxisols*, which are formed largely by the laterization process. They tend to be stained red by iron and aluminum oxides and to be low in organic content. The deserts contain the *aridisols*, light in color from calcification and low in organic matter. The coniferous evergreen forest and the cool, moist climate have combined to produce the *podosols*, a light-colored soil accumulation of mineral and organic matter in the lower horizon. Soils are largely absent on the ice caps because of the severe cold and lack of weathered regolith.

As with climate and vegetation classification, classifying soils in the regions of more variable climate is more difficult. At some locations in midlatitudes, for instance, the laterite process prevails in the summer, and the podzolization process in the winter.

Summary

Climate classification is a difficult task and one with a long history. Generally, however, two main approaches to classification have been used. The empiric approach, typified by the Köppen classification, is based upon observed distributions of the environment; genetic classifications attempt to define climates by their cause. The classification followed here uses air masses and air-mass interaction as its genetic base. Nine primary climates are identified by the system, with other variations introduced in some of the units.

The relationship between climate and other environmental variables is evident, and in many ways vegetation and soil distribution are related to climatic realms.

Map B
World political map.

CIRCLE

RUSSIAN   FEDERATION

ALBAHD TA
(NOR.)

SWEDEN
FINLAND
Oslo
Stockholm
Copenhagen
ESTONIA
LATVIA
LITHUANIA
BELARUS
POLAND
Berlin
Warsaw
UKRAINE
GERMANY
NETH.
Paris
CZECH.
AUS.
Vienna
SWITZ.
HUN.
ROM.
ITALY
ALB.
BUL.
Rome
GREECE
Athens
TURKEY
Ankara
Istanbul
Algiers
TUNISIA
MALTA
CYPRUS
SYRIA
LEBANON
ISRAEL
Cairo

St. Petersburg
Moscow
Samara
Novosibirsk
Irkutsk

KAZAKHSTAN

MANCHURIA
SAKHALIN

ALEUTIAN
ISLANDS
KURIL
ISLANDS

MONGOLIA
Ulaan Baatar
Vladivostok

UZBEKISTAN
Tashkent
KYRGYZSTAN
TURKMENISTAN
TADZHIKISTAN
GEORGIA
ARMENIA
AZERBAIDZHAN
Tehran
Kabul
AFGHANISTAN
PAKISTAN
Baghdad
IRAQ
KUWAIT
BAHRAIN
QATAR
UNITED ARAB
EMIRATES

Beijing
KOREA
Seoul
KOREA
JAPAN
Tōkyō

C H I N A

Nanjing
Shanghai
Chongqing
RYUKYU IS.
(JAPAN)
BONIN IS.
(JAPAN)

NEPAL
BHUTAN
New Delhi
BANGLADESH
Calcutta
MYANMAR
Yangon
LAOS
Hanoi
HONG KONG
Macao (Por)
TAIWAN

WAKE
(U.S.A.)

JORDAN
SAUDI
Riyadh
Mecca
ARABIA
Sana
YEMEN
Aden

LIBYA
EGYPT

NIGER
CHAD
SUDAN
DJIBOUTI
Addis Ababa
ETHIOPIA
SOMALIA

NIGERIA
Lagos
CAMEROON
CENTRAL
AFRICAN
REPUBLIC
UGANDA
KENYA
Nairobi

GUINEA
GABON
CONGO
ZAIRE
Brazzaville
Kinshasa

BENIN
Luanda
ANGOLA
TANZANIA
Zanzibar
RWANDA
BURUNDI

ZAMBIA
MALAWI
ZIMBABWE
NAMIBIA
BOTSWANA
MOZAMBIQUE

MADAGASCAR
Antananarivo
RÉUNION
(FR.)
MAURITIUS

Pretoria
Maputo
SWAZILAND
SOUTH
AFRICA
LESOTHO
Durban
Cape Town

Karachi
Bombay
INDIA
Hyderabad
Madras
ACCADIVE IS.
(INDIA)
SRI LANKA
Colombo
NICOBAR IS.
(INDIA)
MALDIVES

ANDAMAN IS.
(INDIA)
Bangkok
THAILAND
VIETNAM
Manila
CAMBODIA
Ho Chi Minh City
MALAYSIA
BRUNEI
Singapore
Djakarta
INDONESIA

PHILIPPINES

NORTHERN
MARIANAS IS.
(U.S.A.)
GUAM
(U.S.A.)
PALAU
FEDERATED STATES
OF MICRONESIA
CAROLINE
ISLANDS

PACIFIC

OCEAN

MARSHALL
ISLANDS
KIRIBATI
NAURU

PAPUA
NEW
GUINEA
SOLOMON
ISLANDS
TUVALU

SEYCHELLES

INDIAN

OCEAN

AUSTRALIA
Brisbane

Perth
Adelaide
Sydney
Canberra
Melbourne
NEW ZEALAND
Auckland
Wellington

VANUATU
FIJI
Nouméa
NEW
CALEDONIA
(FR.)

IS. DE KERGUÉLEN
(FR.)

CIRCLE

C   A

0   1,500   3,000 MILES
0   1,500   3,000 KILOMETERS

TROPICAL CLIMATES

Chapter Overview

Tropical regions have long held a certain mystique for midlatitude peoples. The rich flora and fauna of certain tropical areas have encouraged some to envision a utopian economic development in the tropics. Diseases such as malaria and yellow fever and fears of the debilitating effects of the tropical climate and ferocious insects, animals, and people have retarded development. Today, many of the diseases are under control and air conditioning has made the climate tolerable. But growth is still restrained, in part because of the very nature of tropical environmental systems.

General Characteristics

Radiation and Temperature

Several attributes characterize tropical climates, the most important being the energy balance. Several aspects of the energy balance distinguish these climates from those closer to the poles. First, the influx of solar energy is high throughout the year in the tropics, although it does vary with the seasons. The great amounts of solar energy create climates without winter. Solar radiation is intense all year because the solar beam is nearly always at a high angle. In addition, the length of the day varies only a little from one part of the year to the next. The photoperiod, the relative length of day to which plants must adjust, varies between 11–13 hours from winter to summer. Contrary to popular belief, though solar radiation is relatively high in the tropics all year, it is sometimes higher in midlatitudes, particularly in summer. In the equatorial lowlands, for instance, clouds often reflect more than half of the total solar radiation. In addition to the large amount that clouds reflect, the high humidity and smoke from widespread burning of vegetation by humans further reduce solar radiation near the ground and filter out most of the ultraviolet radiation. So much solar radiation is reflected and scattered that even a light-skinned person has difficulty getting a tan. Tropical regions never receive as many hours of sunlight as midlatitude locations do in the summer.

Although the length of the photoperiod changes little throughout the year, the variations that do occur are important biologically. Even very small differences in length of day are enough to stimulate flowering or dormancy in plants. Evidence shows that many tropical plants are more sensitive to slight changes in day length than are plants found in higher latitudes. Reduced day length triggers flowering in the common poinsettia; in their natural habitat these shrubs do not bloom at Christmas, but much later in the year. We force them to bloom before Christmas by artificially reducing the exposure to light in early autumn.

Annual temperatures average about the same throughout the tropical regions, although the drier areas have slightly higher temperatures due to more intense radiation at the surface. The annual range in temperature depends on the length of the dry season. Where there is no dry season, the annual range in mean monthly temperatures may be as little as 1° or 2°. Where there is dry weather in the winter months, the mean temperatures drop and the annual range increases.

Another significant aspect of the energy balance of tropical climates is that the primary energy flux is diurnal (Figure 14.1); that is, the variation in temperature from

Time of year

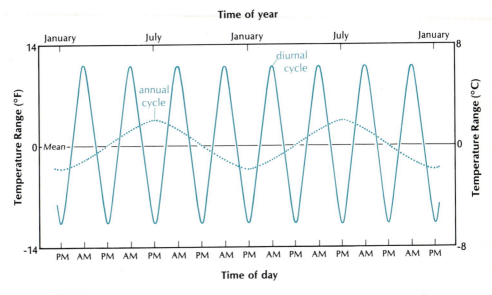

FIGURE 14.1

The relationship between the annual and diurnal energy periodicities in tropical regions.

day to night is greater than the variation from season to season (Figure 14.2). In fact, nights can be rather cool. After arriving in a village in tropical West Africa that was to be home for one of the authors, the first items that his family went looking for were blankets—one night without a cover was enough. The diurnal radiation and temperature cycle are more important than the annual temperature cycle in regulating life cycles.

Precipitation

In tropical areas, rainfall rather than temperature determines the seasons. The amount and timing of rainfall form the chief criteria for distinguishing among the various climates. Contrary to popular belief, only a very small portion of the tropical regions have a year-round rainy season. In Africa, for instance, areas that have heavy rainfall in each month total less than 10% of the land area. Most tropical environments have a marked seasonal regime of rainfall that governs the biological productivity of the system. The remaining areas are deserts, where rainfall is incidental throughout the year. The seasonal pattern of precipitation is certainly the most critical aspect of moisture in this region. This seasonal moisture pattern in fact distinguishes the major tropical environments—the rain forest, savannas, and desert—from one another.

The notion that these are monotonous, seasonless regions is far from the truth. Unlike the midlatitudes, tropical areas do lack snow and ice storms and tornadoes. But change is an attribute of the tropical regions as much as anywhere else, and residents talk about it just as much.

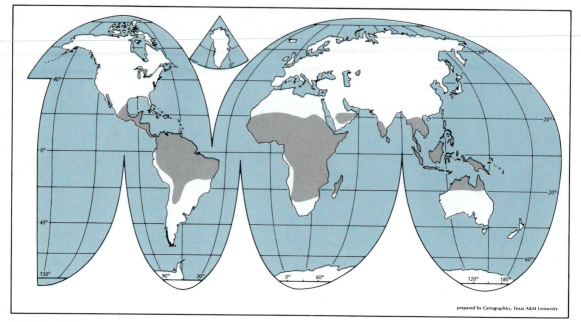

FIGURE 14.2
Land areas (shaded) experience a greater diurnal energy change than annual energy change.

Climatic Types

The Tropical Wet Climate

The term *wet climate* refers to the luxurious evergreen forest typical of the wet tropical lowlands (see Plate 9). The distribution of this climate is more limited than most people imagine because the region that receives abundant precipitation every month is limited. The region has fairly even and high temperatures averaging between 20° and 30°C (68°–86°F), frequent precipitation year-round, and total rainfall of over 200 cm (80 in.). Tropical maritime air masses dominate the climate (see Table 14.1).

The geographic area for wet climates is restricted because the region must lie within the tropical convergence zone throughout the year, and the convergence zone is continually shifting north and south. This migration is due to the passage of the overhead sun as schematically illustrated in Figure 14.3. Figure 14.4 shows the average extreme positions of the ITCZ. Where the migration of the ITCZ is maximum, the rain forest is smallest. The rain forest is also restricted to low elevations, usually below 1000 m (3300 ft),

TABLE 14.1
Climates dominated by mT air.

Köppen type	Af
Thornthwaite types	AA'r
	BA'r
	CA'r

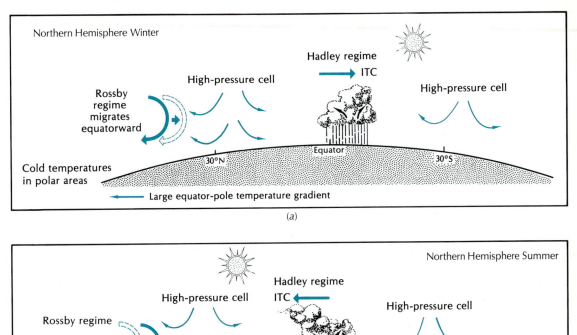

FIGURE 14.3
Schematic diagram showing the extent and migration of the Rossby and Hadley regimes:
(a) winter; (b) summer.

because at higher altitudes temperatures are considerably lower. As altitude increases, the forest changes character. Trees are shorter, the number of species decreases, and highland vegetation replaces the rain forest.

The fairly even temperatures of the rain forest are due primarily to the equal or nearly equal periods of daylight and darkness, the uniformly high intensity of radiation throughout the year, and the constantly high humidity. The monthly average temperatures vary from 24° to 30°C (75°–86°F), with an annual range of only 3°C (5°F) or so. In Belem, Brazil, for instance, the range from the warmest to coolest month is only 1.6°C (3°F). Since radiation varies more within the day than from day to day, the diurnal range may be two to three times the annual range.

Associated with the convergence of tropical maritime air are high humidity and cloudy skies. Since humidity is so high in the daytime, when nocturnal cooling occurs, early morning fogs and heavy dew are common. Dew-point temperatures are in the range of 15°–20°C (59°–68°F).

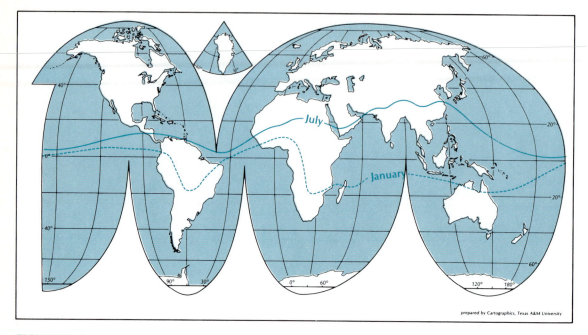

FIGURE 14.4

Mean extreme positions of the ITCZ.

Source: J. Griffiths and D. Driscoll, *Survey of Climatology*. (Columbus: Merrill Publishing Co., 1982), 207. Reprinted by permission of the authors.

The primary circulation determines the regional pattern of atmospheric circulation and the seasonal regime, and thunderstorms and easterly waves account for most of the day-to-day weather in tropical wet climates. Due to the degree of surface heating, local thermals, cumuliform clouds, and thunderstorms predominate. The cumulus clouds build during the daytime hours into towering cumulus clouds and thunderstorms when the air becomes unstable. This process provides a distinctive diurnal rainfall pattern, such as that shown in Figure 14.5. Precipitation occurs on more than half of the days. Duitenzorg, Java, averages 322 days a year with thunderstorms, which are intense and brief. Rainfall amounts of 50–65 mm (2.0–2.5 in.) per hour occur every one or two years, although hourly rainfalls of 95–120 mm (3.7–4.7 in.) have been recorded in the East Indies. Rainfall intensities for hour-long periods are not significantly higher in the tropics than in the midlatitudes, but sustained periods of high-intensity rainfall are significantly more common in tropical regions. The maximum 24-hour rainfall on record occurred at Cilaos, La Reunion, on March 16, 1952, when a total of 1870 mm (73.62 in.) of rain fell (see Table 14.2). In 1922, Mount Cameroon erupted in West Africa, producing a storm that left a snow cover on the summit, and at the foot of the mountain, 14.53 m (572 in.) of rain were recorded during the year. Not all rainfall in tropical wet climates is convectional; sometimes there are periods of prolonged showers.

Average annual rainfall is not as important a factor as is the frequency of precipitation. Annual totals range up to 2 m (6.6 ft). The highest amounts occur on mountain slopes such as Mount Waialeale, HI, where annual rainfall averages 11.8 m (39 ft).

FIGURE 14.5

Diurnal distribution of rainfall at Kuala Lumpur, Malaysia. The midafternoon maximum is typical of many wet tropical locations.

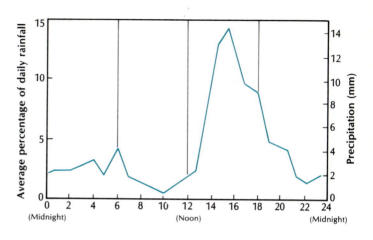

As with most generalizations, there are exceptions to the rules about where certain types of climate fall along the equator. Some extremely dry environments exist where we expect to find humid tropics. Herman Melville wrote of "The Encantadas" (Galapogos Islands):

> But the special curse, as one may call it, of the Encantadas, that which exalts them in desolation above Idumea and the Pole, is that to them change never comes; neither the change of seasons nor of sorrows. Cut by the equator, they know not autumn and they know not spring; while already reduced to the lees of fire, ruin itself can work little more on them. The showers refresh the deserts but in these isles, rain never falls. . . . Nowhere is the wind so light, baffling, and in every way unreliable, and so given to perplexing calms, as at the Encantadas. (p. 182)

A shack inhabited by a woman stranded on one island used a system for collecting dew to provide fresh water:

> . . . and here was a simple apparatus to collect the dews or rather doubly distilled and finest winnowed rains, which in mercy or in mockery, the night skies sometimes drop upon these blighted Encantadas. All along beneath the eave, a spotted sheet, quite-weather-stained, was spread, pinned to short, upright stakes, set in shallow sand. A small clinker, thrown into the cloth, weights the middle down, thereby straining all moisture into a calabash placed below. This vessel supplied each drop of water ever drunk upon the site by the Choles. Hunilla told us the calabash would sometimes, but not often, be half-filled over-night. It held six quarts, perhaps. (p. 231)

TABLE 14.2

Precipitation extremes recorded in the tropical wet climates.

Highest 12-hr total	Belouve, La Reunion Feb. 28–29, 1964	1340 mm (52.76 in.)
Highest 24-hr total	Cilaos, La Reunion Mar. 16, 1952	1870 mm (73.62 in.)
Highest 5-day total	Cilaos, La Reunion Mar. 13–18, 1952	3854 mm (151.73 in.)
Highest number of rain days in a year	Cedral, Costa Rica 1968	355 days

The Tropical Wet-and-Dry Climate

The distinguishing feature of this climate is a very pronounced seasonal moisture pattern. Atmospheric humidity, precipitation, soil moisture, and stream flow change through the year in a rhythmic pattern. These environments are found next to the tropical rain forest environments and in most cases poleward of them. They occupy much of the area from the boundary with the tropical rain forest to the Tropics of Cancer and Capricorn, a much larger area than the rain forest. There are areas with this type of climate north and south of the Amazon Valley in South America, across Africa north and south of the Congo Basin, and in much of southeast Asia and parts of the Pacific islands (see Plate 10). Their location in the general circulation is such that they experience the kinds of weather found in both the equatorial convergence zone and the subtropical divergence zone.

Seasons in the tropical wet-and-dry climates are dominated alternately by maritime tropical and continental tropical air masses (Table 14.3). The seasonal moisture pattern reflects the tropical convergence zone migration: The rainy season coincides with the high sun and the presence of the convergence zone; while the dry season is a product of more stable air from the subsidence in the subtropical highs. Seasonality increases away from the equator. In a traverse away from the equator, the low-sun precipitation begins to decrease first, with the high-sun precipitation remaining as high as in the tropical convergence zone. Between the equatorial convergence zone and the tropical desert, winter precipitation drops to near zero. From that point poleward, high-sun precipitation declines until it is no longer significant and we find the tropical desert.

The tropical wet-and-dry climate has the most pronounced precipitation seasonality of any climatic type, especially in its monsoon variety. Rangoon, Burma, illustrates the contrasts: Precipitation during the three-month winter there averages 25 mm (1 in.); during summer, it averages 1880 mm (74 in.). The extremes increase north at Akyab, where the averages are 3.8 cm and 426 cm (1.5 in. and 167 in.). The greatest extremes in the Asian region are at Cherrapunji, India, where in two winter months the city receives an average of only 2.5 cm (1 in.) of precipitation versus 528 cm (207 in.) of rain in two summer months. Cherrapunji recorded a five-day total of 405 cm (159 in.) of precipitation in August 1841 and a one-month total of 915 cm (30 ft) in July 1861. In one year from August 1860 to July 1861, residents measured a total of 26 m (85 ft) of rain.

The average annual precipitation varies so much from place to place in the tropical wet-and-dry climate that it cannot be used as a criterion for distinguishing the region. On the wet margins, where topography is favorable for orographic increase in rainfall, precipitation can exceed 1000 cm (400 in.) per year, while on the dry margins it drops to less than 25 cm (9.8 in.). Variation in annual rainfall is higher here than in the tropical rain forest. The precipitation total in any year reflects the extent of migration of the general

TABLE 14.3
Climates with seasonal mT and cT air.

Köppen types	Aw
	Am
	BSH
Thornthwaite types	BA′w
	CA′w
	DA′s

TABLE 14.4
Monthly data for tropical wet-and-dry-climates.

	JAN.	FEB.	MAR.	APR.	MAY	JUNE	JULY	AUG.	SEPT.	OCT.	NOV.	DEC.	YR.
				Calcutta 22°32' N 88°22' E: 6 m									
Temp. (°C)	20.2	23.0	27.9	30.1	31.1	30.4	29.1	29.1	29.9	27.9	24.0	20.6	26.94
Precip. (mm)	13	24	27	43	121	259	301	306	290	160	35	3	1582
				Cuiaba, Brazil 15°30' S 56°03' W: 165 m									
Temp. (°C)	27.2	27.2	27.2	26.2	25.5	23.8	24.4	25.5	27.7	27.7	27.7	27.2	26.5
Precip. (mm)	216	198	232	116	52	13	9	12	37	130	165	195	1375
				Dakar, Senegal 14°39' N 17°28' W: 23 m									
Temp. (°C)	21.1	20.4	20.9	21.7	23.0	26.0	27.3	27.3	27.5	27.5	26.0	25.2	24.3
Precip. (mm)	0	2	0	0	1	15	88	249	163	49	5	6	578
				Darwin, Australia 12°26' S 131°00' E: 27 mm									
Temp. (°C)	28.2	27.9	28.3	28.2	26.8	25.4	25.1	25.8	27.7	29.1	29.2	28.7	27.6
Precip. (mm)	341	338	274	121	9	1	2	5	17	66	156	233	1562

circulation and the length of time the zone of convergence stays over an area. The further a site is from the heart of the convergence zone, the lower the annual average precipitation, the shorter the rainy season, and the greater the annual variation. The weather is, of course, very different from season to season. The rainy season is warm, humid, and has frequent rainstorms. During the dry season, more or less desert conditions exist (Table 14.4).

The lag of the rainy seasons behind the sun's migration produces three seasons in many parts of the wet-and-dry tropics: cool, hot, and rainy seasons. The cool season falls during the winter months, when solar radiation is at a minimum. The hot season follows when temperatures rise as the sun moves higher in the sky and solar radiation increases. The onset of the wet season brings an increase in cloud cover, slightly reduced radiation, and cooler temperatures.

Where moist air of oceanic origin exists, precipitation occurs. Where the dry continental air is usually present, precipitation is largely absent. Often a sharp boundary divides the two types of air. The Hadley cells north and south of the equator and the convergence zone between them shift north and south, following the migration of the vertical rays of solar energy. This migration produces the seasonal pattern of precipitation that characterizes so much of the tropical region. The system moves through 20° of latitude through the year from about 5°S to 15°N.

The Tropical Deserts

World deserts cover more than one-fourth of all the land area of Earth (see Plate 11). Table 14.4 provides climatic data for selected cities in tropical deserts, and Table 14.5 lists the major climatic classes in the Köppen and Thornthwaite systems. Deserts exist at all latitudes between 50° N and S, with the largest found between 30° N and S (Table 14.6). The heart of the tropical deserts lies near the Tropics of Cancer and Capricorn, pri-

TABLE 14.5
Climates dominated by cT air.

Köppen type	BWh
Thornthwaite types	EA'h
	EB'd
	EA'd
	DA'd

marily toward the west sides of the continents. They are less common on the east side of the land masses because the trade winds carry considerable amounts of moisture ashore. Several characteristics distinguish the tropical deserts, including

1. low relative humidity and cloud cover
2. low frequency and amount of precipitation
3. high mean annual temperature
4. high mean monthly temperatures
5. high diurnal temperature ranges
6. high wind velocities

The basic climate controls in tropical deserts are upper-air stability and subsidence. Divergence, general stability, and low relative humidity usually characterize the tropical desert. Relative humidity averages 10%–30% in the interior areas, slightly higher in coastal locations; it has been measured as low as 2%. Low relative humidity means cloud cover is also low. The Sahara averages 10% cloud cover in winter and 4% in the summer.

Precipitation is minimal in amount and occurs sporadically in time and space. Low annual precipitation is the norm, but heavy downpours sometimes occur. Arica, Chile, receives an average of 5 mm (0.2 in.) of precipitation per year. Iquique, Chile, experienced a 14-year period with no rainfall, and Wadi Halfa in the Sahara Desert experienced a 19-year period with no rainfall. Iquique, on another occasion, received 100 mm (4 in.) of rain in one day. One rain may bring 125–250 mm (5–10 in.) of precipitation, then no precipitation will fall for years. One station in the Thar Desert where rainfall averages 100 mm annually received 850 mm in two days. At Dakhila, in southern Egypt, where the average is 100 mm, one 11-year period elapsed with no rain. Although infrequent, general rains over large areas of the desert do occur. In February 1980, a widespread rainstorm struck eastern Saudi Arabia with paralyzing results. Streets in the major cities

TABLE 14.6
Major deserts of the world.

DESERT	APPROXIMATE AREA	
	KM² (THOUSANDS)	MI² (THOUSANDS)
Sahara	9100	3500
Central Asia	4510	2200
Australian	3400	1300
North American	1300	500
Patagonian	680	260
Indian	600	230
Kalahari-Namib	570	220
Atacama	360	140

flooded because they have no storm sewers to carry off the water, and water blocked the major highways from Riyadh to Dhahran. Precipitation is more frequent on the equator-ward margins of the deserts in the summertime as the ITCZ moves over the area. On the poleward sides, precipitation falls mainly in the winter because the major mechanism for precipitation is the midlatitude winter cyclone. The precipitation that does fall is primar-ily convective on the equatorward side, with only occasional cyclonic precipitation. As latitude increases, so does the frequency of cyclonic precipitation.

Temperature is a better seasonal indicator than precipitation in the deserts (see Table 14.7). Winter has slightly cooler days and much cooler nights. The highest average annual temperatures on Earth are found in the tropical deserts. They vary between 29° and 35°C (84°–96°F). At Lugh Ferrandi, in Somalia, the temperature averages 31°C (88°F), consid-erably higher than in many of the other tropical climates. The major factor controlling average annual temperature is latitude, and the stations with the highest averages are those closest to the equator. The highest official temperature yet recorded is 58°C (136°F), recorded September 13, 1922 at El Azizia, near Tripoli in northern Africa. In the United States the highest recorded temperature is 56.7°C (134°F) in Death Valley, CA. Winter averages in the deserts dip below those of other parts of the tropics as Earth radiation at night rapidly cools these areas. Averages are as low as 15°–20°C (59°–67°F). The lower winter temperatures give the deserts the highest annual range found in the tropical regions. Aswan, Egypt, located on the Tropic of Cancer, has an annual range of 19°C (34°F).

Diurnal ranges in the deserts are the highest of any of the climates, and, as in other tropical climates, they far exceed the annual range. The average diurnal range is 14°–25°C (25°–45°F), but occasionally it is much greater. At Bir Mirha in the Sahara south of Tripoli, the temperature dropped from 37.2° to −0.6°C (99°F–31°F) in one 24-hr period, a range of 37.2°C (68°F). The largest known 24-hr drop in temperature occurred at the oasis of Salahin in the Sahara Desert on October 13, 1927. The tempera-

TABLE 14.7
Monthly data for tropical desert climates.

	JAN.	FEB.	MAR.	APR.	MAY	JUNE	JULY	AUG.	SEPT.	OCT.	NOV.	DEC.	YR.
Al-Hofuf, Saudi Arabia 25°22′ N 49°35′ E: 145 m													
Temp. (°C)	14.0	15.9	20.6	25.4	30.8	33.5	34.6	33.9	31.3	26.8	20.9	16.1	25.3
Precip. (mm)	23	8	16	16	1	0	0	0	0	1	1	6	49
Marrakesh, Morocco 31°37′ N, 08°00′ W: 458 m													
Temp. (°C)	11.5	13.4	16.1	18.6	21.3	24.8	28.7	28.7	25.4	21.2	16.5	12.5	19.9
Precip. (mm)	28	28	33	30	18	8	3	3	10	20	28	33	242
Alice Springs, Australia 23°38′ S, 133°35′ E: 570 m													
Temp. (°C)	28.6	27.8	24.7	19.7	15.3	12.2	11.7	14.4	18.3	22.8	25.8	27.8	20.8
Precip. (mm)	43	33	28	10	15	13	8	8	8	18	30	38	252

ture dropped from an afternoon high of 52.2° to −3.3°C (126°–26°F) the following morning, a range of 55.6°C (100°F).

Coastal Deserts. Some coastal sections of the tropical deserts contain a distinctive set of atmospheric conditions that distinguishes them from the rest of the tropical deserts. These desert areas have cool temperatures, shallow temperature inversions, cold water offshore, considerable fog and stratus cloud cover, and little rain (Table 14.8). In fact, the areas with the lowest annual rainfall in all the deserts are found here. Baja California, Ecuador, Peru, and the Sahara and Namib deserts of Africa have zones with this type of climate (see Plate 12). The primary factor in producing these conditions is the flow of cool air onto the land. The air is chilled when it crosses a cold current flowing along the coast. One effect of this onshore flow of air is to cool the area during the summer months, dropping the annual mean temperature by 5°–10°C. It also reduces the annual range in temperature. At Iquique, Chile, the relative humidity averages 81% in August, yet the rainfall averages only 28 mm (1.1 in.) per year. During a 20-year period, no measurable precipitation fell. Caloma, Peru, averages 48% relative humidity in August and no measurable precipitation has been recorded there. These coastal deserts have extremely low precipitation, even for desert areas, and a much higher percentage of cloud cover. Humidity is often in the 80%–90% range, and extensive fog is common. Near Lima, Peru, in July and August fog often covers the city for days at a time. When conditions are optimal, the fog will pass over the city and penetrate for miles into the valleys of the Andes Mountains to the east of the city. These are unusual climatic areas that do not cover large areas of Earth's surface.

Desert Storms. Severe storms are not frequent in the deserts, primarily because moisture is insufficient to supply the energy that produces a storm. Deserts are very windy, however, particularly in the afternoon and during the hottest months. The air over deserts commonly has a steep lapse rate and is unstable near the surface. Dust devils are the most visible result of desert turbulence. Developing in clear air with low humidity, they become visible due to the debris they carry. They can reach a height of nearly 2 km. On

TABLE 14.8
Comparison of temperature and relative humidity in tropical deserts and coastal deserts.

| | AVERAGE OF WARMEST MONTH | | | AVERAGE OF COOLEST MONTH | | | MEAN ANNUAL | ANNUAL TEMP. |
| | | R.H. (%) | | | R.H. (%) | | | |
	T(°F)	0700	1400H	T(°F)	0700	1400H	TEMP. (°F)	RANGE (°F)
Tropical deserts								
In-Salah (Sahara)	98	36	25	56	63	37	77	42
Riyadh (Arabian)	93	47	31	58	70	44	75.5	35
Laverton (Gr. Australian)	87	36	24	52	60	43	74.5	35
West coast deserts								
Arica (Atacama)	72	74	61	60	83	74	66	12
Walvis Bay (Kalahari)	66	91	73	58	83	65	62	8
Villa Cisneros (Coastal Sahara)	72	88	63	63	75	51	67	9

occasion they will sustain velocities high enough to blow down shacks or blow screen doors off their hinges. Dust devils are very common in deserts, and occasionally a multitude of them will form at one time under the right atmospheric conditions. The dust devil results when an intense thermal forms at or near the ground surface and surrounding air moves inward to replace the rising air. As the air moves in toward the thermal, the radius of curvature decreases and velocity increases.

Tropical deserts also have dust storms and sandstorms. The dust in these storms occurs as a result of deflation. A readily visible aspect of most deserts is the lack of fine sediments because the desert wind keeps the surface swept clean. Consequently, winds blowing out of deserts are often dust-laden. On the south side of the Sahara they call this dry wind the *harmattan*. Since it is a dry wind, it is cooling and welcome, although it may limit visibility and leave household furnishings covered with dust. These same winds blow out of the north side of the Sahara Desert. Occasionally, they travel across the Mediterranean Sea, devastating crops. They are important enough in the climate of Mediterranean countries to warrant local names such as *sirocco* in Italy and Yugoslavia and *leveche* in Spain.

In the sandy parts of the deserts, the wind's ability to move sand is significant. Sand dunes cover only a small part of the tropical deserts—less than 30% of the Sahara Desert and only about 2% of the Sonoran Desert of North America. Sandstorms occur mainly in areas of sand dunes that have plenty of loose sand and when wind velocities reach a value high enough to move the sand. The velocity at which sand begins to move, called the *threshold velocity*, depends on the size of the sand particles, wetness of the sand, and other less important variables. For medium-sized dry sand (0.25 mm in diameter), the threshold velocity is 5.4 m per sec; that is exceeded some 30% of the time. Sand drift increases rapidly as wind velocities increase above the threshold velocity. We measured an average rate of 12 liters per meter width for winds 5.5–6.4 m per sec (12–13 mph) and a rate of nearly 10 times that for wind velocities of 10 m per sec (22 mph). The increase is rapid because above the threshold velocity the power of the wind increases as the cube of the wind velocity.

Two kinds of sandstorms develop in tropical deserts. One is the result of surface heating and the resultant turbulence. It is not unusual for sand temperatures at the surface to reach 85°C (180°F). This type of sandstorm happens chiefly during the daytime and in the hottest months. At Al-Hofuf, Saudi Arabia, the percentage of hours in which sand drift occurs between 6:00 A.M. and 6:00 P.M. ranges from 76% in February to 91% in June. The total hours per day when winds exceed the threshold velocity also follows a seasonal pattern, increasing from 3 hr per day in February to 9.4 hr per day in June. Sand movement is greater in summer, when high-velocity winds are more frequent. In the Nafud it increases from 116 liters per meter per day in February to 406 liters per meter per day in June. The high frequency of afternoon sandstorms makes travel and other outdoor activities particularly difficult in summer.

The second type of sandstorm results from low-pressure disturbances passing through the area. These storms often generate higher-velocity winds than does diurnal heating. They also last longer. One such sandstorm in the Nafud lasted for 43 hr, during which sand blew constantly. The wind averaged 10.7 m per sec (24 mph) and exceeded 15.6 m per sec (35 mph) for hours at a time. This kind of regional sandstorm halted the American attempt to rescue the hostages from Iran in 1980.

Both types of sandstorms can move huge amounts of sand. In the Nafud, the wind moves an estimated 80 m^3 of sand across each meter width of the sand field each year. The dunes themselves move at fairly high speed. In the Nafud, dunes with an average height of 10 m (33 ft) move at a rate of 15–19 m each year. Both the highway across Saudi Arabia and the only railroad go through the Nafud; both close sporadically as dunes encroach upon the right of way.

■ Applied Study

Desertification

In many parts of the world, particularly on the desert margins, a process of environmental degradation called **desertification** is taking place. The process consists of the breakdown in the vegetal cover and soil until the land is no longer productive and erosion by water and wind produce relatively sterile land. Desertification does not involve the outward expansion of the climate conditions responsible for the core areas of the world deserts. Rather, it is human activity transforming the land to desert-like character.

Lands on the fringes of the desert are particularly susceptible to these processes because of the seasonal nature of precipitation. Precipitation in the wet-and-dry lands fluctuates considerably from year to year, as illustrated by the rainfall data for Khartoum, Sudan, shown in Figure 1. Precipitation from 1941 to 1974 varied from less than 50 mm (2 in) to more than 300 mm (12 in.). The amount of summer rainfall determines how much forage there will be on the grazing land and how big the yields of crops will be. This in turn determines how many people and animals can be fed (Figure 2). Since rainfall varies so much from year to year, the number of people the grassland can adequately support varies as well.

Even total annual precipitation is often misleading, because the intraseasonal distribution varies from year to year. In central Sudan, most of the rainfall occurs in May through September. In 1973, 175 mm (7 in.) of rain fell during the growing season, but none in the months of June and August (Figure 3). In the following year, precipitation fell in each of the four months from June to September, but the total was only 50 mm (2 in.). Both years were disastrous for farmers, especially for those engaged in crop cultivation, since the two dry moths came at critical times in the growing season.

The nature of the ecosystems in much of the wet-and-dry lands is such that they will not support many people on a subsistence basis. Dry-land cultivation practices are mostly primitive and damaging to the land. Farmers clear areas and farm them for several years, during which the soil gradually deteriorates and loses fertility. When the soil is exhausted, they abandon it and assume that it will rejuvenate with time. If the amount of land cultivated in any given year is small enough, and the livestock on the grassland does not exceed the carrying capacity, the ecosystem will suffer little long-range damage.

Desertification happens when the numbers of people and livestock exceed the capacity of the precipitation to support them. It is not a new problem, but it is happening at an unprecedented rate. Before colonization of Africa, intermittent periods of environmental degradation occurred. Whenever a dry cycle began, excessive grazing and cultivation

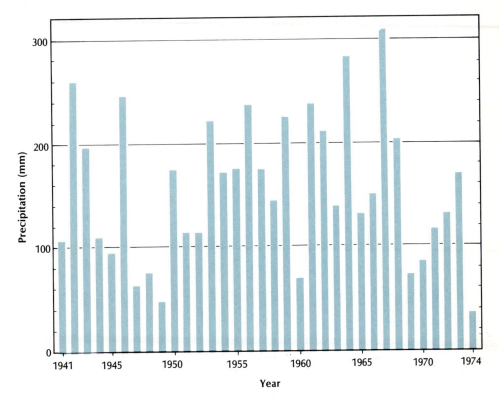

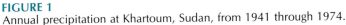

FIGURE 1
Annual precipitation at Khartoum, Sudan, from 1941 through 1974.

preceded loss of vegetation and soils. Population increased during wet periods and decreased during drought (Figure 4).

   If an area became unproductive, the population either died off or moved away. The introduction of colonial administration brought some measures that altered the balance between the population and environment, reducing intertribal warfare and introducing public health practices. These measures resulted in rapid population growth. In 1990, sev-

FIGURE 2
Precipitation and carrying capacity. The precipitation over much of sub-Saharan Africa is subject to marked cycles. Since most of the population engages in subsistence agriculture or herding, precipitation translates into carrying capacity.

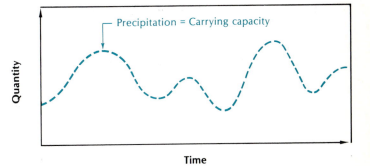

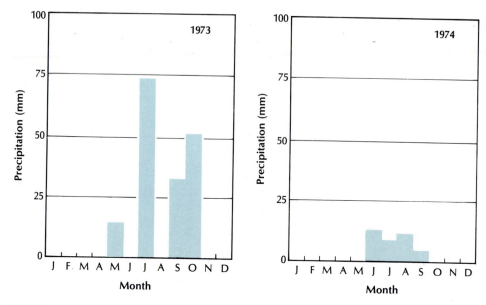

FIGURE 3

Seasonal rainfall in Khartoum, Sudan, in 1973 and 1974. In 1973 the total was above average, but no measurable rain fell in the two crucial months of June and August. In 1974, the total was far below normal.

eral of the wet-and-dry land countries had the highest population growth rates in the world. Along with the human population, animal populations have also grown.

As the population grows rapidly in the face of varying rainfall, the demands on the ecosystem exceed the carrying capacity more frequently and for longer periods. As Figure 5 shows, the demand ultimately exceeds the carrying capacity on the continuing basis.

The increasing pressure on the land has caused the Sahara Desert to expand southward into more-humid areas. Human activity may not be entirely responsible for this phenomenon, but it certainly accelerates it. Degeneration into desert usually occurs in scattered patches of bare ground from a few meters to several kilometers across. Because it is not an even front of desert surface advancing across the landscape, it is not easy to mea-

FIGURE 4

Population and carrying capacity in precolonial Africa. Population size followed the precipitation cycles. During favorable periods, population expanded; when dry periods followed, the population died back. The shaded areas represent times of major environmental degradation.

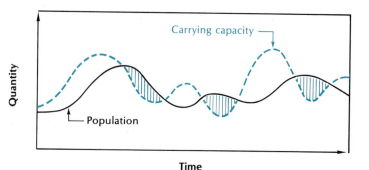

FIGURE 5
Population and carrying capacity following African colonization, when conditions changed to greatly increase population growth rates. At point A, the carrying capacity begins to decline and various forms of aid to the distressed areas control the death rate. Thus when precipitation increases again at point B, a much larger population remains from which to continue growth. The result is a population greater than the environment can support, even under more favorable conditions. Desertification accelerates.

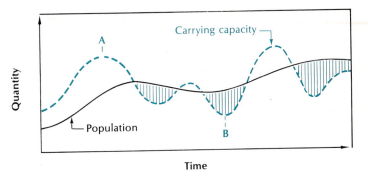

sure. Desertification is taking place at a steady rate in many areas of Earth's grasslands and occurs on every continent except Antarctica. It accelerates during periods of major drought like the one that struck sub-Saharan Africa in the early 1970s.

When plant cover falls below the minimum required to protect the soil against erosion, desertification becomes irreversible. Soil holds less water and dries out. The temperature increases, speeding the collapse. Bare soil appears, and wind and water remove the fine solid particles. This loss of topsoil is significant: The total drift westward off the African coast may be as much as 60 million metric tons each year. It is extensive enough to be measurable in the Caribbean Sea.

Wind erosion does not affect only dust-sized particles. When large patches of bare ground form, it exposes the soil to erosion of the larger particles such as sand, organic matter, and mineral salts. During the height of a recent drought, satellite images showed thousands of square kilometers of surface obscured by blowing sand. Blowing sand and silt further destroy the standing vegetation by stripping leaves or by burying plants under dunes.

The only long-range solution to the problem is to decrease the intensity of land use below historical levels. Animal population levels must remain below the maximum carrying capacity represented by the wetter-than-normal years. This is the only mechanism that will provide immediate relief from the drought hazard and allow the pastures a chance to survive. Determining these critical levels requires a thorough understanding of the regional environment as well as close monitoring of the natural changes in precipitation. If land use intensifies, desertification and famine will continue, and if production doesn't increase to keep up with the growing population, famine will also be certain. ■

Summary

Tropical climates cover a large portion of the Earth's surface between latitudes 25° N and S. In most of this region, temperatures change at least as much from day to night as from summer to winter. Radiation intensity is relatively high all year. Precipitation defines the climatic regions within the tropics. Some places have a high chance of precipitation every day of the year, and others have an equally low chance of precipitation on any given day of the year. Between these extremes lie the vast areas with a rainy season and a dry season. The wet-and-dry tropics have wide range in both the length of the rainy season and in total precipitation received during the rainy season. The monsoon regions of Asia represent the extreme example of seasonal precipitation; some of the highest amounts of annual precipitation occur there. The driest areas of the world are also in tropical regions: The coastal margins of the tropical deserts include weather stations with the lowest known annual total precipitation.

THE MIDLATITUDE CLIMATES

Chapter Overview

General Characteristics
- Midlatitude Circulation
- Summer
- Winter
- Water Balance

Climatic Types
- Midlatitude Wet Climates
- Midlatitude Winter-Dry Climates
- Midlatitude Summer-Dry Climates
- Midlatitude Deserts

Climates found in the midlatitudes differ from those of tropical regions in two major respects. In tropical regions, the variation in radiation and temperature is greater from day to night than it is from season to season. In the midlatitudes, seasonal variation is greater than diurnal (Figure 15.1) because midlatitude winters have much less solar radiation than summers do (see Chapter 2). The primary circulation in the midlatitudes is also different from that of the tropics. The subtropical high over the oceans and the polar high act as major sources of air moving toward the convergence zone centered in the latitudinal range of 50°–60°. These two source areas are very different in character—one is very warm and the other quite cold. While that difference is enough to produce different air streams in the midlatitudes, varying moisture content further distinguishes the air masses.

General Characteristics

Midlatitude Circulation

Westerly upper-air flows of the Rossby regime dominate the circulation in the midlatitudes. The equatorward limit of this regime is about at the axis of the zone of divergence associated with semipermanent high-pressure cells. These are located beneath the subtropical jet stream. The Hadley circulation operates between this axis and the equator.

The temperature gradient between the equator and the poles determines the location of the subtropical jet stream. As polar areas warm in the summer, the gradient lessens and the strength of the westerlies diminishes. The jet migrates poleward to attain the average position shown in Figure 15.2*a*. When a very steep gradient exists, the westerly circulation

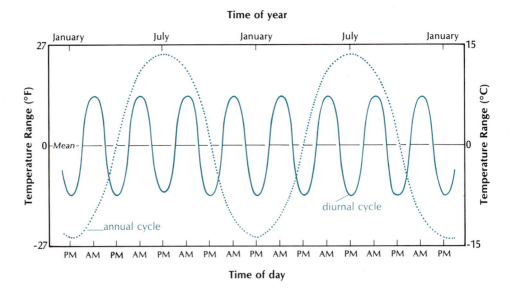

FIGURE 15.1
The relationship between the annual and diurnal energy cycles in the midlatitudes.

is strong and the jet migrates closer to the equator. Figure 15.2b shows the mean jet stream for the Northern Hemisphere in winter. Surface circulation patterns are closely related to the flow of the upper-air westerlies in the midlatitudes. The migration of the subtropical jet and the weakening of the westerly circulation provide very different con-

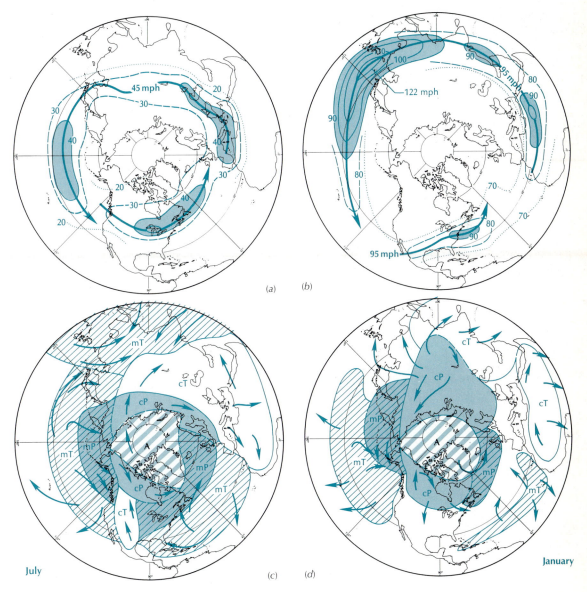

(a)   (b)

(c)   (d)

July

January

FIGURE 15.2

The mean jet stream in (a) summer and (b) winter. Principal Northern Hemisphere air mass source regions in (c) summer and (d) winter.

Source: (a) and (b) from S. Peterssen, *Introduction to Meterology*, 3d ed. (New York: McGraw-Hill, 1969), 189. Used with permission of McGraw-Hill Book Company.

ditions from season to season. The movement and relative dominance of air masses varies, as does the frequency and path of traveling low-pressure systems. Figures 15.2c and 15.2d show some of the changes that occur in air-mass dominance over the Northern Hemisphere. Note, for example, the difference in extent of cP air masses in winter and summer.

The difference between summer and winter is one of the basic attributes of midlatitude climates.

Summer

Summer in the midlatitudes has many of the characteristics of the tropical climates. At the time of the summer solstice, radiation intensity at 47° N and S is as high as it is at the equator. Not only is the radiation intensity high, but the duration of radiation is longer than it is at the equator on this date (see Figure 15.3). The inequalities in length of day and night increase with latitude. When added to the variation in solar intensity, the longer or shorter days compound the seasonal differences. The result is that in the summer, more radiation reaches the surface in the midlatitudes than in the equatorial zone. During summer, temperatures average in the 20°–25°C (68°–77°F) range at most midlatitude weather

FIGURE 15.3
The length of daylight in summer and winter and the relative intensity of solar radiation compared to a vertical solar beam.

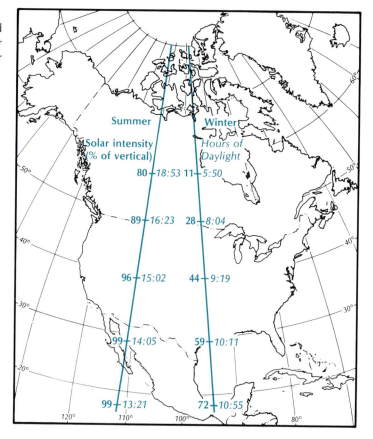

stations. Along the margin facing the equator, the warm-month temperatures may average 26°–30°C (79°–86°F), comparable to rainy tropics. The warm-month means drop slowly toward the poles. At latitudes greater than 45°, the summer averages can drop below 20°C (68°F). In July, averages differ only slightly from the Gulf Coast to the Canadian border (Figure 15.4).

Precipitation decreases toward the poles and toward the interior of the continents. The primary reason for this decrease is the increasing distance from the source of mT air. Areas on the west sides of the continents that get precipitation from mP air normally have less total precipitation than areas receiving precipitation from mT air. This is because the mP air is cooler and contains less water vapor. The frequency of the thunderstorms decreases rapidly from south to north. Cyclonic precipitation is most frequent at the polar margins and is more frequent in all sections of the midlatitudes in the winter. Hurricanes provide another precipitation mechanism; some of the heaviest rains along the coastal areas result from hurricanes.

Winter

Winter truly sets the midlatitude zone apart from the tropics. All areas in the midlatitudes experience a range of 47° in the angle of the solar beam between the summer solstice and the winter solstice. This varying angle produces changes in radiation intensity at the surface. In winter, solar intensity is low and the hours of sunlight are short; temperatures reflect these conditions. Mean winter temperatures also decrease markedly as latitude increases. January temperatures average above 10°C (50°F) closer to the equator and drop to below −18°C (0°F) toward the poles. The range between mean July and mean January temperature increases from about 8°C (14°F) in the warmer areas to about 40°C (72°F) in the Canadian Arctic. The coldest winter temperatures have a steep poleward gradient also, but not as steep as the average temperatures. Temperatures as low as −15°C (5°F) occur along the equatorward margin, and lows from −31°C to −46°C (−25°F to −50°F) characterize the more northerly locations.

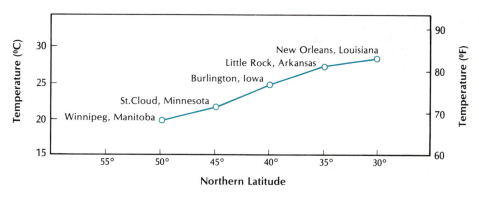

FIGURE 15.4
Mean July temperature along a north-south transect from New Orleans, LA, to Winnipeg, Manitoba.

Diurnal ranges in winter are slightly greater than the summer ranges because the relative and absolute humidity are somewhat lower. The winter controls are, in essence, low solar radiation, short days, and moderate atmospheric humidity.

Water Balance

Climates in tropical regions are distinguished on the basis of the seasonal water balance; the same three patterns occur in midlatitudes. Some regions have frequent precipitation year-round, some have pronounced wet and dry seasons, and desert areas have only occasional precipitation. In tropical regions, rainfall may vary from month to month, but temperatures tend to stay the same through the year. Some midlatitude regions have seasonal changes in both solar energy and moisture (Figure 15.5).

The effect of high summer radiation is apparent in water balances of midlatitude stations (Figure 15.6), where potential evapotranspiration peaks in summer. In some

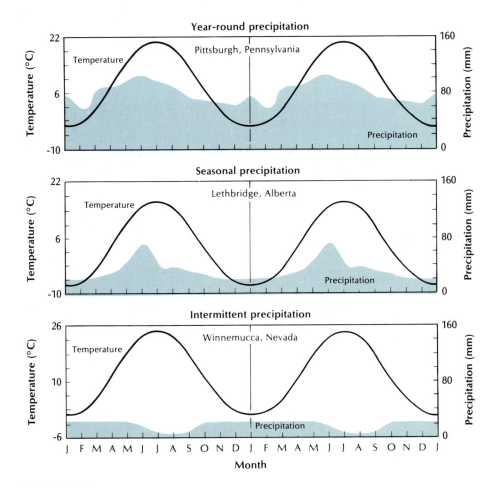

FIGURE 15.5
The three major types of midlatitude climates based upon the seasonal pattern of precipitation.

cases, such as in the summer-dry climates, this summer maximum of potential evaporation is out of phase with seasonal rainfall. Here precipitation occurs in the cool season when rainfall is more effective, with less water lost to evaporation.

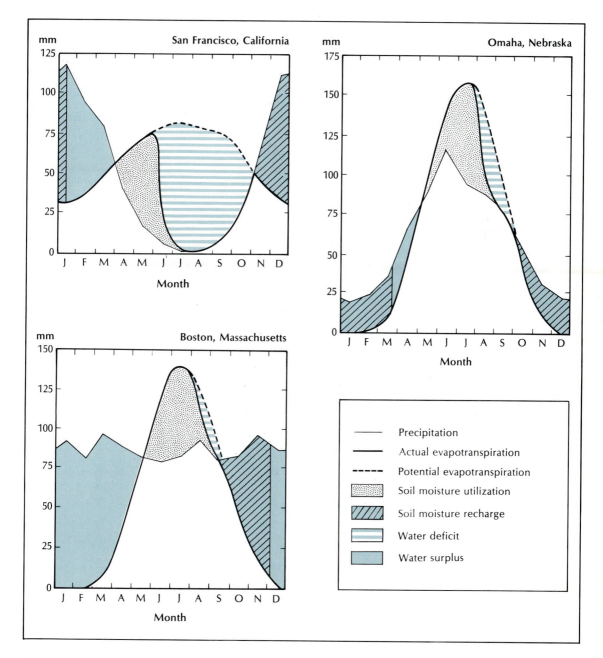

FIGURE 15.6
Selected water-balance graphs for North American stations.

In the interior areas of the midlatitudes, the summer is the time of maximum rainfall with the period of highest solar radiation and the time when plants are actively growing. Irrigation is thus an integral part of agriculture. In the humid midlatitude areas, precipitation may well exceed evapotranspiration for much of the year, except during droughts. Most water shortages occur in summer.

An important aspect of the water balance in the midlatitudes is periodic freezing of streams, lakes, ponds, and soil moisture, which occurs when winter temperatures drop well below freezing. Most parts of the midlatitudes experience freezing temperatures for varying periods during the winter. These temperatures freeze soil moisture and surface water, stopping the flow of runoff and cutting the water supply available to plants. Moisture accumulates in the environment. Even though winter precipitation is generally less than summer precipitation over the Northern Hemisphere land masses, demands on the water supply are also lower. Evapotranspiration drops in winter because of lower temperatures and reduced plant growth. Over the land masses, some 70% of all precipitation goes directly back to the atmosphere through evaporation. In winter, this may drop to 40% or less, allowing moisture to accumulate on and in the soil. Removing water from circulation by freezing shapes the annual pattern of stream flow. Reflecting that, the U.S. Geological Survey uses a water year in all its calculations that differs from the calendar year. For most of North America, streams reach their low point in September or October following the summer, when evapotranspiration uses a big share of the water. As temperatures cool, the vegetation uses less water, and moisture begins to collect. During winter, moisture accumulates either as soil moisture or as snow. The spring thaw then causes streams to rise. Flooding may occur if a large amount of snow covers the land surface. Spring floods due to snowmelt generally characterize poleward locations and mountain regions, but floods also affect other areas.

Climatic Types

Tropical climates were distinguished from each other on the basis of the seasonal pattern of precipitation. All midlatitude locations have a seasonal temperature regime that is stronger in some places than others. Midlatitude climates are also distinguished from one another on the basis of the seasonal pattern of precipitation.

Midlatitude Wet Climates

The midlatitude wet climates are associated with the subpolar lows over continental areas that spread toward the equator on the eastern sides of the continents. They tend to stretch nearly across the continents near 60° N and S and reach equatorward as far as 25° or 30° along the eastern margins of land masses. In North America, this type of climate extends from the Pacific coast of Canada eastward in a crescent to the Atlantic coast (see Plate 13). In the United States, it includes the area lying east of the Mississippi River and the first row of states west of the Mississippi. In Eurasia, this climate extends from the offshore islands of Great Britain and Ireland eastward into the Soviet Union. It also occurs along the Pacific coast of Asia in latitudes 30°–60°. The climate also exists in South America,

TABLE 15.1
Climates with seasonal mT and mP air.

Köppen type	Cf, Df
Thornthwaite types	BB′, BC′

Africa, and Australia. South America has two such areas: southern Chile, including the southern end of the central valley; and the Pampas of Argentina and Uruguay, extending southward to include part of Patagonia. The area in Africa is very limited, consisting of a small region on the very southeast tip of the continent. Both Australia and New Zealand include sectors with this type of climate. Table 15.1 lists the climatic types, and Table 15.2 shows sample climatic data.

Summer midlatitude–wet climate conditions are very similar to those of the rainy tropics, with two primary exceptions: cyclonic storms that move across the region and temperatures that average from 21° to 26°C (70°–79°F). The tropical margins reach as high as 29°C (84°F), somewhat warmer than the humid tropics.

The summer temperature extremes in the midlatitude wet climates exceed those of the tropical wet climate. Few weather situations in the humid tropics have recorded temperatures in excess of 38°C (100°F), while most stations in the midlatitude climate have recorded maximums at least this high. The summer diurnal range is typically 8°–11°C (15°–20°F), comparable to the tropical humid climates. In the summer months of June, July, and August, temperatures in the southeastern United States average 1°–2°C (2°–3°F) higher than in Belem, Brazil, and rainfall is about the same. Nighttime temperatures are often within a few degrees of the daytime temperatures, because the nights are

TABLE 15.2
Monthly data for midlatitude wet climates.

	JAN.	FEB.	MAR.	APR.	MAY	JUNE	JULY	AUG.	SEPT.	OCT.	NOV.	DEC.	YR.
New Orleans, LA 29° 59′ N, 90°04′ W: 1 m													
Temp. (°C)	12	13	16	20	24	27	28	28	26	21	16	13	20
Precip. (mm)	98	101	136	116	111	113	171	136	128	72	85	104	1371
Montreal, Quebec 45° 30′ N, 73° 35′ W: 60 m													
Temp. (°C)	−11	−9	−4	5	13	18	21	19	15	8	1	−7	5
Precip. (mm)	83	81	78	72	72	85	89	77	82	78	85	89	971
Buenos Aires, Argentina 34° 20′ S, 58° 30′ W: 27 m													
Temp. (°C)	23	23	21	17	13	9	10	11	13	15	19	22	16
Precip. (mm)	103	82	122	90	79	68	61	68	80	100	90	83	1027
London, England 51° 28′ N, 0°00′ : 5 m													
Temp. (°C)	4	4	7	9	12	16	18	17	15	11	7	5	10
Precip. (mm)	54	40	37	38	46	46	56	59	50	57	64	48	595

short. Summer temperatures in midlatitude humid climates depend mainly on the high solar intensity, long days, and high atmospheric humidity.

Precipitation is frequent throughout the year. The annual total is quite variable, depending upon the latitude and continental position. It varies from as little as 510 mm (20 in.) upward to 1780 mm (70 in.). Mobile, AL, is among the wettest of U.S. cities with a mean annual precipitation of 1.73 (68 in.). The variability of annual precipitation is similar to that of the humid tropics—normally less than 20%. Seasonal distribution is fairly even, but some areas have more in one season. Frequency of precipitation varies. Bahia Felix, Chile, averages 325 days a year with rain, and rain has a 90% probability on every day of the year. This is not typical of midlatitude humid climates, of course, but it does show the extent to which the precipitation mechanisms can persist.

The precipitation varies more in form than it does in the tropics. During summer and closer to the equator, convectional rainfall is the most common type of precipitation in midlatitude wet climates. The southeastern United States, for example, averages 40–60 days per year of thunderstorms. The frequency of the thunderstorms decreases rapidly toward the poles, where cyclonic precipitation is common. This type of precipitation also is more frequent throughout the region in the winter. Hurricanes provide another mechanism for inducing precipitation, and some of the heaviest rains along coastal areas result from hurricanes.

While most of the annual precipitation takes the form of rain, snow is a factor to some extent in all places. In the United States, mean snowfall varies from a trace along the Gulf of Mexico to an average of about 1 m (40 in.) through the Great Lakes district. Even the southern areas sometimes have snowfall.

Humidity is generally high. On occasion it will drop quite low but usually only for short periods. High humidity and high temperatures make the sensible temperatures of summer quite high. As in the tropics, radiation fogs are common on clear nights in summer and fall.

Summer in the humid midlatitudes is much like the rainy tropics. Temperatures and humidity are high, and convectional showers are common. Summer storms such as tornadoes, thunderstorms, hurricanes, heat waves, and cold waves bring variety to summer. Cold waves result from the equatorward flow of summer continental polar air from the interior of high-latitude land masses. These cold waves may bring temperature above 10°C (50°F), much cooler air than is normal.

Within the wet climate area, differences in temperature provide further divisions in the climate. Köppen distinguished between areas that have hot summers and mild winters (Cfa) and those with cool summers and mild winters (Cfb). The latter region is mainly on the western side of the continents, where summer temperatures are cooler due to the onshore winds from the oceans.

The midlatitude wet climates are perhaps best developed on the west sides of the continents from about 45° to 65°. Here the westerlies from the oceans bring mP air masses onshore, increasing cloud cover, humidity, and the frequency of precipitation. The largest area with this type of climate is in Europe, where it spreads inland a fair distance because no north–south mountain range exists to block the flow of the westerlies from the Atlantic Ocean. In North America, the zone extends from northern Oregon to Alaska. In South America, it extends from 40° to the tip of the land mass. Mountain ranges parallel to the coast restrict this type of climate in both North and South America. Small areas with this type of climate exist in South Africa, Australia, and New Zealand.

Winter is fairly long but mild, and summer is cool in the wet midlatitudes. Winter maritime air masses are usually at least as warm as the offshore ocean. Winter air is damp and chilling, and fog is common because of the high humidity. The storm systems often become occluded, bringing extensive periods of rain and drizzle. London offers a good example of this phenomenon. The city averages 164 days each year with precipitation, translating into a 45% probability of rain every day. On the other hand, the total annual rainfall in London is only 625 mm (25 in.), meaning the amount of rain received on each rainy day averages less than 4 mm (0.15 in.). Cloudy weather, fog, and drizzle are indeed typical.

Very cold weather is not common along these coasts. Cold cP air masses develop over land masses, and the westerlies carry them south and east most of the time. Only when an extremely large mass of cold air spreads out so it expands in a westerly direction do these cold waves affect coastal areas. Warm ocean currents also increase winter temperatures. Figure 15.7 shows how the North Atlantic drift produces milder winter conditions in northwestern Europe.

In summer, precipitation is less frequent because of the stabilizing effect of the subtropical highs as they move northward. Sites near sea level do not receive high amounts of precipitation, primarily because of the coolness of the air and cyclonic activity that produce the precipitation. Summers are subject to cool weather as the result of the influence of the adjacent oceans. The oceans in these latitudes never warm very much, and a cold current flows equatorward along most of these areas. Where highlands exist along

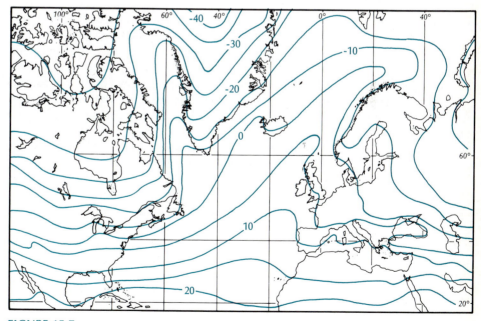

FIGURE 15.7
The effect of the northward movement of warm water in the Gulf stream. Note the shape of the isotherms (°C).

these coasts, even summer precipitation is frequent, and rain forests have evolved in which huge trees and dense surface vegetation are interspersed. The famous redwoods of California and Oregon typify these forests. The Olympic Peninsula of Washington contains a magnificent forest of Douglas fir. These forests extend well into Canada. The green meadows of Ireland are also due to the frequency of the moist Atlantic air masses.

Midlatitude Winter-Dry Climates

The midlatitude wet-and-dry climates (Table 15.3) follow a strong seasonal pattern of both temperature and moisture (Table 15.4). The general location of this climate is the interior of the continents in the midlatitudes. This continental location provides the characteristic feature of a large average annual temperature range. Figure 15.8a shows the increase in summer temperatures and decrease in those of winter for selected Eurasian stations. Figure 15.8b shows how this continental effect varies in Eurasia. The values given in the diagram are based upon the Index of Continentality devised by Gorczynski using the formula $C = 1.7 \times A - 20.4/\sin \Phi$. This formula assesses the continental effect of a location (C) taking into account the annual range of temperature (A) and latitude (Φ). As the map in Figure 15.8 shows, Asia is a region of maximum continental effect on climate.

Specific areas with this type of climate are the Great Plains of the United States and Canada, the grasslands of Eurasia, and small areas of Australia, Africa, and South America (see Plates 14 and 15). These are continental locations where wind systems reverse seasonally. Convergence occurs frequently during the summer, causing frequent precipitation. In winter, divergence is more common, and the atmosphere is drier. This is the most variable climate. These areas host maritime tropical air masses and occasional tropical continental air masses from the adjacent deserts in summer. In winter, they experience frequent outbreaks of polar continental air masses and an occasional maritime polar air mass. Each type of airstream brings its particular variety of weather.

A major difference in the response to solar input in the midlatitude winter-dry region is the result of lower atmospheric humidity. Clear days are more frequent, so incoming and outgoing radiation are higher. Both summer and winter extremes are greater as the result of more frequent dry air masses. Extreme high temperatures range from 38° to 44°C (100°–110°F) in the wet and dry regions. They range up to 49°C (120°F) in the warmer sections as they did in the Great Plains in 1980.

Winter is the dry season. Dry weather increases land cooling during the long nights, resulting in mean winter temperatures usually being a few degrees lower than in the humid climates. Extreme low temperatures are 3°–5°C (5°–9°F) cooler. This climatic region experiences the lowest temperatures in the Northern Hemisphere. In North America, the temperature dropped to −52°C (−62.8°F) at Snag, Yukon, on February 3,

TABLE 15.3

Climates with seasonal mT and cP air.

Köppen type	Cw	Dw
Thornthwaite types	BB′w	CB′d
	CB′w	CC′d
	CC′w	DC′d
	DB′w	

TABLE 15.4

Monthly data for dry climates.

	JAN.	FEB.	MAR.	APR.	MAY	JUNE	JULY	AUG.	SEPT.	OCT.	NOV.	DEC.	YR.
Denver, CO 39°22' N, 104°59' W: 1588 m													
Temp. (°C)	0	1	4	9	14	20	24	23	18	12	5	2	11
Precip. (mm)	12	16	27	47	61	32	31	28	23	24	16	10	327
Bismarck, ND 46°46' N, 100°45' W: 507 m													
Temp. (°C)	−13	−10	−5	6	13	18	22	21	14	8	−2	−8	5
Precip. (mm)	11	11	20	31	50	86	56	44	30	22	15	9	385
Ulan Bator, Mongolia 47°55' N, 106°50' E: 1311 m													
Temp. (°C)	−26	−21	−12	−1	6	14	16	14	9	−1	−13	−22	−3
Precip. (mm)	1	2	3	5	10	28	76	51	23	7	4	3	213

1947. At Tanana, AK, a −51°C was recorded in January 1886. The lowest official temperature of the Northern Hemisphere is −57°C at Verkoyansk in the former U.S.S.R. (−61°C was unofficially recorded at Oimekon, U.S.S.R.). As a result of less-humid conditions and excessive Earth radiation, the average annual range is up to 34°C (67°F) in extreme cases and the absolute range up to 100°C (186°F). A winter characteristic of these areas is rapidly changing temperatures. At Browning, MT, on January 23–24, 1916, the temperature dropped from 6.7° to −48.9°C (44° to −45°F)—a total of 55.6°C in 24 hours. Such drastic changes are usually the result of the inflow of warm air, but adiabatically heated air will cause equally sharp rises in temperature.

The midlatitude winter-dry climates are subject to the effects of a monsoon. During summer, the excessive heating of land in the Northern Hemisphere causes the subpolar low to expand equatorward. This forms a low trough, or cell, in the middle of the continent. Converging air brings moisture to the continent, producing a summer rainy season. During winter cooling of the land mass, the subpolar trough splits into cells over the oceans, and a ridge of high pressure develops between the polar high and the subtropical high. This reduces the probability of precipitation in midlatitude climates.

Winds are a significant factor in the weather. A seasonal change in wind direction accompanies the seasonal change in pressure. During the high-sun season, onshore winds prevail. During the winter, prevailing wind direction changes 90°–180°. The degree of wind shift depends to a large extent on the continental position. The more inland the weather station, the greater the wind shift. Wind velocities are high and persistent. In the United States, Oklahoma City has the highest average wind velocity of any city, with a mean of 22 km per hr (14 mph).

Atmospheric humidity and precipitation are seasonal in character, and cloud cover varies with humidity. Precipitation decreases with distance from the equator and the sea. Although it is difficult to delineate annual precipitation, most areas in the winter-dry midlatitudes receive between 300 and 900 mm (12–35 in.). Of the annual total, about 75%

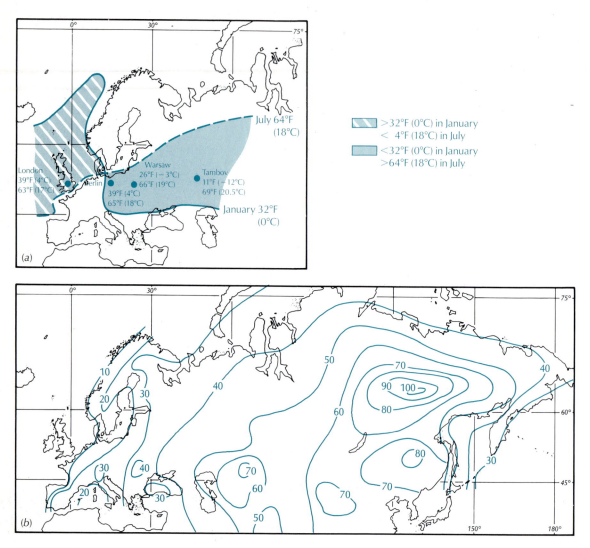

FIGURE 15.8

The influence of continentality: (*a*) In January, continental areas are colder than neighboring oceans; in summer, the reverse is true. (*b*) An index of continentality applied to the Eurasian land mass.

Source: K. Boucher, *Global Climates*, (Seven Oaks, Kent, England: Hodder & Stoughton, 1975), 224. Used by permission of Hodder & Stoughton, Ltd.

falls during the summer half-year. The annual total is highly variable and subject to long-term cycles as yet unexplained. Several factors influence the amount of rainfall. Nearly all rainfall east of the Rocky Mountains comes from midlatitude cyclones. Winds from the Gulf of Mexico take a curved path across the eastern United States, moving east of the Great Plains. The dry, tropical air masses that move northward across the Great Plains develop in Mexico and contain little moisture.

In the Great Plains, cold fronts are often very steep as they move to the southeast, and extremely severe rainstorms develop with them. The heaviest unofficial 12-hr rain fell at Thrall, TX, on September 9, 1921, when 810 mm (32 in.) of rain fell. At Holt, MO, an unofficial rain of 42 minutes' duration totaled 300 mm (12 in.) on June 22, 1947.

The winter-dry climate is well-defined over the Eurasian land mass. The largest contrasts between summer and winter weather conditions occur here, with the possible exception of the two polar regions. Much of Asia is subject to seasonal reversal of pressure, wind direction, and precipitation.

Summers in Eurasia are hot and humid, with intense summer convectional storms. Along the equatorward fringe, as much 70% of the annual precipitation occurs in the summer months; toward the interior, this increases to as much as 80%. The summer precipitation results from convectional instability, and severe rainstorms are frequent. A single storm may produce a significant portion of the annual rainfall. Total annual rainfall may be as much as a meter along the coastal areas, but this decreases inland to less than 675 mm (25 in.).

In winter, cold, dry, continental air from the Siberian high-pressure zone brings a majority of clear days, with short periods of cyclonic disturbances. In the interior winter temperatures drop below −34°C (−30°F). The frost-free growing season is short—less than 100 days in most years—particularly on the northern fringes. Since most of the precipitation occurs in the summer, winter snowfall is not particularly abundant. Snow stays on the ground for as much as 200 days in the northern interior.

The extreme of the midlatitude winter-dry climate occurs in northeastern Siberia. Summers can be warm and winters severely cold. At Oimekon, the range in temperature between January and July is 67°C (115°F). The range between the highest and lowest temperature recorded is 104°C (187°F). Winter is at least seven months long, with snow often remaining on the ground until May, when temperatures rise rapidly.

Permafrost appears in the coldest parts of this climatic zone. Since the soil is often covered by only a little snow, the ground freezes to a depth of 50 m (150 ft). In summer, the top few centimeters thaw, leaving wet, soggy soil and extensive swamps.

Midlatitude Summer-Dry Climates

One area with a seasonal rainfall regime differs from the rest. In this climate, the rainy season happens in the winter rather than in the summer as illustrated by data in Table 15.5. Table 15.6 provides the major types of summer-dry climates. This type of climate occurs on the western margins of the continents between 30° and 40° latitude. It normally does not spread into the continents very far, so its extent is limited. Summer-dry climates describe three main areas. The largest is around the Mediterranean Sea and extends eastward into Iran. The second is the west coast of North America from near the Mexico–United States boundary northward into the state of Washington, between the Pacific Ocean and the Sierra Nevada and Cascade Mountains. The third major area is

TABLE 15.5
Types of midlatitude summer-dry climates.

Köppen type	Cs, Ds
Thornthwaite types	BB's, CB's, DB's

TABLE 15.6

Monthly data for summer-dry climates.

	JAN.	FEB.	MAR.	APR.	MAY	JUNE	JULY	AUG.	SEPT.	OCT.	NOV.	DEC.	YR.
Santiago, Chile 33° 27' S, 70° 40' W: 512 m													
Temp. (°C)	19	19	17	13	11	8	8	9	11	13	16	19	14
Precip. (mm)	3	3	5	13	64	84	76	56	30	13	8	5	360
Los Angeles, CA 33° 56' N, 118° 23' W: 37 m													
Temp. (°C)	13	14	15	17	18	20	23	23	22	18	17	15	18
Precip. (mm)	78	85	57	30	4	2	T	1	6	10	27	73	373
Rome, Italy 41° 48' N, 12° 36' E: 131 m													
Temp. (°C)	8	8	10	13	17	22	24	24	21	16	12	9	15
Precip. (mm)	65	65	56	65	51	30	25	17	66	78	96	97	711

across Southern Australia. Smaller areas are in the Capetown district of South Africa and in central Chile. The climate is often called the Mediterranean climate after the area associated with the Mediterranean Sea. In total, only about 2% of Earth's land area experiences this type of climate.

While much of Earth's land area receives the bulk of its precipitation in the summer or spread over the year, the summer-dry climates have more precipitation in winter. At Perth, Australia, 85% of the annual precipitation occurs in the winter six months. At Istanbul, Turkey, some 70% of the yearly total falls in the winter half-year. San Diego receives 90% of its annual total from November through April. Santa Monica, CA, has recorded only traces of precipitation in the three months of June, July, and August, and the April through September period averages only 25 mm (1 in.) out of an annual total of 381 mm (12 in.). The occurrence of the precipitation in winter and the relatively low intensity of the cyclonic precipitation permit more efficient plant growth than is the case in climates where summer is the rainy season. Although mean winter temperatures average above freezing, snow occasionally falls, even along the equatorward margins of the summer-dry regions. It rarely stays on the ground very long, as daytime temperatures rise high enough to melt the snow. Total annual precipitation averages less than 750 mm (30 in.) and tends to increase poleward. Table 15.7 provides the annual rainfall for several cities along the west coast of North America. The data show the rapid increase in annual total that happens closer to the poles.

TABLE 15.7

Latitudinal variation in annual precipitation in the summer-dry climates

San Diego	269 mm
Los Angeles	380 mm
San Francisco	510 mm
Portland	900 mm

Winter temperatures are mild, with a lot of sunny weather and short periods of rain. The subsidence associated with the subtropical high keeps the sky relatively clear so it has an unusually bright blue color. The Mediterranean Sea is noted for its brilliant blue color. This is partly due to sky color and partly due to the low organic content of the water.

Like all climates that have a rainy season, these environments have two distinct seasons, each associated with a different air mass and a different part of the general circulation. During summer, these areas are under the influence of stable oceanic subtropical highs, giving them essentially tropical desert weather conditions. In winter, the anti-cyclonic circulation moves equatorward, allowing the westerlies to bring moisture to the region.

Midlatitude Deserts

The midlatitude deserts generally occur in the interiors of continents, although they merge with tropical deserts on the west sides of the land masses (Table 15.8). In North America, desert extends from southern Arizona northward into British Columbia between the Sierra Nevada and Rocky Mountains (see Plate 16). In Eurasia, deserts lie either in the trans-Eurasia cordillera or on the flanks of these mountain masses. All of the central Asian countries have desert areas within their boundaries. The Southern Hemisphere has a much smaller desert area in midlatitude locations because the hemisphere lacks land area in the latitudes where deserts are most extensive.

Midlatitude deserts have the highest percentage of possible sunshine of any of the midlatitude climates. Arizona averages 85% possible sunshine through the year, with an average of 94% in June and 76% in January. Temperatures reflect the clear skies that dominate the weather. Summer temperatures are the highest of the midlatitude climates (Table 15.9). Summer averages are 5° –8°C (9°–14°F) higher than in humid areas, and maximum temperatures are also higher. During summer, when the days are long and the sun reaches high in the sky, temperatures can exceed 50°C (122°F); Death Valley in California recorded an official shade temperature of 56.7°C (134°F), only 1°C below the highest sea-level temperature ever recorded. Winter temperatures compare with those of the humid climates at the same latitude, primarily because the coldest temperatures anywhere in the midlatitudes occur with outbreaks of cold, arctic air.

This climatic type is the most difficult to place in the pattern of the general circulation. Several factors contribute to the aridity of these regions. Each is not enough in itself to produce extreme drought, but coupled with other elements, they result in aridity. Lying between the subpolar low and the subtropical highs, these deserts maintain dry conditions due to low moisture supply and intermittent anticyclonic circulation.

The midlatitude deserts are not as arid as the tropical deserts. During the winter months, the general circulation shifts equatorward far enough to allow occasional passage of weak cyclones associated with the polar front. Precipitation totals are, of course, very low. Phoenix averages 180 mm (7 in.) of precipitation per year, and Ellensburg, WA, 230

TABLE 15.8
Midlatitude desert climates with seasonal cT and cP air.

Köppen type	BWk
Thornthwaite types	Eb', Ec'

TABLE 15.9
Monthly data for midlatitude deserts.

	JAN.	FEB.	MAR.	APR.	MAY	JUNE	JULY	AUG.	SEPT.	OCT.	NOV.	DEC.	YR.
Yuma, AZ 32°40′ N, 114°36′ W: 62 m													
Temp. (°C)	13	15	19	22	26	31	35	34	31	25	18	14	24
Precip. (mm)	10	9	6	2	T	T	6	13	10	10	3	8	77
Winnemucca, NE 40°50′ N, 117°43′ W: 1312 m													
Temp. (°C)	−3	0	3	8	12	16	22	20	15	9	2	−1	9
Precip. (mm)	27	24	21	21	24	19	7	4	9	21	20	24	219
Alice Springs, NT, Australia 23°38′ S, 133°35′ E: 546 m													
Temp. (°C)	28	27	25	20	16	12	12	14	18	23	25	28	21
Precip. (mm)	27	45	18	10	18	15	14	10	6	25	23	29	250
Ashkhabad, U.S.S.R. 39°45′ N, 37°57′ E: 230 m													
Temp. (°C)	2	5	9	16	23	29	31	29	24	16	8	3	16
Precip. (mm)	22	21	44	38	28	6	2	1	3	11	15	19	210

mm (9 in.). Precipitation does not seem to follow a seasonal pattern. Phoenix receives a trace of snow and Ellensburg averages 780 mm (31 in.) of snow each winter. Moisture efficiency is very low throughout these deserts. In some areas, potential evaporation exceeds precipitation as much as tenfold.

Summary

Midlatitude climates have marked differences in temperature through the year in response to changes in the intensity and duration of insolation. Winter, accentuated by the incursion of cold air from the polar regions, distinguishes these climates from the tropics. The convergence of very different airstreams results in the development of actively moving low-pressure systems, which often produce severe storms. These changing atmospheric conditions provide a distinct pattern of summer, fall, winter, and spring. Like the tropics, midlatitude climates can be subdivided on the basis of the distribution of precipitation through the year. In midlatitudes, the wet-and-dry climates are further subdivided into summer-dry and winter-dry regimes. Having the rainy season in the winter in the one case and the summer in the other results in different types of climate.

CLIMATES OF
NORTH AMERICA

Chapter Overview

Chapter 15 outlined the kinds of weather that typify the midlatitudes and identified the major variations in climate in these regions. This chapter examines the climate of North America in greater detail.

Perhaps the most significant aspect of the climate is that the polar front and the westerlies cross the continent. The traveling cyclones that move with the westerlies bring extremely varied weather to much of the continent. Equally significant is the Rocky Mountain cordillera, which extends north and south, nearly at right angles to the westerlies. The coastal ranges and interior ranges reach elevations more than one-third the height of the troposphere. These ranges heavily influence the flow of air over the continent.

Table 16.1 shows the air masses that influence North American weather and their seasonal characteristics, and Figure 16.1 shows the source regions for these airstreams. The data in the table and the location of the source regions point up two important attributes: One, different types of air masses dominate the areas east and west of the continental divide; and two, airflow changes seasonally. Airstreams from the Pacific Ocean dominate the western part of the continent, and airstreams from the Canadian Arctic and the Atlantic Ocean dominate the continent east of the Rocky Mountains.

FIGURE 16.1

Source regions and paths of air masses affecting North America.

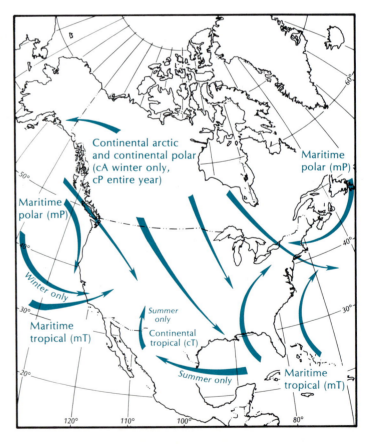

TABLE 16.1

Weather characteristics of North American air masses.

AIR MASS	SOURCE REGION	TEMPERATURE AND MOISTURE CHARACTERISTICS IN SOURCE REGION	STABILITY IN SOURCE REGION	ASSOCIATED WEATHER
cA	Arctic basin and Greenland ice cap	Bitterly cold and very dry in winter	Stable	Cold waves in winter
cP	Interior Canada and Alaska	Very cold and dry in winter Cool and dry in summer	Stable entire year	a. Cold Waves in winter b. Modified to cPk in winter over Great Lakes bringing "lake-effect" snow to leeward shores
mP	North Pacific	Mild (cool) and humid entire year	Unstable in winter Stable in summer	a. Low clouds and showers in winter b. Heavy orographic precipitation on windward side of western mountains in winter c. Low stratus and fog along coast in summer; modified to cP inland
mP	Northwestern Atlantic	Cold and humid in winter Cool and humid in summer	Unstable in winter Stable in summer	a. Occasional "northeaster" in winter b. Occasional periods of clear, cool weather in summer
cT	Northern interior Mexico and southwestern U.S. (summer only)	Hot and dry	Unstable	a. Hot, dry, and clear, rarely influencing areas outside source region b. Occasional drought to southern Great Plains
mT	Gulf of Mexico, Caribbean Sea, western Atlantic	Warm and humid entire year	Unstable entire year	a. In winter it usually becomes mTw moving northward and brings occasional widespread precipitation or advection fog b. In summer, hot and humid conditions, frequent cumulus development and showers or thunderstorms
mT	Subtropical Pacific	Warm and humid entire year	Stable entire year	a. In winter it brings fog, drizzle, and occasional moderate precipitation to N.W. Mexico and S.W. United States b. In summer it occasionally reaches western U.S., providing moisture for infrequent conventional thunderstorms

Source: Frederick K. Lutgens, Edward J. Tarbuck, *The Atmosphere*, 2nd edition, ©1982, p. 203. Reprinted by permission of Prentice-Hall, Inc., Englewood Cliffs, New Jersey.

Climate East of the Continental Divide

Day-to-day and month-to-month changes characterize weather east of the Rocky Mountains. Shifts in the general circulation and the passage of midlatitude cyclones cause most of the changes.

Summer

During summer, the north–south temperature gradient is at its weakest. The westerlies are less intense and shift northward. Figure 16.2 shows the approximate boundary between the airstreams of Pacific origin and those from the Arctic and Atlantic Oceans during July. The southern boundary of the continental polar air stretches across Alaska and northern Canada. The boundary between the air from the Pacific and the Atlantic Oceans extends through the Great Basin of the United States and eastward to the vicinity of Hudson Bay.

East of the continental divide, maritime tropical air flows northward over the continent, bringing warm, humid weather conditions. Monthly data obscure the varied weather systems that occur in each season, and the differences from one year to the next. Some

FIGURE 16.2

July circulation over North America (inches).

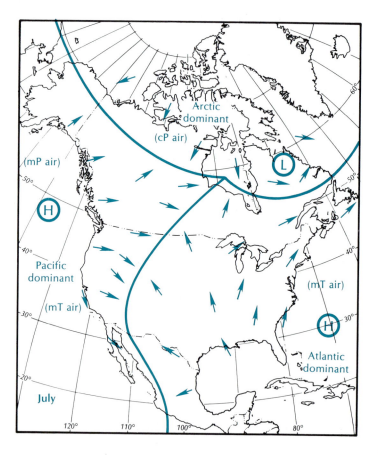

summers are hotter than normal, others are cooler, some are rainier than normal, and some are drier. In any summer, the probability that record high or low temperatures will be set is high. Figures 16.3 and 16.4 provide additional information on the distribution of precipitation and temperature on the continent.

Heat Waves.   Heat waves are periods of clear, dry, and unusually hot weather. They develop when the westerlies move unusually far north, permitting the subtropical high to

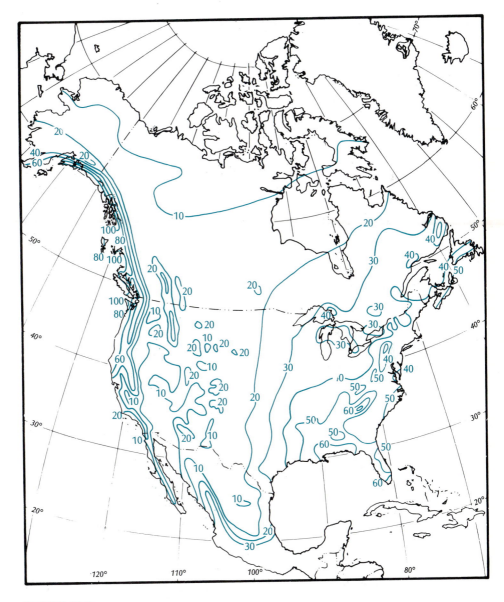

FIGURE 16.3
Distribution of precipitation in North America. Precipitation in inches.

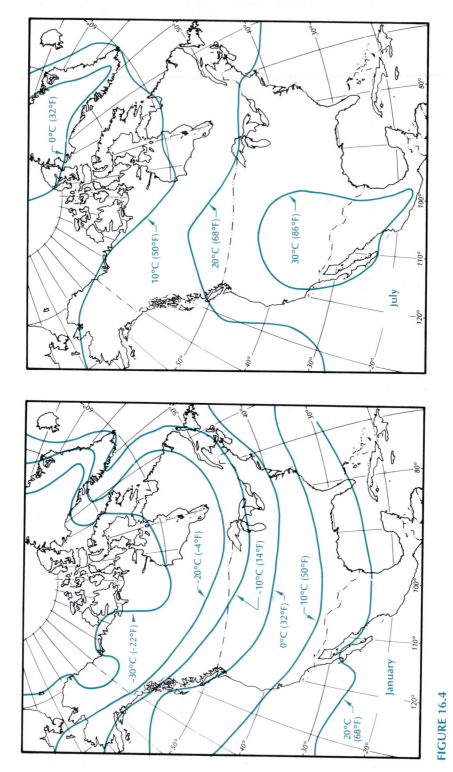

FIGURE 16.4
Mean January and July temperatures over North America.

expand and cT air to displace the mT air. Temperature rises and humidity drops. Heat waves occur nearly every summer someplace in North America. Major heat waves took place in 1830, 1860, and 1901. The worst period for heat waves was during the drought years of the 1930s. The hottest of those years, and the worst for fatalities, was 1936. Temperatures reached 43°C (109°F) or higher in Nebraska, Louisiana, Minnesota, Wisconsin, Michigan, Indiana, Pennsylvania, West Virginia, New Jersey, and Maryland. Highs of 49°C (120°F) or more occurred in the Great Plains states of North Dakota, Kansas, Oklahoma, Arkansas, and Texas. Heat-related fatalities approached 15,000 in that summer.

A heat wave in the summer of 1980 took at least 1265 lives, nearly seven times the number of an average year. Most of those who died were the elderly or poor who lived in non-air conditioned housing. Over 25% of the deaths occurred in Missouri. The heat wave began in southwest Texas when temperatures topped the 38°C (100°F) mark. By the middle of July, most of the central third of the nation was experiencing afternoon temperatures in excess of 38°C (100°F). The hot, dry weather spread over all of the eastern United States until the first week in September. Dallas had temperatures of 38°C (100°F) or more every day from June 23 through August 3. On July 13, several cities reported their highest recorded temperatures: Augusta, GA, had 42°C (107°F); Atlanta had 41°C (105°F), and Memphis recorded 42.2°C (108°F). The heat wave caused hundreds of kilometers of concrete pavement to buckle, often exploding violently along seams and cracks. Asphalt roads softened as the material reached temperatures of more than 66°C (150°F). Trucks moving on the soft pavement destroyed many kilometers of road surface.

Drought in the Great Plains. The Great Plains area of North America suffers recurring droughts. Evidence shows that between 1825 and 1865, rainfall was low in many parts of North America. During this period the Great Plains were being explored but were not yet settled. Pioneer wagon trains ground tracks into the sediments on the floor of some western lakes in the 1840s when the lakes were dry. The lakes refilled and covered the route until the turn of the century. Probably the driest year in the plains states since settlement began was 1860, when Kansas, Missouri, Iowa, Minnesota, Wisconsin, and Indiana were all affected. A less-severe drought struck the same area again in 1863–64.

A severe drought peaked in the fall of 1881, spreading over all of the United States and southern Canada east of the Mississippi River system. Many of the wells, cisterns, and springs that had never been dry before now failed. Lack of water for steam delayed freight trains. The water supply of New York City dried up. Again, in the middle of the 1880s, rainfall diminished, culminating in the drought of 1894 and 1895. This was perhaps the most-intense and widespread drought experienced since colonization. Many of the settlers of the Great Plains left their land. Most moved westward, looking for more favorable farming sites. Wetter years returned to the plains and people quickly forgot the drought problem, with new settlers moving into the plains states in large numbers. Another drought around 1910 drove off many of these later settlers.

Following World War I, a third wave of farmers moved into the Great Plains on the heels of some wetter-than-normal years. The dry years of 1930s spelled real disaster for the land and for the socioeconomic structure of the Great Plains. The drought of the 1930s was not of equal intensity over the entire plains or through the decade. It was a dry period for most of the United States and southern Canada, with the drought spread over the northern and central plains and the northern Rockies. The core areas were in Kansas and the Dakotas. The first period, from mid-1933 to early 1935, was the worst. The dust began to blow during this time, with topsoil blowing or washing away by the billions of tons. The dust storms of 1934 and 1935 demonstrated to people as far east as the Atlantic coast that severe problems existed in the West. By 1935, an estimated 80% of the land in the Great Plains was suffering from erosion. An estimated 150,000 people moved out of the plains states. John Steinbeck's novel, *The Grapes of Wrath*, dramatized the drought's impact. Two more periods of intense drought occurred in 1936 and 1939–40. The 1936 period was very intense, but short; the later spell was long, but less intense.

Drought struck the Great Plains again in the 1950s. The flow of some perennial streams decreased to nearly half their normal volume, and salinity tripled. Wind erosion was again a problem, but not so severe as in the 1930s. In 1957, an estimated 13,000 km^2 (3.2 million acres) of land was damaged in the Great Plains. Another 118,000 km^2 (29.2 million acres) was susceptible to blowing.

Winter

During winter, both the Pacific and Atlantic high-pressure systems move equatorward. This considerably reduces the influence of the subtropical high over the Atlantic from what it is in the summer. The flow of air from the Pacific Ocean penetrates further into the continent, creating a broad zone of mixing cP, mT, and mP air through the Great Lakes (Figure 16.5)

Most areas in the Northern Hemisphere north of 30° experience snow at one time or another. In North America, snow covers the majority of the land for 2–6 months each winter. Higher elevations sustain snow longer, sometimes for the entire year. In general, snowfall frequency increases with latitude, although the amount of snow that falls and latitude have no direct correlation. In the United States, the mean annual snowfall varies from a trace along the Gulf of Mexico to an average of 1 m (3.3 ft) or more near the Great Lakes (Table 16.2). Total annual snowfall does not increase north of the latitude of the Great Lakes. Areas with the coldest weather have less precipitation and thus less snowfall. Snow is possible at all temperatures found in the lower atmosphere, but below −20°C (4°F) the chance of getting much snow is very low because the atmosphere simply does not hold much moisture. The heaviest snowfalls occur when temperatures are between 4° and −4°C (39° and 25°F). To produce heavy snowfall, temperatures must be as warm as possible to have abundant precipitable water, but cold enough for the precipitation to be in the form of snow. Those conditions explain why very heavy snowfalls often occur near the beginning and end of winter. In the fall, measurable amounts of snowfall begin in late November. In the Great Lakes region it often snows during the Thanksgiving holiday, when the probability for heavy snowfall is high. The cP air masses are becoming quite cold and more frequent, and yet warm, moist mT air is still flowing northward. One of the earliest heavy snowstorms to move across the United States was November 17–18, 1980. The early storm blocked interstate highways in New Mexico and Texas with drifts up 1.25 m (4 ft) high. The storm dumped snow in a band all the way from New Mexico to New England. Snow occurring this early in the year does not remain on the ground very long. Soil temperatures are still high and the snow quickly

FIGURE 16.5
January circulation over North America.

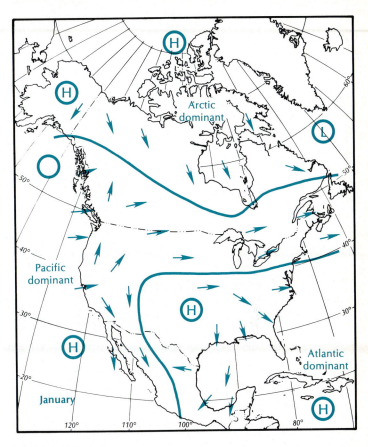

melts. These fall storms have brought disaster to shipping on the Great Lakes. On November 8, 1913, an early fall storm crossed lakes Superior, Michigan, and Huron, sinking nearly a dozen ships and badly damaging more than 50 others. On November 10, 1975, the Edmund Fitzgerald went down in an unexpected storm on Lake Superior.

The same kind of phenomenon occurs again in the spring. Warm maritime tropical air begins to move northward, and very cold masses of continental polar air move south; when mixed, heavy snowfalls can result. Macon, GA, received 587 mm (23 in.) of snow from a spring storm in 1973. The blizzard of 1888 in New England took place on March 11–14. In 1978, the major blizzard of the winter in the midwest was on March 9 and 10.

Winter storms produce disasters someplace almost every year. The two most severe winter storms are the blizzard and the cold wave, and it is not uncommon for a cold wave to follow a blizzard. Most winter storms develop from disturbances along the polar front and the activity resulting from mixing cold dry air (cP) and warm moist air (mT). Many low-pressure systems develop or redevelop along the east side of the Rocky Mountains anywhere from New Mexico north into Alberta. They move eastward, with most traveling over the Great Lakes and out over New England or the St. Lawrence depression.

Blizzards.   The term **blizzard** likely comes from the German word *blitz*, meaning lightning. In North America the term has come to mean any sudden, violent force or

TABLE 16.2

Snowfall records in the Great Lakes Basin.

STORM TOTALS

1 Day		
	Bennett Bridge, NY	1.3 m (51 in.) Jan. 17, 1959
	Buffalo, NY	1.2 m (48 in.) Dec. 10, 1937
5 Day		
	Oswego, NY	2.56 m (101 in.) Jan. 27–31, 1966

ANNUAL TOTALS

Old Forge, NY	10.37 m (408 in.) 1976–77
Steep Hill Falls, Ontario	7.65 m (307 in.) 1939–40
Herman, MI	7.83 m (308 in.) 1975–76
Tahquamenon Falls, MI	8.45 m (333 in.) 1976–77
Bennett Bridge, NY	8.94 m (352 in.) 1946–47
Hooker, NY	11.86 m (467 in.) 1976–77

onslaught. During the Civil War it was used to describe a heavy enemy volley of musketry. In the spring of 1870, a newspaper in Estherville, IA, used it to describe a severe winter snowstorm accompanied by high winds. The use of the term spread, and by the severe winter of 1880–81 it had gained general usage throughout the continent. Today, NOAA defines a blizzard as a storm with winds of at least 56 km per hr (35 mph) and temperatures below −6.7°C (20°F), combined with enough blowing or falling snow to reduce visibility to less than 0.4 km (0.25 mi). A severe blizzard is one in which wind velocities equal or exceed 72.5 km per hr (45 mph), temperatures drop below −12.2°C (10°F), and visibility is near zero. Many severe blizzards have occurred over the years. In 1888, two substantial blizzards affected large but different areas of the continent. The first took place on January 11–13 and paralyzed the Great Plains from Alberta south to Texas. The second was a late storm, striking the eastern seaboard from Maine to Chesapeake Bay March 11–14. An average of 1.02 m (40 in.) of snow fell over New England and southeastern New York State. In the space of just a few hours, the storm crippled New York City and other cities in the region. Snow drifts in Herald Square in New York climbed 9 m (29.7 ft). Transportation ground to a halt: Horse-drawn vehicles couldn't move, and railroads suspended operations because trains couldn't penetrate the drifts. Telegraph communications were knocked out, and mail to and from the city nearly stopped. A food panic threatened to develop in some areas. Fires in New York City became virtual firestorms, fed by high winds; firemen could not respond because they could not get equipment through the snow. The accompanying cold was so intense with the wind chill that the East River froze. Many people crossed back and forth over the ice between Manhattan and Brooklyn. Sparrows died by the tens of thousands, many of them frozen to tree limbs. Cattle froze to death, some of them freezing solid while standing. Human deaths from the storm reached 400, with 200 deaths in New York City alone.

Lake-Effect Snow.   As noted in Chapter 10, regions that receive unusually large amounts of snowfall are the lee sides of the Great Lakes, which contain distinct snow belts associated with the lakes. These snow belts are the result of the winter winds picking up moisture as they cross the lakes. The lakes remain unfrozen far into the winter, and in some winters none of them freeze over. Whenever the air flowing over the lakes is colder than the water beneath, large amounts of water evaporate into the air. As the air streams over the land on the downwind side of the lake, it cools, and the moisture precipitates out as snow. The greater the difference in the temperature of the air and the water, the greater the evaporation will be and the more likely a lake-effect snowfall will result. When a lake freezes, it shuts off the supply of water and lake-effect snow declines. Lake-effect areas receive much more snow than the surrounding region: The Keweenaw Peninsula of Upper Michigan receives snowfall averaging more than 5 m (195 in.) a year, for example, and several locations have recorded more than 10 m (390 in.) of snow in a given winter (Table 16.2). The most snow officially recorded east of the Rocky Mountains was at Hooker, NY, during the 1976–77 season, when 11.86 m (467 in.) of snow fell. These snow belts only extend about 45–50 km (25–30 mi) away from the lakes.

Climate of the West Coast

Different air masses dominate the weather of western North America than do eastern North America. Relatively dry subsiding air from the subtropical high and maritime polar air from the Pacific Ocean dominate the weather in the west.

Summer

During summer, the westerlies are farther north and relatively weak. The coastal states of California and Oregon are under the influence of the subtropical high and experience clear, dry weather. During some years, no measurable precipitation has fallen from California to Washington in June, July, or August. Even Vancouver, British Columbia, has experienced a July with no measurable precipitation. Along the immediate coast, fog is common even though rain is not. The cause of the fog is discussed in more detail later in this chapter.

Summer days are long, solar radiation is intense, and temperatures soar. On July 18, 1982, for example, temperatures topped 32°C (90°F) in each of the 17 western states, exceeding 38°C (100°F) in Nevada, Arizona, and New Mexico.

In California, heat waves take one of two forms. The first happens when the subtropical high is displaced northwestward, producing tropical desert conditions for long periods. The second type comes from the hot, dry Santa Ana winds, which flow down from the high plateaus in the summer. Also called the red winds for the large amounts of dust they carry, the Santa Anas result from the development of a high-pressure system over the Nevada desert. The air, heated by subsidence and the hot desert surface, flows outward from the desert. As air currents descend into the Los Angeles Basin, they heat still more by compression. Most frequent in September and October, they bring temperatures into the 38°–43°C (100°–110°F) range. Often the winds are dust-laden, as in November of 1969, when winds reached velocities of 122 km per hr (78 mph) through the passes. During one Santa Ana, the air temperature was 13.9°C (57°F) and the dew point was −23°C (−30°F). Such a desiccating wind can raise the fire hazard in mountain forests to a dangerous level. The sudden change in weather and the extreme dryness of the air often make these weather events—and the people who experience them—highly disagreeable.

Fall weather often alternates between fires and floods. The natural vegetation is a scrub grassland, which is explosively dry after the long, hot summer. Dry thunderstorms at the onset of the rainy season, human carelessness, and arson trigger wild fires that destroy large areas of vegetal cover every year. The first fall rains are often torrential. When the cyclonic storms first move onto the land surface, the land is extremely warm. Additional heat from the ground makes the air more unstable as it moves onshore. Where fires have raged, the unprotected soil turns into mud. Landslides and mudflows destroy much valuable property and leave behind sterile hill slopes. Nearly every year in California, residential areas are either incinerated, buried in sediment, or simply carried off down the hillsides in a mass of mud, kindling, and plasterboard.

Winter

A winter concentration of precipitation distinguishes the west coast of the continent, from southern California into Canada. Cities in California typically receive an average of 84% of the annual precipitation in the winter, and more than half of the annual total falls during December, January, and February. Farther north along the coast, the precipitation increases. Most of the increase occurs in the spring and fall and is due to the longer season in which the westerlies flow over the coast. Oregon averages only 73% of its annual total in the six winter months. Even Prince Rupert, British Columbia, at 54° N, receives 62% of its annual total of 2400 mm (100 in.) in the six winter months. This winter maximum extends inland into the Great Basin of the United States and into the interior valleys of western Canada. While the annual total tends to increase poleward to about 50°, the actual amount received at any given location depends on site characteristics. At some places, the totals run fairly high. Monumental, CA, receives an average of 3.91 m (154 in.) per year. A single heavy rain can produce 24-hour totals of 250 mm (10 in.) or more, enough to produce major flooding.

Since most precipitation falls in the winter, much of it takes the form of snow. The amount varies from a trace at San Diego to 11.5 m (449 in.) at Tamarack, in Alpine County, CA. The heaviest snowfalls occur on the mountains' southwestern slopes. Like precipitation, the amount of snowfall varies from one winter to the next, particularly in the southern mountains. Throughout this region snow is a major source (if not *the* major source) of water for irrigation and power.

In North America, the areas with most snowfall are the mountain areas of the far west. The mean permanent snowline in mountains in the humid tropics is about 4700 m (15,400 ft). The snowline increases slightly in a poleward direction to 5200 m (17,000 ft) near 30° N. From there it drops to near 3000 m (9800 ft) at 45° N latitude, and about 1400 m (4500 ft) at 60° N. Actual height of the snowline on a given mountain depends on its orientation to the sun and to the wind direction. The snowiest location in North America is Paradise Ranger Station on Mount Rainier in Washington. At an elevation of 1692 m (5550 ft) and facing the Pacific winds, snowfall there averages more than 15 m (49.5 ft) a year. In the 1971–72 season, a phenomenal 28.5 m (1122 in.) of snow fell.

When the winter rains and snows fail, the west experiences drought. Such winters occur intermittently. One such winter happened in 1976–77, when a ridge of high pressure extended northward along the west coast of North America, interrupting the usual zonal flow of air across the continent and having drastic effects on the precipitation of the west. Figure 16.6 shows the circulation at the 700-mb level for January in the four years from 1975 to 1978. Between 1975 and 1977, the pattern of upper-air circulation had shifted over the continent in such a way that a high-pressure ridge extended from California to Alaska. This pattern shows on the charts as the poleward bend in the height

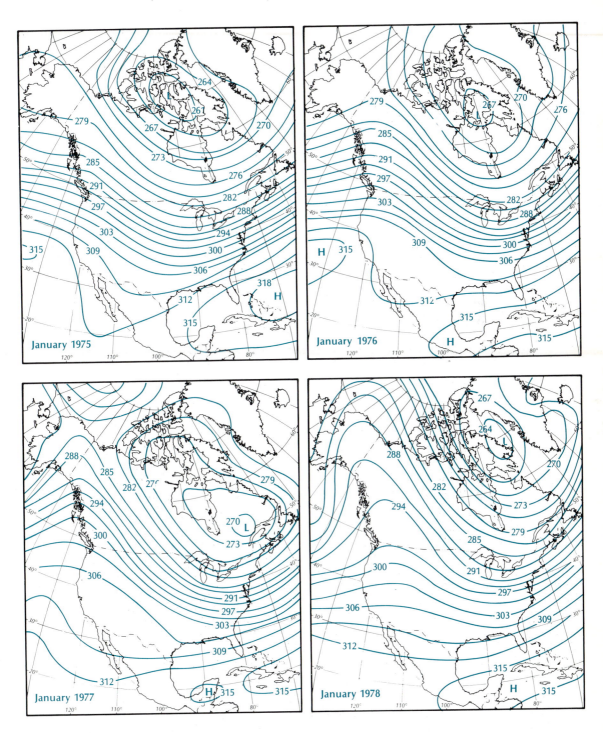

FIGURE 16.6
Mean 700-mb height contours for January 1975–78. Heights in meters.

of the 700-mb level over the west coast. This ridge blocked the onshore movement of mT air enough to bring severe drought to California and southern Oregon. In January 1978, the circulation reverted to a more zonal flow over the continent.

Effects of the Mountains on Local Climate

The coastal states and provinces in the west have a wide range of local climatic conditions resulting from the influence of the mountains on temperature and rainfall. The westerly winds from the Pacific Ocean produce a truly marine climate, but the coastal ranges restrict the marine climate to a very narrow zone. The climate on the lee sides of the coast ranges is different from that on the windward sides. Because of the mountains, some of the wettest *and* driest areas of North America occur in British Columbia and the Pacific states.

Temperatures depend on an area's proximity to the open ocean. Coastal sites have cooler temperatures in the summer and warmer temperatures in the winter than do sites inland. The cold currents that flow offshore help keep the summers cool. San Francisco has a July average of 17°C (63°F); Sacramento, farther inland in the Central Valley, has a July average of 25°C (77°F). The extremes of temperature at the coastal stations are also moderate for the latitude, reflecting the influence of the onshore flow of air. Fog, a frequent phenomenon along the coasts, results from the cold current stabilizing the moist, oceanic air. The fog of San Francisco Bay is well-known, but the phenomenon is widespread along the west coast of North America as well as in other areas with this type of climate.

Coastal California and Oregon have small daily and annual ranges in temperature, and subfreezing weather is infrequent. The frost-free growing season—365 days along the southern coast—declines northward. Growth of citrus fruit as far north as 40° attests to the mildness of the winters on the coast. The leeward locations are drier and hence have a greater range in temperature. The range between January and July at North Head, WA, is 8.4°C (15°F). Inland a few kilometers at Vancouver, WA, it is 15.5°C (28°F); and at Kennewick, WA, it is 23.9°C (43°F). Because of the ocean, January averages from 45° to 50° N are up to 15°C (27°F) warmer on the west coast than on the east coast of the continent. Port Hardy, British Columbia, averages 2.4°C (36°F) in January, and September Isles on the St. Lawrence River averages −12.7°C (10°F).

West of the coast ranges, and to a large extent west of the Sierra Nevada and Cascade Mountains, severe thunderstorms and tornadoes are infrequent. The primary factor in the development of these storms, the cold front, has less steep a gradient than farther inland, since the mP air is relatively warm. Thunderstorms increase away from the coast and at higher elevations.

■ Applied Study

Cold and Human Physiology

Cold climates pose a distinctive problem to humans, placing undue stress on the body. Humans are warm-blooded animals whose mean body temperature is 37°C (98.6°F). Maintaining this body temperature requires a balance between heat losses and heat gains. Body *metabolism* relates the intake of food (for production of energy) to energy expended by bodily activities. The heat produced by a resting person is about 50 kcal per hr per cm^2 of body surface. This amount of energy is designated the MET. Varying activities are expressed in terms of MET:

Sleeping: 0.0 METs

Awake, resting: 1.0 METs

Standing: 1.5 METs

Walking, on level: 3.0 METs

Short sprint: 10.0 METs

Besides the heat produced by such activities, the body also gains heat through absorption and conduction of energy from the environment. Losses of energy occur through outward radiation, conduction, and evaporation of moisture from the skin. In normal body function, the losses and gains balance.

$$M + R + C - E = 0$$
where
M = metabolic heat
R = radiation
C = conduction
E = evaporation

If the body loses more heat than it gains, the equation is no longer balanced and body temperature drops. If it gains more than it loses, body temperature rises.

The heat lost from the body depends on a variety of factors, such as the amount of the body covered by clothing, the thickness of the clothing, and the amount of physical activity. Normally in cold weather only the face or the face and hands are exposed, representing about some 3%–10% of the body's surface area. When the legs are uncovered, however, the exposed surface increases to 30% or more. An average layer of winter clothing has a resistance to heat loss of about 1 cal per m^2 per sec compared to .5 cal per m^2 per sec for gloves or shoes. Bare skin, of course, has no resistance. An additional source of heat loss is through the lungs: it is very difficult to prevent this kind of heat loss. In fact, as much as 20% of the body's total heat loss may occur through respiration. Some of the heat loss is caused by the evaporation of water in the lungs and some by direct heating of the air. The cold air that is breathed in is heated to body temperature and the humidity is increased to near the saturation level.

A combination of factors influences the exchanges that occur. For example, low temperature and air movement affect the loss of body heat. The *wind chill factor* expresses this relationship. This chilling effect is the result of the removal of a microlayer of warm air generated by body metabolism. If wind removes the layer, more body heat is required to replace it. If the process of removal continues, the loss of the heated layer causes the sensation of cold air at the skin surface. The more rapid the heat loss, the colder it feels. The rate of heat loss is a function of wind speed. This heat loss can be expressed scientifically as calories per unit area per unit time. Because the public does not easily understand chilling expressed in energy units, it is expressed in terms of equivalent temperatures. Equivalent temperatures are the temperature that one would perceive under

Wind speed (mph)	Actual thermometer reading (°F)											
	50	40	30	20	10	0	-10	-20	-30	-40	-50	-60
	Equivalent temperature (°F)											
	50	40	30	20	10	0	-10	-20	-30	-40	-50	-60
5	48	37	27	16	6	-5	-15	-26	-36	-47	-57	-68
10	40	28	16	4	-9	-21	-33	-46	-58	-70	-83	-95
15	36	22	9	-5	-18	-36	-45	-58	-72	-85	-99	-112
20	32	18	4	-10	-25	-39	-53	-67	-82	-90	-110	-124
25	30	16	0	-15	-29	-44	-59	-74	-88	-104	-118	-133
30	28	13	-2	-18	-33	-48	-63	-79	-94	-109	-125	-140
35	27	11	-4	-20	-35	-49	-67	-82	-98	-113	-129	-145
40	26	10	-6	-21	-37	-53	-69	-85	-100	-116	-132	-148
Wind speeds greater than 40 mph have little additional effect	**Little danger** (for properly clothed person)				**Increasing danger**				**Great danger**			

FIGURE 1

Wind-chill equivalent temperatures.

a given temperature and wind speed. For example, if the temperature is 8°C (25°F) and the wind speed is 40 kph (25 mph), the result is the equivalent to what would occur at a still air temperature of 19°C (−2°F) (Figure 1).

In calculating the effective temperature due to wind chill, it is necessary to take into account that the wind speeds people normally experience are less than those reported at weather stations. Wind velocity is measured at a height of 10 m (33 ft) at North American weather stations. Wind velocity increases rapidly from the ground surface upward, nearly doubling at 10 m. At head height for a person 168 cm (5.5 ft) tall, the wind velocity is only about 57% of reported velocity. An exception to this rule is when wind is reported calm. A person walking 5 km per hr (3 mph) has an effective wind speed of 5 km per hr.

Response mechanisms of the body come into effect when an imbalance occurs. Cold conditions reduce blood flow to the hands and feet to minimize heat loss. However, reduced blood flow to the surface can result in *frostbite*, the freezing of tissue and cell destruction. Shivering is another response to cold. Its function is to increase the metabolic rate of the body. When shivering occurs, the transfer of blood to the surface layers of the body increases the loss of heat by conduction and radiation. This reaction leads to more shivering, and the body temperature falls rapidly.

When the body temperature falls to a level where its mechanisms have no further warming effect, hypothermia occurs. Failure of the thermoregulatory system occurs at or about 33°C (91°F) and death at near 25°C (77°F). The writings of Arctic explorers include many accounts of suffering from the cold as do accounts from people of more temperate lands who had the misfortune to be exposed to severely cold weather. ■

Summary

The climate over North America varies from place to place, depending upon the frequency with which mT, mP, cT, or cP air is present. West of the Rocky Mountains, the weather is that of either mP air or stable mT air. East of the Rocky Mountains, the most frequent air streams are cP and mT. The result is that west of the mountain ranges, most precipitation comes from mP air and falls in the winter. East of the mountains, most precipitation comes from mT air in the summer or throughout the year. Seasons change as one type of airstream replaces another, and seasonal changes in the different airstreams increase the variety of weather. The region that experiences the most extreme temperature and precipitation is the winter-dry region of the Great Plains. Here, no mountains obstruct the flow of air between the Canadian Arctic and the Gulf of Mexico. In winter, extremely cold, arctic air occasionally flows south as far as Mexico. In summer, warm air from the Gulf of Mexico or the high plains travels north well into Canada. The weather changes so rapidly in the Great Plains that an oft-repeated phrase is: "If you don't like the weather, just wait for a moment."

POLAR AND HIGHLAND CLIMATES

Chapter Overview

The distinguishing feature of polar climates is cold weather brought about by low levels of solar radiation. The radiation balance differs from that of the tropics and midlatitudes in both the diurnal and annual pattern of solar energy. In the tropics, the major energy change is from day to night, and daily changes in radiant energy and temperature are greater than from season to season. The midlatitudes have large fluctuations in both the diurnal and annual periods. The closer a location is to one of the poles, the larger the seasonal fluctuation in radiation and temperature relative to the diurnal changes. Like the tropics, polar climates are dominated by a single periodic fluctuation in energy. In polar areas, the annual fluctuation, not the diurnal, is most important (Figure 17.1).

Each place on Earth's surface receives nearly the same number of hours of sunlight for the entire year. The distribution throughout the year, however, varies a great deal from place to place. In the tropics, the radiation is evenly spread throughout the year, with about 12 hours of insolation each day. At latitudes higher than the Arctic and Antarctic circles, most of the energy comes in one continuous dose during the summer half-year. These areas receive very little energy during the other half-year. At the time of the summer solstice in the Northern Hemisphere (June 22), the north polar axis tilts toward the sun 23.5°. This is the longest day of the year in the Northern Hemisphere. On this date the sun does not drop below the horizon in the zone north of 66.5° N. At Murmansk, Russia (69° N), the continual daylight of summer lasts for 70 days. Spitzbergen, Norway, (80° N) has 163 days of continuous light, and the North Pole has 189 days of continuous daylight.

At the time of the fall equinox, days are 12 hours long everywhere. The vertical rays of the sun are over the equator, and when sunset is occurring at the North Pole, sunrise is occurring at the South Pole. The equinoxes are far more important to the polar regions than are the solstices, because they mark major times of change in energy. The equinoxes represent times when solar radiation either becomes significant or drops to near zero.

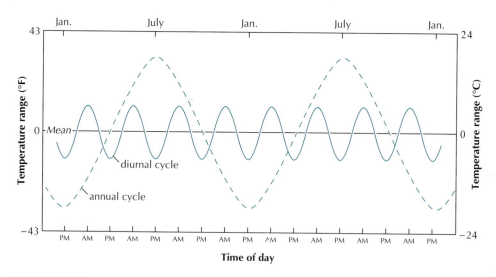

FIGURE 17.1
Relationship between the annual and diurnal energy cycles in polar climates.

At the time of the winter solstice, the days in the Northern Hemisphere are the shortest. The North Pole is in the midst of a single period of darkness lasting 176 days. At Spitzbergen, the night is 150 days long, and at Murmansk, continuous darkness lasts for 55 days, from November 26 through January 20.

The winter months in polar regions may have no direct solar radiation. Some radiation, including some in the visible range, does continue to find its way to the surface so that total darkness seldom exists. When the sun is below the horizon, refraction and scattering of the rays bring some sunlight to the surface until the sun is 18° below the horizon (astronomical twilight). Thus, some latitudes may have several months with no direct sunlight but almost continuous twilight. The stars, moon, and auroras are also sources of light for the poles, so the darkness is not as intense as it might be.

Another important element in the polar climates is that the intensity of radiation is never very high for any length of time when compared to the midlatitude or tropical regions. Even on the equatorward margins, the sun rarely reaches more than 45° above the horizon. In the summer months, however, although its intensity is low, the duration of daylight is long, with more than 20 hours of daylight near the summer solstice. Although summer insolation persists for many hours a day, temperatures do not rise very much because radiation intensity is low and much of the energy goes to melt snow and ice.

While the two polar areas have some similarities, several basic differences influence the climate of each. One important and obvious difference between the Arctic and the Antarctic: The Antarctic is a relatively large, high land mass surrounded by water, and the Arctic is a sea surrounded by land. The relationship of the land–sea interface reverses at the two poles, which is very significant to the climate of the two areas. The three macroclimates of the polar regions are polar wet, wet-and-dry, and dry (Figure 17.2). The polar wet-and-dry climate exists primarily in the Northern Hemisphere, and the other two climates occur mainly in the Southern Hemisphere.

The Arctic Basin

The Arctic Ocean consists of all the sea surface north of the Arctic Circle, whether covered by ice or open water (Figure 17.3). The central part of the Arctic Ocean is frozen most of the year but does have occasional open leads (Figure 17.4). The central ice pack covers 11.7 million km² (4.5 million mi²) in winter and melts back to 7.8 million km² (3 million mi²) in the warmest months. The ice drifts from east to west. In the process of moving, leads open and close, so some unfrozen areas are always present. The ice ranges in age up to 20 years and in thickness up to 5 m (16.5 ft), with the oldest ice being the thickest. In the summer, the ice melts away from the edge of North America and Eurasia so the ocean may be mainly open water with patches of drifting ice for up to two months. Because of heat from the Gulf stream, the Norwegian Sea and Barents Sea remain open all year at latitudes where the sea is frozen in other areas.

The earth–space energy balance in the polar areas makes the poles major energy sinks. The polar regions emit much more energy than they receive from solar radiation. In the Arctic Basin, radiation reflected and radiated away as long-wave radiation is 60% greater than direct solar radiation. For this to happen, a supplementary heat source must make up

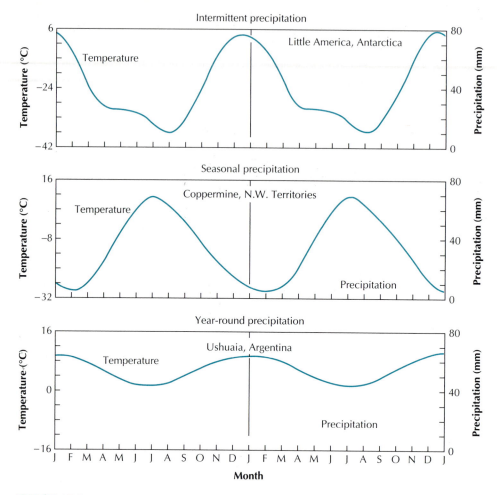

FIGURE 17.2

The polar environments: All three environments have a low solar energy input year-round. The heat from atmospheric water and the sea keeps the average temperature at Ushuaia above freezing.

the difference. This energy source is heat carried poleward by air and water currents. This steady flow of energy poleward also maintains Arctic temperatures at a level much higher than insolation would allow (Figure 17.5). The primary source of heat for the Arctic Basin is inflow from the sea and the atmosphere. This energy is nearly double that of solar radiation absorbed at the surface. Warm ocean currents move much of this heat into the Arctic, releasing heat to the atmosphere by radiation, conduction, and evaporation. The amount of open water compared to ice is of extreme importance, since the water surface transfers over 100 times more heat to the atmosphere than an ice surface does.

Seawater acts differently from fresh water as it cools toward the freezing point. In fresh water, the maximum density occurs at 4°C (39°F) and fusion takes place at 0°C (32°F). In seawater, the maximum density happens just near the fusing point—about

FIGURE 17.3
Map of the Arctic Basin showing the average location of the ice pack in March and August.

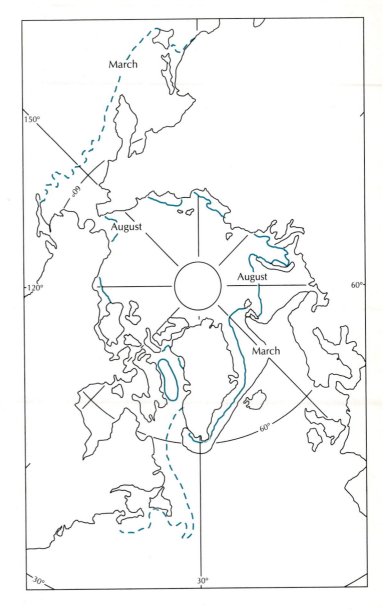

−2°C (28°F), depending on the salinity. One result of this phenomenon is that surface water in polar seas may cool up to 6°C (11°F) more than it would if it were fresh water. The colder polar seawater is denser than either fresh water or warmer seawater. As it chills to near the freezing point, it sinks to the bottom and spreads out, settling in the deeper parts of the ocean basins. This settling cold water helps promote circulation in the ocean, which in turn helps equalize the energy between the equator and poles. This fact also explains why most ocean water is so cold, averaging about 3.5°C (38°F). Warm surface water in tropical regions and along beaches in the midlatitudes is the exception, not the rule.

FIGURE 17.4
Arctic ice pack showing open leads.
Source: Stephen J. Krasemann/DRK Photo.

The atmosphere over the Arctic Basin never becomes really cold because the water beneath the ice and in open areas does not drop below −1.6°C (29°F). The sea, even when frozen, modifies atmospheric temperatures. Whenever the air temperature over the ice drops below −1.6°C (29°F), heat is transferred through the ice and from the open

FIGURE 17.5
Energy balance of the Arctic Basin.

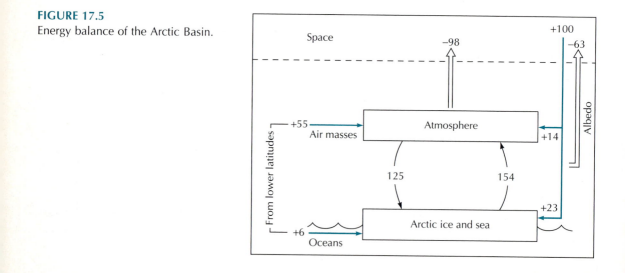

water. Over much of the Arctic Basin, temperatures over the ice range between −20°C (4°F) and −40°C (−40°F), with the lowest temperature yet recorded near the surface being −50°C (−58°F). The Arctic Ocean's heat causes an abrupt atmospheric change along the coastline.

Polar Lows and Arctic Hurricanes

Intense, short-lived low-pressure systems have long been observed along the edges of the continents in the North Pacific and North Atlantic Oceans. Tor Bergeron used the term *extratropical hurricane* in 1954 to describe severe storms in the Arctic. These polar lows are small compared to midlatitude lows. The polar lows are less than 1000 km in diameter; sometimes, a fully developed system is only 300 km across. Wind speeds may be sustained at velocities of 40 knots, with gusts to 60 knots. The central pressure drops to 970 mb or less. The same type of low forms around the Antarctic continent year-round.

In some respects these storms are similar to tropical hurricanes. They are circular, have a relatively clear eye, and sustain high wind velocities. They also have symmetrical spiral bands of deep cumulus clouds. Like tropical cyclones, polar hurricanes produce an outward flow of air aloft and dissipate rapidly when they reach land. Like tropical hurricanes, polar hurricanes reach to the tropopause. The outflow of moist air aloft produces a broad, thick layer of cirrus clouds.

Polar hurricanes differ from tropical cyclones in important respects as well. They develop to full strength in 12–24 hours and may have a total life span of only 36–48 hours before they strike land. They form under very different conditions than tropical lows, in cold air north of the polar front. They travel at speeds up to 30 knots, nearly twice as fast as the average tropical hurricane, and bring gale-force winds, heavy snow, and sleet. In the Northern Hemisphere, they form near the ice pack and move eastward toward land. They form in places of extreme differences in the merging airstreams along an arctic front, and where a low-pressure disturbance exists along the edge of the ice pack. This low draws in the cold air from the ice sheet and relatively warm, moist air from the subpolar ocean. Very cold air develops over the pack ice in winter, with temperatures as low as −40°C (−40°F). The temperature contrast between the air masses may be more than 40°C (72°F). The sea-surface flux of heat plays a major role in intensifying and maintaining the mature storm.

The Polar Wet-and-Dry Climate

The polar wet-and-dry climate, found primarily along the shores of the Arctic Ocean, has cold winters, cool summers, and a summer rainfall regime. Specific areas experiencing this climate are the North American Arctic coast, Iceland, Spitzbergen, coastal Greenland, the Arctic coast of Eurasia, and the Southern Hemisphere islands of Macquarie, Kerguelen, and South Georgia. The land areas with this climate represent only about 5% of Earth's land surface (Table 17.1).

A sample of five tundra weather stations in the Northern Hemisphere shows a January average temperature of −30°C (−22°F). Mean annual temperatures are below freezing, but summer temperatures go above freezing at least for a few weeks. The summer period, if measured by above-freezing temperatures, averages 2–3 months but may be as long as

TABLE 17.1
Climates with seasonal mP and cP air.

| Köppen type | ET |
| Thornthwaite type | E |

5 months. Mean temperatures in the warmer months seldom go above 5°C (41°F). High temperatures range from 21° to 26°C (70°–79°F).

Pressure gradients are weak in the tundra, so it is not a stormy region. In fact, it is one of the least-stormy areas of Earth's surface. High-pressure systems with cold, still, dry air are frequent, particularly in winter. The North American Arctic is even less stormy than other areas. Because of the mountainous character of Alaska, the cyclones of the Pacific Ocean do not move inland very far or very often. Spring and fall are the seasons with the most-active weather, as in most high-latitude locations. Winds are light in velocity and variable in direction. The average velocity is less than 8 knots—less than 5 knots in the Canadian tundra.

Atmospheric humidity is high during the summer months, averaging 40–80%. During the winter it is less, averaging 40–60%. Annual precipitation averages less than 250 mm (10 in.) for most stations; more than three-fourths of the annual total falls during the summer half-year. Winter snowfall is only about 50 mm (2 in.) of water, because the snow is dry and compact. A deep snow cover is absent over most of the tundra, as winter snow is not excessive and tends to drift easily. Thus, exposed areas are often free of snow, although the sheltered areas along streams may have a deep snow cover. Cloudiness varies through the year in the tundra. In the summer months, cloud cover may average 80%, but during the winter it drops to around 40%.

The Antarctic Continent

The Antarctic is not only the coldest area on Earth's surface, but the least-known area. Ice covers nearly 98% of the Antarctic land mass, yet it receives so little precipitation that it qualifies as a desert (Table 17.2). The ice sheet is comparable to the tropical deserts in amount of precipitation. The extremely cold air is incapable of holding much water vapor. Average annual precipitation probably does not exceed 200 mm (8 in.) any place on the ice cap (see Plate 17).

Antarctica is about 14.2 million km^2 (5.5 million mi^2) in size—nearly twice the size of the United States and nearly three times greater than the Arctic Basin. The environment around this land mass is so harsh that even the existence of the continent was not known until extremely late in history. Throughout much of historic time it showed on maps as *Terra Australis Incognito* ("unknown southern land"). Nathaniel Palmer first discovered it on November 17, 1820. The first expedition commissioned by the United States set out in 1838 under Lieutenant Charles Wilkes. The U.S. National Science Foundation currently spends more than $100 million a year on research in the Antarctic.

TABLE 17.2
Climates dominated by cP air.

| Köppen type | EF |
| Thornthwaite type | F |

The outstanding feature of the Antarctic land mass is, of course, the huge mass of ice surmounting and surrounding the continent. A sheet of ice more than a mile thick covers 98% of the land mass. This ice sheet is divided into two separate pieces, the West Antarctic ice sheet and the East Antarctic ice sheet. The eastern ice sheet is firmly attached to the ground beneath, but the western sheet is unstable. Some 86% of the planetary ice is on this land mass. The pole of inaccessibility, or the center of the ice mass, is at about 4000 m (13,200 ft) elevation and 670 km (400 mi) away from the geographic pole. The polar ice cap is not thickest where it is coldest; instead, the ice reaches its maximum depth where air temperatures average a little below freezing. Ice accumulates very slowly in some areas: Near Byrd Station, at a depth of about 300 m (1000 ft), the ice first formed from snow that fell some 1600 years ago. The rate at which the ice moves seaward varies, in some cases moving as much as a kilometer per year and in some locations becoming buoyant and extending far out to sea. In the Ross Ice Shelf, glacial ice extends almost 800 km out to sea. This floating ice is as much as 270 m (890 ft) thick, with roughly 40 m (130 ft) above water. Icebergs breaking off from this shelf ice may be very large in area compared to their depth. A huge iceberg broke away from the Ross Ice shelf in early October 1987. It was about 154 km long and 36 km wide (92.4 by 21.6 mi). The surface area was about 4750 km^2 (1834 mi^2). It is known to be at least 250 m (825 ft) thick.

The oldest known drifting iceberg is one known as C-2. It was first spotted by satellite in 1978 and broke up in 1990. It was a fairly large iceberg. It was about 30 km in diameter and more than 60 m (190 ft) thick. For over four years it was grounded in 60 m of water. It traveled nearly around Antarctica for a distance of about 10,000 km (6214 mi).

The Polar Dry Climate

The Antarctic region is a major energy sink. Almost all of the weather stations on the continent have an annual net radiation loss. In fact, most have a net radiation loss even in the summer. Winter lasts six months and summer, two months. The heat loss to space is greatest over the ocean surrounding the land mass. In July in the midst of the Southern Hemisphere winter, radiative heat is lost to space over the entire area from the pole halfway to the equator. In January, the area of net loss shrinks southward to as far as 85° in some places.

Antarctic temperatures average 16°C (30°F) below those of the Arctic, for several reasons. First, Antarctica is mainly a land mass rather than water. Also, average elevation is high, with parts of the ice sheet 3000 m (10,000 ft) above sea level. The South Pole is 3 million mi farther from the sun during its winter than is the Arctic during its winter, reducing solar radiation 7% from that of the Arctic winter. Temperatures reflect both elevation and proximity to the ocean. Mean annual temperatures range from −19°C (−2°F) to as low as −57°C (−70°F) at Vostok. Minimum temperatures in the winter drop to −57°C along the coast. They decrease with elevation and distance from open water. Vostok, a Russian Antarctic station located inland at an elevation of 3950 m (13,000 ft), has recorded the coldest temperature at Earth's surface, −88.3°C (−127°F). The geographic South Pole is at 3000 m (10,000 ft) elevation. Table 17.3 lists some temperature extremes.

TABLE 17.3

Climatic data for Amundsen-Scott Station (South Pole).

Mean annual temperature	20 years (1957–76)	−49.3°C
Coldest year	1976	−50.0°C
Coldest day of record	July 22, 1965	−80.6°C
Coldest month of record	August 1976	−65.1°C
Coldest half-year	1976 (Apr.–Sept.)	−60.7°C
Warmest day of record	Jan. 12, 1958	−15.0°C

In Antarctica, the coldest temperatures lag far behind the solstice as a result of the long absence of direct solar radiation. At Vostok, the coldest days occur after the sun has risen above the horizon for a day or two. The sun normally rises above the horizon there on August 22, and the coldest temperature yet observed occurred two days later, on August 24.

As stated earlier in this chapter, the intensity of radiation reaching the surface in summer is low, although some periods have continuous sunshine and the total radiation received is high. In fact, weather stations in Antarctica hold the record for the most solar energy received in a monthly period. These stations also hold records for the longest continuous period of direct sunlight and the most hours of daily sunshine in a month. Despite that radiation, the coldest temperatures on the planet are also found here.

In summers the temperatures range upward toward freezing but rarely reach it. Most of Antarctica does not get warm enough in summer to melt the surface snow, although a few areas do shed their snow cover. Temperatures above the ice cannot rise much above the freezing point because the surface doesn't get above freezing. If the ice at the surface does melt, the additional energy used in melting more ice keeps the temperature at freezing.

The equinoxes are times of great change in the Antarctic. Shortly after the spring equinox (September), the polar atmosphere receives a burst of solar energy. This burst of energy begins the mixing process that causes the chlorine compounds to interact with the other gases to remove ozone. At the fall equinox, temperatures drop rapidly over the ice sheet. This results in very strong temperature, pressure, and moisture gradients around the edge of the continent. Intense cyclones develop and move around the perimeter (the arctic lows discussed earlier in the chapter). These storms send large quantities of relatively warm, moist air inland above the temperature inversion. This influx both retards cooling in the fall and heats the surface throughout the winter.

By far the dominant feature of the Antarctic is the persistent cold. Second to the cold is a temperature inversion that prevails above the ice over the entire continent in the winter and much of the continent in summer. The inversion ranges up to 25°C (45°F) in the lower 4 km (2.4 mi). On July 2, 1960, Vostok had an inversion of 30°C (54°F). The amount of the inversion increases away from the coast in winter, reaching 25°C (45°F) or more at the cold pole. The high altitude and clear atmospheric conditions, which permit rapid radiation loss from the surface, cause the strong inversion.

The absence of the normal pattern of day and night and the high heat capacity of the snow and ice reduce diurnal temperature fluctuation to less than 5°C (9°F). Table 17.4 shows the extreme nature of the temperature of the Antarctic.

TABLE 17.4
World climatic extremes recorded on the Antarctic continent.

CONDITION		RECORD
SOLAR RADIATION		
Longest continuous period of direct sunlight	Dec. 9–12, 1911	60 hr
Highest monthly mean (hours)	Syowa Base (69° S 40° E)	14 hr/day
Highest mean monthly intensity	South Pole (Dec.)	955 ly/day
TEMPERATURE		
Lowest mean monthly diurnal range	South Pole (Dec.)	2°C (3.5°F)
Lowest average temperature (warmest month)	Vostok	−32.5°C (−27.6°F)
Lowest mean annual temperature	Cold Pole (78° S 96° E)	−58°C (−72°F)
Lowest mean monthly maximum	Vostok	−66.5°C (−88°F)
Lowest monthly mean	Plateau Station (Aug.)	−71.7°C (−97°F)
Lowest mean monthly minimum	Vostok	−75°C (−103°F)
Lowest daily maximum	Vostok	−83°C (−117°F)
Lowest absolute temperature	Vostok	−88°C (−126.9·F)
HUMIDITY		
Lowest dew point	South Pole	−101°C (−150°F)
WIND VELOCITY		
Highest annual mean	Cape Denison	19.3 m per sec (43 mph)
Highest monthly mean	Cape Denison (July 1913)	24.5 m per sec (55 mph)

The energy balance of the Antarctic consists of rapid radiative heat loss from the ice- and snow-covered surface, partly offset by advection of heat in the warm, moist air from the surrounding ocean. The advected air contains both sensible and latent heat, which is carried downward by vertical motion. The airflow at the surface is largely downslope and offshore.

The extreme cold of Antarctica influences the climate far from the ice cap itself. Cities that lie on the southern margins of the Southern Hemisphere land mass have average temperatures 3°C (5°F) cooler than cities at the same latitude in the Northern Hemisphere. Several of the islands in the South Seas support glaciers, including Kerguelen and Heard Islands. In South America, mountain glaciers reach the ocean at latitudes nearer the equator than do those in the Northern Hemisphere.

The Windy Continent

Winds are a predominant factor in the weather of the polar deserts. Westerly winds are common at the surface poleward to around 65°, where they give way to low-level easterly winds (Figure 17.6) that extend to about 75°. Cyclonic storms develop over the oceans in the westerlies and move around the Antarctic from west to east. They normally do not penetrate far inland, and they account for much of the precipitation and weather along the coast. Winds of a high enough velocity to move snow and produce blizzard conditions occur at Byrd Station about 65% of the time. They have enough velocity to produce zero visibility about 30% of the time. A slight increase in wind velocity brings a large increase in blowing snow. The amount of snow moved by the wind varies as the third power of the velocity. Ironically, the most-accessible locations are also the windiest. Mawson's Base at Commonwealth Bay experiences winds above gale force (44 km per hour or 28 mph) more than 340 days a year. At Cape Denison, the mean wind velocity is 19.3 m per sec (43 mph), and during July of 1913 it averaged 24.5 m per sec (55 mph).

FIGURE 17.6

Average surface winds and January surface pressure over Antarctica.

Source: After Schweratfeger, Mather, and Miller, from K. Boucher, *Global Climate* (Seven Oaks, Kent, England: Hodder and Stoughton, Ltd., 1975), 294. Used by permission of Hodder and Stoughton, Ltd.

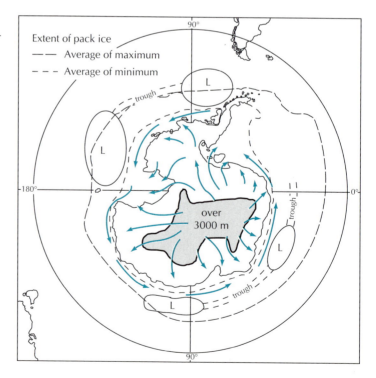

Strong gravity-winds develop and blow off the land masses. These katabatic winds are the flow of cold, dense air down a topographic slope due to gravity's force. They result from the extreme cooling of the air over the ice. A topographic slope of 0.2% is about equal to the force of the average pressure gradient over the Antarctic. Where the slopes are steep, katabatic winds may regularly flow in a direction other than that of the pressure gradient. The strongest winds occur when the pressure gradient and topographic slope coincide. The katabatic winds are the prevailing winds over much of the Antarctic, particularly where the interior highlands drop steeply to the coast. Where these winds are fairly constant, they will produce formations of waves in the snow and ice surface called **sastrugi.** These frozen waves grow as high as 2 m (6.5 ft) and make surface travel extremely difficult. Strong winds, blizzards, and rising temperatures often occur together when the usual temperature inversion disappears and warmer maritime air moves inland.

Even the interior can be very stormy, particularly in winter. On July 27, 1989, a six-member international team of men and dogs set out to cross Antarctica by dog sled and skis. It took them seven months to make the crossing, the first on foot. They encountered 150 km-per-hr (90 mph) winds and temperatures as low as $-45°C$ ($-49°F$). One man got lost in a blizzard one evening only a matter of meters from the camp. He dug himself a little trench in the snow and lay down and snow soon buried him. When morning came, other members of the party resumed their search for him. When he heard them calling, he crawled out of the snow—cold, but alive and unharmed. The party had only rare interludes without howling wind and blowing snow. Visibility was often zero, preventing any travel. Sometimes the group had to stay put for as much as three days at a time. One blizzard lasted for 17 days, and another stormy period lasted 60 days.

Summers are not so stormy. Temperatures rise to $-29°C$ ($-20°F$) and winds drop to the 45 m/s^{-1} (20-mph) range.

Precipitation

Moisture data for the polar deserts are more difficult to obtain than are temperature data. Data show that relative humidity, absolute humidity, and cloud cover decrease inland. Relative humidity often drops as low as 1%. Humidity and cloud cover are also lower in winter as a result of the stronger subsidence and the surface inversion. The cloud decks that appear over the ice caps produce warmer temperatures as they provide a heat source that radiates heat to the ground. Precipitation may be frequent, but it is light. The atmosphere contains little moisture for precipitation because of the low temperatures. The depth of precipitable water probably never exceeds 10 mm (0.4 in.) at any time. Annual precipitation ranges from as little as 51 mm (2 in.) on the cold plateau to as much as 510 mm (20 in.) at some peninsular locations. Most of the precipitation occurs as snowfall, averaging 300–600 mm (12–24 in.) per year. At Halley Bay, snow falls about 200 days per year. Because of the high winds, blizzards and drifting snow are common. Drifting snow also occurs 200 days a year. Blowing snow reduces visibility to less than 1 km about 170 days a year. Without wind, loose dry snow will pile up to as much as a meter deep. Some of the precipitation is in the form of ice pellets, and some moisture is deposited directly as condensation. Along the coasts, steam fogs have drifted inland when temperatures were as low as $-30°C$ ($-34°F$). These steam fogs add to the accumulation of moisture.

The Polar Wet Climate

Surrounding the Antarctic is the planetary ocean. It is uninterrupted by land masses north to a latitude of 40°S except at the tip of South America. Over this ocean the atmosphere has year-round high humidity, frequent precipitation, and low temperatures. The only land areas are a series of islands scattered around Antarctica (Table 17.5). Table 17.6 contains data representative of this climate type.

Humidity is very high, averaging over 50% all year. Cloud cover is also extensive, averaging 80% during the winter and slightly less in the summer. Precipitation frequency is high in all areas, with at least a 25% chance of precipitation on any day of the year. The annual total varies, however, as the intensity of precipitation varies with latitude and distance from the ocean. The totals vary from 370 mm to 2.92 m (15–117 in.). The annual variation in precipitation is among the lowest found anywhere. Snow, a common form of precipitation in this climate, is generally very wet, and snowfalls last a short time.

The high humidity and cloud cover have a considerable degree of control over insolation and temperature. The summer averages range up to 10°C (50°F), and winter averages are between −6.7° and 0°C (19°–32°F), so the annual range in temperature is not very great. Winter lows are exceptionally warm for a polar climate. The mild winter temperatures and low annual range reflect the marine location of these areas. Because the ocean does not freeze, it provides a constant source of heat for the atmosphere. The diurnal ranges are also quite low because of the high humidity. No season is frost-free in the polar wet climate. Frost can occur any day of the year, and these areas average over 100 days annually with frost.

In winter, storms with high winds occur much more frequently. This is the area named "the roaring 40s, the furious 50s, and the shrieking 60s" by the sailors of the sixteenth century. They gave these names to the area when they started sailing around Cape Horn in South America.

This climate is distinctive in several ways. It is the cloudiest of all climates, with most of the area experiencing few clear days. No place can equal the South Atlantic for its frequency and severity of storms.

Highland Climates

The tropical, midlatitude, and polar systems are distinguished from each other on the basis of periodic aspects of the energy balance. We subdivide each of these latitudinal regimes in turn on the basis of the seasonal pattern in the hydrologic cycle. *Highlands*, or mountain areas, exist in all of these different climatic regions and hence are subject to the same diurnal and seasonal patterns of solar energy and moisture as the surrounding lowlands. For example, the coast ranges of California are subject to a strong seasonal pattern of energy and a seasonal moisture regime consisting of a dry summer and wet winter.

TABLE 17.5
Climates dominated by mP air.

Köppen type	EM
Thornthwaite type	AE

TABLE 17.6
World climatic extremes recorded in the polar wet climate.

Lowest mean annual diurnal temperature range	Heard Island, Indian Ocean (53°10' S 74°35' E)	3.3°C (–0.5°–2.8°C) 6°F (31°–37°F)
	Macquarie Island Indian Ocean (54°36' S 185°45' E)	3.3°C (2.8°–6.1°F) 6°F (37°–43°F)
Lowest mean annual hours of sunshine	Laurie Island (66°44' S 44°44' W)	500 hr
Lowest monthly percentage of possible sunshine	Argentine Island (June) (61°15' S 64°15' W)	5%
Highest mean annual relative humidity	Deception Island (63° S 61°W)	90%

The primary characteristic of the climate in highland areas is the rapid change that takes place with elevation and orientation of the slopes to the sun or prevailing wind. Radiation, temperature, humidity, precipitation, and atmospheric pressure all change rapidly with height. Radiation intensity increases with height because of a steady decrease in thickness of the atmosphere. La Quiaca, Argentina, for instance, has the highest average annual radiation level of any known surface location (Table 17.7). The increase in radiation intensity with height is responsible for the sunburn that skiers or hikers often get in mountain resort areas; ultraviolet radiation increases sharply as the atmosphere thins. Ambient air temperature decreases with height above sea level at about 1°C per 100 m (390 ft). As the density of the atmosphere decreases, the concentration of water vapor and CO_2 decrease. In the absence of these gases, the greenhouse effect of the atmosphere drops sharply, and the ground surface heats and cools rapidly, resulting in higher diurnal temperature ranges. El Alton, Bolivia, at an elevation of 4.081 m (13,468 ft), has a mean annual temperature of 9°C (48°F) and a mean daily range of 13°C (23°F).

The microclimate of a site in a highlands area depends also on slope orientation. Orientation to the sun is significant in determining local heating characteristics. A slope facing the sun has much higher surface temperatures (and, consequently, higher air temperatures) than one facing away from the sun. This is a result of more-intense solar radiation (Figure 17.7).

Relative humidity increases with elevation up to a point, since the ability to hold moisture is a function of atmospheric temperature, and temperature decreases with height. This explains why fog is more prevalent at higher elevations (Table 17.8). Moose Peak in Maine averages 1580 hr and Mount Washington in New Hampshire, averages 318 days of fog annually. The highlands' seasonal pattern of precipitation is similar to that of surrounding lowlands. The amount of precipitation changes rapidly with elevation and orientation to the wind. Since highlands provide a mechanical lifting mechanism, precipitation tends to increase up to a point, then decrease as moisture precipitates out. Orientation of slopes to the prevailing winds is just as significant as elevation in determining local precipitation. Windward slopes receive up to five times the amount of precipitation on lee-

TABLE 17.7
World climatic extremes recorded in highland climates.

EXTREME		AMOUNT RECORDED
	RAINFALL	
Highest monthly mean	Cherrapunji, India (July)	2692 mm (106 in.)
Highest monthly total	Cherrapunji, India (July 1861)	9296 mm (366 in.)
Highest annual mean	Mt. Waialeale (Kauai, HI)	11.68 m (460 in.)
Highest total (4-month period)	Cherrapunji, India (April–July 1861)	18.74 m (737.7 in.)
Highest annual total	Cherrapunji, India (Aug. 1, 1860–July 31, 1861)	26.47 m (1042 in.)
	SNOWFALL	
Highest daily total	Silver Lake, CO (April 14–15, 1921)	1930 mm (76 in.)
Highest 6-day total	Thompson Pass, CO (Dec. 26–31, 1955)	4420 mm (174 in.)
Highest 12-day total	Norden Summit, CA (Feb. 1–12, 1938)	7722 mm (304 in.)
Highest monthly total	Tamarack, CA (Feb. 1911)	9906 mm (390 in.)
Greatest depth on ground	Tamarack, CA (Mar. 9, 1911)	11.53 m (454 in.)
Highest annual mean	Paradise Ranger Station (Mt. Rainier, WA)	14.78 m (582 in.)
Highest annual total	Paradise Ranger Station (Mt. Rainier, WA)	25.83 m (1017 in.)
	WIND VELOCITY	
Highest 1-hr mean	Mt. Washington, NH	77.3 m per sec (173 mph)
Highest 24-hr mean	Mt. Washington, NH	57.7 m per sec (129 mph)
	RADIATION	
Highest mean annual	La Quiaca, Argentina (22° S, 3459 m elev.)	667 ly/day

ward slopes. The rainiest areas on Earth are where there is orographic lifting of onshore winds. Table 17.9 provides sample data for the effects of elevation on precipitation.

Since temperature decreases with elevation, a greater percentage of total precipitation occurs as snow in highland areas. The mean snowline for Earth is about 4550 m (15,000

FIGURE 17.7
Slope orientation and radiation intensity.

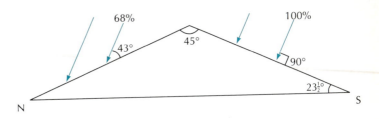

ft). In the tropical regions, it rarely goes above 5450 m (18,000 ft). The snowline is usually higher on the leeward sides of mountain ranges because less snow falls on that side, insolation is higher, and Foehn winds melt the snow. Mountains in Africa, South America, and Asia all exceed this elevation, so snowcapped mountains occur there, even along the equator. Extremely heavy snowfalls are common in some highlands. The 25-m snowfall at Paradise Ranger Station on Mount Rainier in the winter of 1970–71 serves as an extreme example.

The chinook is a downslope breeze or wind found along the east side of the Cascade and Rocky Mountains of North America. The name comes from the Chinook Indians, who lived along the lower reaches of the Columbia River. The chinook is a relatively warm, dry wind. It develops with the uplift of relatively mild, stable air ascending windward slopes of mountains. The wind develops quickly with a sharp rise in temperature and drop in relative humidity. These winds are most often identified with the sudden changes in temperature they effect. On January 22, 1943, the temperature rose from −20°C (−4°F) at 7:30 A.M. to 7°C (45°F) at 7:32 A.M. at Spearfish, SD. This was a total rise of 27°C (49°F) in two minutes. These winds occur most often in a narrow zone about 300 km east of the crest of the Rockies from New Mexico north into Canada. They form on the leeward sides of mountain ranges.

The primary source of heat in the Chinook comes from compression. Chinooks form when a strong westerly flow of air blows across the mountains. During a strong wind, a trough or cell of low pressure forms on the east side of the mountains. Because the air is stable, the low-pressure trough pulls the wind down the eastern slope

TABLE 17.8
Comparison of number of days per year with fog at neighboring mountain and valley stations.

STATION	ELEVATION (FT)	FOG DAYS/YEAR
Mt. Washington, NH	6280	318
Pinkham Notch	2000	28
Pikes Peak, CO	14,140	119
Colorado Springs	6072	14
Mt. Weather, VA	1725	95
Washington, DC	110	11
Zupspitze, Germany	9715	270
Garmish, Germany	2300	10
Taunis, Germany	2627	230
Frankfurt, Germany	360	30

Source: H. L. Landsberg, *Physical Climatology*, 2d ed. (Dubois, PA: Gray Printing Co., 1958), 140.

TABLE 17.9
Temperature and precipitation variation with height at Ecuador weather stations.

STATION	LATITUDE	ELEVATION	MEAN TEMPERATURE	ANNUAL PRECIPITATION
Cotopoxi	0°38' S	3308 m	8°C	1113 mm
Banos	1°23' S	1833 m	16°C	1519 mm
Mera	1°30' S	1063 m	21°C	4770 mm
Santo Dominga de los Colorados	0°15' S	497 m	22°C	3383 mm

of the mountains to its original altitude. The subsiding air heats at the dry adiabatic rate of 10°C per km.

The chinook picks up strength if precipitation takes place in the airstream as it rises over the windward side of the mountains. Condensation adds heat to the air, so when the air descends on the lee side of the mountains, it is warmer than at the same height on the west side.

A cloud called a chinook wall cloud forms over the Rockies when a chinook is blowing. The cloud forms and remains over the mountain crests, providing a visible sign of the wind system. As the air descends the eastern flanks of the mountains, the clouds evaporate in the warmer air and a clear sky may exist above the chinook.

Chinook winds are gusty and can reach velocities as high as 160 km per hr (100 mph). They can also absorb much water. Because they are warm and dry, in the winter they may sublimate large amounts of snow. It is not unusual for these winds to remove 150 mm (6 in.) of snow a day. They have been known to remove half a meter of snow in a single day. No wonder these winds are called "snow eaters."

The same wind occurs in other parts of the world—as the Föhn wind in the Alps Mountains of Europe and the Zonda in Argentina, for example.

■ Applied Study

Altitude and Human Physiology

The metabolism of the human organism depends, in part, upon oxygen consumption. Changing the amount of oxygen available to the body has marked physiological effects. The atmospheric gas mixture remains the same up to a height of about 80 km (48 mi), but its density and pressure decrease. As a result of the pressure decrease, the partial pressure due to oxygen decreases from 212 mb at sea level to only 55 mb at a height of 10 km (6 mi).

A person's need for oxygen does not change to any significant extent when moving from one altitude to another. Any rapid change in altitude can thus cause severe body stress. Even at moderate altitudes, the effect of hypoxia (oxygen deficiency) can cause nausea. Above 6 km (3.6 mi), the lack of oxygen can seriously affect the human brain.

To make use of the high-altitude environment of mountain regions or for air travel, humans have learned to operate in artificial habitats. Aircraft flying at high altitudes have pressurized cabins. While they do not maintain sea-level pressures, they maintain pressures high enough to eliminate stress for most travelers. Commercial jet aircraft also include emergency oxygen supplies. Oxygen decreases so rapidly with height that even mountain climbers often need supplemental oxygen at the highest altitudes. Visitors to mountain sites such as Cusco, Peru, may suffer substantial stress from the reduced oxygen.

A difference in air pressure can itself be stressful. On takeoff and landing involving substantial altitude changes, the ears may suffer pain as the pressure inside the eardrum and that outside change suddenly. This pressure change is often aggravated if a person has congestion due to a cold or flu. The space program has produced a significant amount of research into the effects of reduced pressure on the human body. Artificial environments such as vacuum chambers have been widely used in research about low-pressure environments.

Studies of how the human body adjusts to living conditions at different altitudes have examined people living and working at high altitudes. Acclimatization to altitude changes results as the physiological functions of the body adjust to a different set of climatic conditions. Full acclimatization is reached when the body is as efficient at high altitudes as it is at the pressure in its former environment. Investigations in the highlands of South America have shown that people accustomed to sea-level conditions may take months or even years to acclimate to high altitudes. Some people may never fully adjust to the difference. Common symptoms that occur during acclimatization are nausea, weight loss, intestinal problems, and emotional problems.

People who have lived long periods at high altitudes may suffer similar consequences when they descend to lower altitudes. The Spanish conquerors of the South American Indians became aware of this in the sixteenth century. In 1588, Philip II of Spain decreed that the Indians should not be forcibly moved to places where the climate was substantially different from their native region. The Spaniards suffered from the effects of living at high elevations. Among those living above 10,000 ft, it was 53 years before a child was born who lived beyond a few months.

Long-term acclimatization to reduced oxygen is well demonstrated in many of the Andean Indians. They tend to be relatively short and stout, with disproportionately large chests. The compact nature of the body provides the best geometric shape to minimize the ratio of surface area to volume, thereby reducing heat loss, while the deep chest seems well-developed for taking in large quantities of air at high altitudes. ■

Summary

Cold is the distinguishing characteristic of the polar climates. Solar radiation of very low intensity is the primary factor in producing the cold in both polar regions. The Antarctic is colder than the Arctic because the latter is an elevated land mass. The Arctic Ocean continually supplies heat to the atmosphere, even when it is covered with ice. Because of this, the Northern Hemisphere cold pole is some distance away from the geographic pole. The major continental ice sheets existing on Earth are in Greenland and the Antarctic. Both are related to moderate snowfall and cold temperature. The primary circulation of the atmosphere carries heat into the polar areas to raise the temperature above what it would otherwise be.

Mountain climates are unique in that elevation and orientation to the winds are the primary variables associated with temperature and precipitation amounts. The seasonal distribution of precipitation and insolation is the same in the mountains as in the surrounding lowlands.

Three# Part

CLIMATES OF THE
PAST AND FUTURE

RECONSTRUCTING
THE PAST

Chapter Overview

Using the instruments of today, minor changes or trends in weather are routinely monitored and analyzed. However, the period for which weather instruments have been available is but a tiny fraction of time in Earth's history. To understand current climates and predict future ones, we must consider them within the framework of climatic change over geologic time. To achieve this, we must first reconstruct the climates of the past. This process requires painstaking, detective-like research, because much of the evidence is based upon the relationship of climate to other environmental processes and to living organisms. Prior to presenting an account of how climate has varied, this chapter deals first with the way in which we deduce climates.

The reconstruction of climates over time is a fascinating puzzle. Instrumental records of meaningful spatial extent have become available only in very recent times, and information about earlier climates requires the use of *proxy data*, observations of other variables that serve as a substitute or proxy for the actual climatic record. Proxies are paleoclimatological archives.

Surface Features

Evidence from Ice

Study of the processes that modify Earth's surface was the first indication that climates have varied over time, particularly in relation to the existence of ice ages. Perhaps the best-known early researcher in formulating these ideas was Swiss scientist Louis Agassiz, whose work began in the early 1800s.

Beginning his research in Switzerland, Agassiz noted that the valleys had a U-shape, rather than the typical V-shape of river valleys. Other researchers had previously noted the same phenomenon and commented upon the fact that the U-shaped valley contained a relatively small river that was out of proportion to the valley's size. The other researchers explained this occurrence by referring to the biblical flood so vividly recorded in the Old Testament. It was assumed that the huge valleys must have been carved by much larger streams when the flood occurred.

Agassiz was also impressed by the presence of large boulders set amid assorted finer sediments that had obviously been transported from an area from where they landed. These boulders, called *erratics*, aroused much curiosity, and their presence was again attributed to the great biblical flood.

Other geologists disagreed with this view of a catastrophic cause for the erratics. These geologists maintained that the forces that caused the U-shaped valleys and the erratics were the same forces that now operated. They maintained that present processes provided the key to what happened in the past. Accordingly, they believed that the large erratics scattered over many areas of the world had been dumped by ice and, basing their ideas upon observed facts, they concluded that the erratics had been deposited by icebergs that floated on an extensive sea that formerly covered large areas of Europe. Using the same reasoning, the geologists decided that the finer sediments had come from melting icebergs; they called these sediments *drift*.

At first, Agassiz agreed that the iceberg theory best explained the erratics, but, in his work he met other scholars who had different ideas. Scientists like Jean de Charpentier and Ignace Venetz had studied the landscape of the Swiss Alps and were convinced that the glaciers they saw had once been much more extensive and were responsible for the valley shapes, the drift, the erratics, and the parallel striations found on hard rock surfaces.

The accumulation of this kind of evidence prompted Agassiz to formulate a comprehensive theory of extensive glaciation. He suggested that a great ice sheet had extended from the North Pole to the Mediterranean and that the moraines, striations, erratics, and drift seen in Switzerland had resulted from the action of glaciers. The idea of an ice age was born. Nowadays it seems astonishing that the idea of major glaciation covering much of Earth was not accepted until the nineteenth century.

The results of ice erosion and deposition are characteristic of large parts of the Northern Hemisphere. Figure 18.1 shows a typical view of an area that has undergone

FIGURE 18.1
The town of Revelstoke lies on the floodplain of the Columbia River. The valley was carved by glaciation, and the floor is now filled with sediment.
Source: John J. Hidore

mountain (alpine) glaciation. The resulting features—the U-shaped valleys, the hanging troughs, the aretes, and the tarns—are well known and clearly indicate ice activity. Features associated with continental glaciation differ markedly from those of alpine glaciation. Studying the position of terminal moraines and the erratic boulders in detail allows reconstruction of the extent and movement of the ice. The erosional features, glacial striations, ice-gouged lakes, and so on, further differentiate the two types of glaciation. Using such findings, we can reconstruct conditions that existed in North America during ice advance. Figure 18.2 provides one interpretation.

While the results of the work of ice age are an important interpretive device, glaciers themselves provide keys to temperature and precipitation conditions. The advance or retreat of glaciers has been used in interpreting climatic change over historic periods. Evidence indicates that glaciers shrank about 3000 B.C. and that the alpine snow level was at least 1000 ft (330 m) higher than today. By 500 B.C., the snow level had again advanced markedly, then retreated. Between the seventeenth and nineteenth centuries, a general resurgence was observed in the Alps and Scandinavia. Glaciers have tended to retreat in the twentieth century, although we seem to be in an in-between stage, with some glaciers growing and others disappearing.

FIGURE 18.2

The maximum extent of the Pleistocene ice advance in North America. Note the reconstructed regions equatorward of the ice.

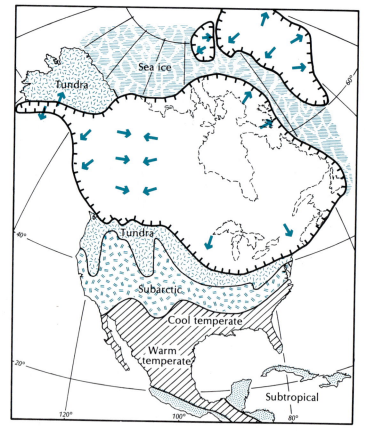

Ice Cores

The great ice caps of Greenland and Antarctica contain a wealth of climatic information. Ice sheets are formed layer by layer from the snowfall of each year; with time, the snow is compressed into ice, often filled with bubbles of trapped air. Drilling into the ice shows us a great deal about the atmosphere in which original snow formed (Figure 18.3).

The water that comprises ice contains oxygen, which has atoms with atomic weights varying from the normal atom; that is, there are isotopes of oxygen, the two most common of which are oxygen 18, the heavy isotope, and the lighter oxygen 16. By analyzing the ratio of one isotope to another in the ice deposit, we can determine the existing environmental temperature, then derive a pattern of temperature over time, using the layers of ice in the core. The trapped air bubbles provide additional significant information: Minute samples can be used to determine, for example, the level of CO_2 in the atmosphere. Layers of dust may give insight into the atmosphere's storminess or the extent of volcanic activity.

Drilling through the ice has become a key tool in climate-reconstruction. Until recently, the deepest cores have been derived by Russian and French scientists at the Vostok base in Antarctica. Ambitious programs are underway on the Greenland ice cap, where U.S. and European scientists are drilling two cores—the Greenland Ice Sheet Project 2 (GISP 2) and the Greenland Ice Core Project (GRIP). GISP 2 plans a core 3000 m (10,000 ft) deep.

Periglacial Evidence

Features associated with the ice itself do not supply the only geomorphic evidence regarding ice ages. Areas not directly affected by the great ice sheets experienced a cli-

FIGURE 18.3
Examining ice cores taken from polar ice caps.
Source: Richard Monastersky.

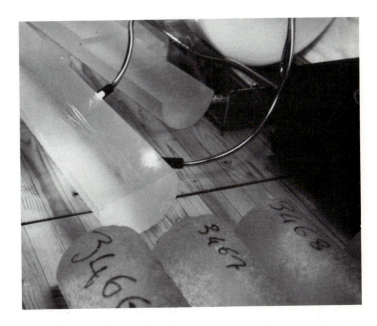

mate very different from today's. Some areas that are now dry experienced much wetter climates as a result of modified circulation patterns. Many inland basins, for example, have been occupied by large lakes. Such **pluvial** lakes, so named because they resulted from higher precipitation in earlier times, have been widely identified.

The western part of the United States, particularly in the Great Basin area, has fine examples of the extent of such lakes. Two glacial lakes in this area were enormous. At its maximum, Utah's Lake Bonneville occupied some 50,000 km^2 (20,000 mi^2) an area approximately the size of Lake Michigan. Evidence of the extent of the this great lake is found in the present Bonneville salt flats and in the *strand lines*, the shore areas indicative of the former level of the lake. Such lakes would necessarily modify the drainage systems, and the formation of a lake and its eventual overflow might lead to a totally new drainage direction.

Sediments

Sediments deposited during geologic time offer evidence of the climatic environment in which they were formed. One example is offered by *evaporite* (salt) deposits. Salt deposits form when, on a long-term basis, evaporation exceeds precipitation in an area where water flows in from other areas. Water evaporates to leave the salts, which were formerly in solution, as sediments. Eventually, these may be buried and turned into sedimentary rock. The large evaporite beds of the western United States, Germany, and central Asia are similar deposits being formed at the present time.

Wind-deposited materials also provide a guide to prevailing winds in earlier times. During the last ice age, sand dunes formed around glacial lakes and along shorelines. The dunes' structure and location have been used to estimate local wind conditions. Similarly, deposits of loess occurred over wide areas. Loess is made up of silt picked up from the edge of the glacier by the wind and scattered over a broad area many miles from the glacier.

Sea-Level Changes

Changes in the ocean level can occur either when the volume of water or the volume of the ocean basins decreases or increases. Many factors can cause either of these to occur, but most of them, such as the accumulation of sediments on the ocean floor or the extrusion of igneous rocks into the oceans, require many years. Rapid changes result mostly from ice alternately accumulating and melting. If the present Antarctic ice sheet were to melt, some researchers calculate that the additional water would cause a sea level rise of 60 m (200 ft). The removal of ice from a large land area would, however, be accompanied by an upward readjustment of the land. The ocean floors would sink further into the crust under the weight of additional water. Even allowing for this, melting of the Antarctic ice would still cause a rise of 40 m (135 ft), sufficient to flood most of the world's major ports.

The rise and fall of sea level during the Pleistocene is an important guide to glacial and interglacial periods. Submergence and emergence of coastal areas and marine terraces point to the amount of water tied up as ice; using appropriate dating methods, such evidence provides a guide to glacial advance and retreat.

Past Life

Studying species of organisms that lived in the past provides much information about the climate of the time. All plants and animals have a preferred set of physical conditions in which they live. Plant and animal fossils can be used in the reconstruction of past ecological conditions, including climates.

Faunal Evidence

Invertebrate Fossils. Fossils without backbones are widely used in establishing the geologic sequence of rock; they have also helped in reconstructing past climates. The value of invertebrate fossils in climate studies is limited, however, by the fact that many of these fossil species lived in fresh or salt water and thus were not directly affected by the atmospheric climate. The sea, lake, or river in which they lived acted as a buffer to direct exposure.

While the physiology of fossil animals compared to modern species helps determine paleoclimates, the *chemistry* of invertebrates has also provided much information recently. Isotopes have been particularly useful, especially as part of cooperative research.

CLIMAP (Climate: Long Range Investigation, Mapping, and Prediction) is one of the more important multidisciplinary projects funded by the National Science Foundation to study and understand the climate of the past. It was part of a larger program entitled the International Decade of Ocean Exploration. CLIMAP investigated ocean-atmosphere exchanges to assist in reconstructing the earth's climate. One of the key forms of evidence obtained from CLIMAP came from cores taken from the layers of sediment on the ocean floors. CLIMAP was a ten-year program that ended in 1980. Another current project, COHMAP (Cooperative Holocene Mapping Project) has similar goals, but focuses on the last 10,000 years.

The layers of sediment contain billions of microscopic skeletal remains of plankton. When the creatures die, their remains fall to the ocean depths and are incorporated in the mud layers. Since the tiny organisms are temperature specific, each species lives within a certain ocean climate; if that climate changes, they drift away and are replaced by plankton better adapted to the new conditions. The fossils found in the mud layers thus provide a record of temperature changes in the oceans and, consequently, in the atmosphere above the ocean. To obtain a sample of the undisturbed layers, researchers remove cores up to 30 m (100 ft) long from the ocean floor. Analysis of the cores is a lengthy task, since 20–50 species and up to 500 individuals exist in each few centimeters of core.

Carbon-14 analysis determines the date when the organism was deposited. The basis of this technique is that skeletons contain both ordinary carbon and a minute trace of the isotope carbon-14. The proportion of carbon-14 to carbon-12 remains fixed while the organism is alive. After it dies, the carbon-14 begins to decay; the ratio of carbon-12 to carbon-14 determines the age of the shell.

Vertebrate Fossils. These fossils provide important clues to past climates, through their distribution and their physiology as related to environment. The great changes in vertebrate life over geologic time have generated a number of interpretations. For example, the extinction of the dinosaurs in Cretaceous time has promoted much discussion; many believe that progressive and increasing world aridity might have caused their extinction. Others question such an interpretation for many reasons, not the least of which is that while other animals became extinct (for example belemnites and ammonites), related reptiles (notably crocodilians) did not. Note, too, that many plants, which are usually more responsive to changes of climatic regime, did not disappear from Earth. A number of other theories explain the demise of the dinosaurs. One, for example, blames temperature rather than moisture change, suggesting that changing temperatures might have influenced sperm production in the dinosaurs; because they could not reproduce, they became extinct. Another theory involves climatic change resulting from the impact of an asteroid on Earth.

Floral Evidence

Plant distribution provides an important guide to the distribution of today's climate, and the same is true of paleoclimates. Identifying vegetation patterns and their changes over time indicates the character of past climates. Often, plant evidence is combined with other environmental features. For example, it has already been noted that the extent of mountain glaciers varies. Mountain vegetation, particularly the elevation of the tree line on the mountain, reflects changes in glaciers. Tracing the tree line over time gives clues to climatic trends in selected areas.

The physiology of plants, like that of animals, also provides much information. The development of drip leaves on plants, for instance, indicates they have existed under very moist conditions; the fossil remains of plants with thick, fleshy leaves probably reflect arid or semiarid climates.

Interpretations of the climate during the Carboniferous, a time of prolific vegetation that gave rise to great thicknesses of coal, show how we can learn from plant physiology. Many of the fossil plants in the Carboniferous appear to be related to the horsetail and the club mosses, both representative of a marsh or swamp environment. Such an interpretation gets support from fossils that suggest plants with layered roots such as those found in modern bog plants and by many minor structures that appear to indicate that some of the plants actually floated on water. Trees lacked growth rings, indicating a climate without marked seasonal differences; the dominance of trees over herbaceous plants would further indicate a swamp environment. In all, the vegetation suggests a warm, moist climate that favored luxurious, if wet, plant cover.

Similar evidence has helped reconstruct the climates of late Paleozoic and Mesozoic rocks. More-recent deposits can be interpreted by pollen analysis; tree-ring analysis is used for much more recent plant distributions (Figure 18.4).

Pollen Analysis.   The study of pollen grains, or spores, is termed *palynology*. Its success depends upon the fact that many plants produce pollen grains in great numbers (e.g., a single green sorrel may produce 393 million grains; a single plant of rye, 21 million grains) and that pollen is widely distributed in the area in which plants are found. Most important, the outer wall of the pollen grain is one of the most durable organic substances known. Even when heated to high temperatures or treated with acid, it is visibly unchanged. This is important, for pollen possesses morphological characteristics that allow it to aid in identification of groups above the species level.

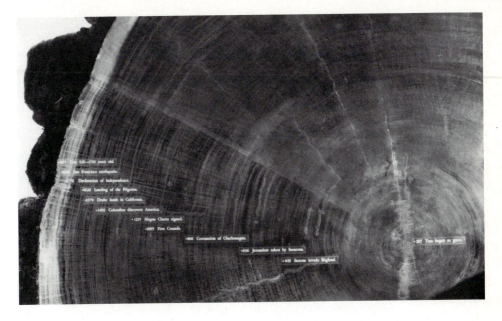

FIGURE 18.4
Tree ring analysis of the trunk of a 1710-year-old Sequoia.
Source: Tom McHugh/California Academy of Sciences/Photo Researchers, Inc.

To use pollen in interpreting the past distribution of vegetation and hence inferring past climate conditions, we must obtain a layered sequence of the pollen. As shown in Figure 18.5, this often occurs in ancient lakes or peat bogs, where seasonal pollen deposits would be covered by sediments. Cores taken would show a sequenced pattern.

Pollens from the cores are identified and a frequency distribution of plant types derived. A high proportion of spruce pollen in the lower core, for example, might give way to oak pollen at higher levels. This variation would indicate a vegetation change over time and suggest that the difference could be related to a passage from cool to warmer climatic conditions. Even a relatively crude classification of pollen type (for example, pollen from trees versus from other plants) could provide a rough guide to changing climatic conditions. A change from pollen associated with the treeless climates of the cold tundra to climates with trees might indicate an amelioration of climatic conditions. In-depth statistical counts obviously provide a good deal more detail. Much of this work has been completed in Scandinavia, where the first palynological stratigraphy occurred.

Despite the important progress using this method, it does have shortcomings. Vegetation cover only attains maturity after a lengthy period, and it is feasible that the type of vegetation established through pollen analysis reflects a successional stage that does not truly represent the prevailing climate. In some areas where vegetation cover is mixed, it becomes difficult to establish any dominant type that can be related to climate. In addition, from Neolithic times people have interfered extensively with the forest cover, and human-induced changes might well give misleading results.

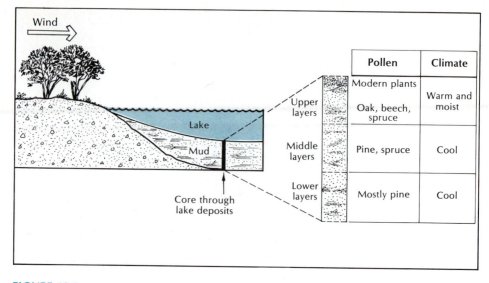

FIGURE 18.5

Simplified diagram showing the method of reconstructing past climates using pollen analysis.

Source: J. E. Oliver, *Climatology: Selected Applications* (Silver Spring, MD: V. H. Winston & Son, Inc.), 223.

Dendrochronology.   A. E. Douglas and his colleagues at the University of Arizona pioneered tree-ring study. Initial studies attempted to relate seasonal growth of trees to sunspot cycles, and a great deal of significant work was completed. In the quest for ancient living trees, the *Pinus aristata* was found to be 4000 years old. Analysis of such ancient trees permitted reconstruction of climates of the American southwest during various settlement periods.

Tree-ring analysis assumes that growth rings reflect significant events that happened during a tree's life. Growth rings form in the xylem wood of the trees. Early in the season, xylem cells are smaller and darker. An abrupt change from light to dark-colored rings delineates the annual increments of growth. Studying the size and variations in these rings provides information about the varying environmental conditions the tree has endured and is the purpose of **dendrochronology**. The method is most valuable in determining conditions that existed during relatively recent geologic history and is widely used in archeological research.

Climatic Change Over Geologic Time

Table 18.1 shows the geologic time divisions, naming the eras and periods and identifying the times at which Earth underwent ice ages. Perhaps the most significant idea expressed in the table is that average global climates have been much warmer than they are today throughout much of geologic time. Periodically, ice ages have interrupted that warmth.

TABLE 18.1
Geologic time divisions.

ERA	PERIOD	BEGINNING (MBP)*	ICE AGES
Cenozoic	Quaternary	2–3	Pleistocene ice age
	Tertiary	65	
Mesozoic	Cretaceous	135	
	Jurassic	190	
	Triassic	225	
Paleozoic	Permian	280	Ice age at approximately 300 mbp
	Carboniferous	345	
	Devonian	400	
	Silurian	440	
	Ordovician	500	Ice age at approximately 450–430 mbp
	Cambrian	570	
Precambrian		>570	Ice age at approximately 850–600 mbp

*mbp = millions of years before present.

Rocks of the Precambrian period, extending back to Earth's origin, provide but few details of the climates that existed, and only the late Precambrian times can be reconstructed with any degree of confidence. We do know, however, that during the late Precambrian, much of Earth was glaciated, perhaps from 950 million years before present (mbp) to 650 mbp. During the Paleozoic (570–225 mbp) period, fossil and sedimentary evidence became more widespread. Organisms and rocks suggest that the Cambrian and early Ordovician times were largely warm, although glaciation may have happened toward the end of the Ordovician. The withdrawal of ice and deglaciation at the end of this period made the climates of the Silurian and Devonian periods similar to those of today. The Pennsylvanian/Mississippian periods were dominated by widespread humid climates with intermittent glaciation. Considerable evidence exists to indicate a major glaciation in the late Paleozoic.

The Mesozoic (225–65 mbp) was essentially a time of widespread warmth and aridity. Climates of the Triassic appear similar to those of the upper Paleozoic—cool and humid—but they gave way to a long period of warmth, especially marked during the Cretaceous. This time of history saw the world in its "greenhouse mode" when climate was predominantly warm, polar ice caps were nonexistent, and sea level was high. The change from this to an eventual "icehouse mode" may not have been smooth but rather episodic.

During the early Tertiary (65–22.5 mbp), the warm temperatures of the Mesozoic began to decline. Long episodes of relatively warm climates were punctuated by abrupt drops in temperature. During this time, the first glaciers since the Paleozoic began to form in Antarctica. By the later Tertiary (22.5–2 mbp), wide temperature swings occurred until in the Pliocene when the downward swings produced glaciations such as those associated with the Pleistocene.

The climates of the Pleistocene consisted of glacial and interglacial times, with polar ice advancing and retreating from numerous sources. The most recent full glacial, known as the Wisconsin in North America, lasted from perhaps 30,000 to 12,000 years before the present. The coldest temperature was 4°–6°C (7°–11°F) lower than present and occurred about 18,000 years ago. Huge ice sheets extended as far south as 50° N in Scandinavia and 40° in North America. Frigid polar water extended in the North Atlantic to 45° N.

Climate Since the Ice Retreat

The period from 18,000 to 5500 years ago corresponds to the deglaciation of Earth. By 12,000 years ago, only scattered areas of ice sheets remained in western North America, with the main ice sheet confined to eastern Canada.

About 10,200 years ago, a strange event occurred that affected Scandinavia and Scotland particularly: The margins of the remaining ice sheet expanded and some small ice sheets reappeared. This time, known as the Younger Dryas (named for a small flower found in cold climates), did not last long and shortly after climatic conditions in the Northern Hemisphere resembled those of the present day. But the rapid temperature decline of the Younger Dryas illustrates that climatic change need not be the long, deliberate process that most people think of it as. Figure 18.6 shows the pattern of temperature change over the past 18,000 years.

After the cooling associated with the Younger Dryas, the global climate warmed. By 7000 years ago, conditions had improved such that only remnants of ice remained. The warm period peaked about 5500 years ago, and most ice disappeared, leaving only the Greenland Ice Sheet and the Arctic Ice that we have today. All evidence points to this being a time when the mean atmospheric temperature of the midlatitudes was about 2.5°C (4.5°F) above that of the present. This time has been described as the *Climatic Optimum*, a term originally applied to Scandinavia when temperatures were warm enough to favor more varied flora and fauna.

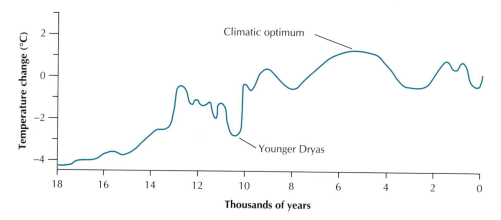

FIGURE 18.6
Patterns of surface temperatures of the last 18,000 years based largely upon Greenland isotope temperatures. Vertical axis provides temperature change from current global average temperature.

Many researchers have used historical records to establish climatic changes that have occurred during the brief existence of humans on Earth. Their findings have allowed fairly detailed reconstruction of climates over the past 6000 years. Medieval chronicles contain many references to prevailing weather conditions. Unfortunately, although not surprisingly, these pertain to exceptional weather events rather than day-to-day conditions. These events include such unusual phenomena as the freezing of the Tiber River in the ninth century and the formation of ice on the Nile River. While these sources do not supply a continuous record, we can construct an overall view of the usual conditions by assessing the number of times given events are recorded. The freezing of the Thames River provides one such example. Between 800 and 1500, only one or two freezings per century were recorded. In the sixteenth century, the river froze at least four times, in the next century it froze eight times, and six freezing periods were recorded in the eighteenth century. One can only suppose that progressive cooling increased the frequency.

The Last 1000 Years

Contemporary literature has aided in reconstruction of Iceland's climate, using one of the most complete sequences of climate over the last 1000 years (Figure 18.7). The data for Iceland are essentially based upon the following sources:

1846 to present: actual meteorological instruments

1781–1845: a reconstruction of weather conditions as derived from the relative severity and frequency of drift ice in the vicinity of Iceland

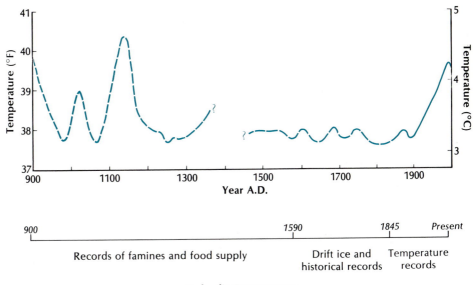

Icelandic Temperatures

FIGURE 18.7
The derived sequence of Icelandic temperatures over the last 1000 years.

1591–1780: historical records combined with incomplete drift-ice data

900–1590: information from Icelandic sagas indicating times of severe weather and related famines

Such a complete record as this can be used as a base guide to the climate of the entire North Atlantic area.

The time extending from about 950 to 1250 is known as the Little Climatic Optimum. Evidence of agriculture and other indicators have been used to reconstruct the climates that existed at the time Greenland was settled by the Vikings. Under the leadership of Eric the Red, the Vikings passed from Iceland, which they had settled in the ninth century, to Greenland. While an icy land, it supported sufficient vegetation (dwarf willow, birch, bush berries, pasture land) for settlement. Two colonies were established, and farming was begun. The outposts thrived and regular communications were established with Iceland.

Between 1250 and 1450 A.D., climate deteriorated over wide areas (Figure 18.8). Iceland's population declined and grains, which had been grown in the tenth century, were no longer produced there. Greenland was practically isolated from outside contact, with extensive drift ice preventing ships from reaching the settlements. In Europe, storminess resulted in the formation of the Zuider Zee and the excessively wet, damp conditions led to a high incidence of the horrifying disease St. Anthony's Fire (ergotism).

Conditions continued to worsen and from about 1450 to 1880 the time known as the Little Ice Age occurred. During the period, glaciers enlarged to the extent that some Alpine villages were overwhelmed by ice.

The Little Ice Age marked the end of the Norse settlements in Greenland that had begun in the tenth century. In fact, in 1492 the Pope complained that none of his bishops had visited the Greenland outpost for 80 years because of ice in the northern seas. (He was not aware of the fact that the settlements were gone.) By 1516 the settlements had

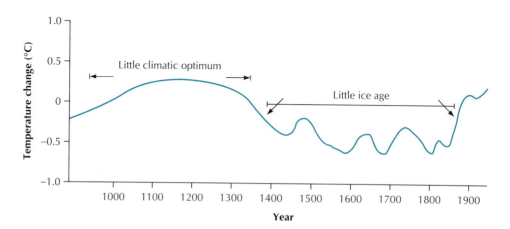

FIGURE 18.8

Variations from current average surface temperature over the last 1000 years. The Little Climatic Optimum was followed by a deterioration in the Little Ice Age.

FIGURE 18.9
The Hunters in the Snow by Pieter Brueghel the Elder.
Source: Collection of Kunsthistorischen Museum GG 1838.

practically been forgotten, and in 1540 a voyager reported seeing signs of the settlements, but no signs of inhabitants. The settlers had perished.

Whether this was due to the deteriorating climate or to invasion by other groups is not known, although a Danish archeological expedition to the sites in 1921 found evidence that deteriorating climate must have played a role in the population's demise. Graves were found in permafrost that had formed since the time of burial. Tree roots entangled in the coffins indicated that the graves were not originally in frozen ground and that the permafrost had moved progressively higher. Examination of skeletons showed that food supplies had been insufficient; most remains were deformed or dwarfed, and evidence of rickets and malnutrition was clear. All the evidence points to a climate that grew progressively cooler, leading eventually to the settlers' isolation and extinction.

Later in the cold spell, the colonies in the eastern United States suffered as well. The soldiers of the American Revolution suffered in the cold weather, although the unusual ice sometimes served as a useful tool. British troops, for example, were able to slide their cannons across a frozen river from Manhattan to Staten Island.

From North America comes a well-known account of life during the last years of the Little Ice Age, a description of the year 1816, known as "the year without a summer." The year began with excessively low temperatures across much of the eastern

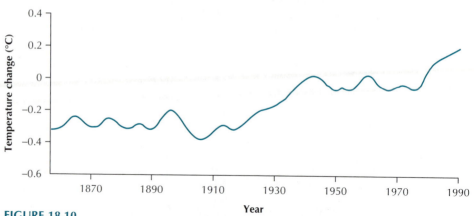

FIGURE 18.10

Globally average temperatures since about 1860 shown as a change from the 1951–80 average.

seaboard. As spring came, the weather seemed to be cool, but not excessively so. In May, however, the temperatures plunged; Indiana had snow or sleet for 17 days, which killed off seedlings before they had a chance to grow. The cold weather continued in June, when snow again fell, devastating any remaining budding crops totally. No crops grew north of a line between the Ohio and Potomac Rivers, and returns were scanty south of this line. In the pioneer areas of Indiana and Illinois, the lack of crops meant that the settlers had to rely on fishing and hunting for their food. Reports suggest that raccoons, groundhogs, and the easily trapped passenger pigeons were major sources of food. The settlers also collected many edible wild plants, which proved hardier than cultivated crops.

The image of the period shown by artists of the time is very different from that of today. Paintings of scenery and activities in the low countries show winter scenes in which ice and snow are central to the theme. One famous painting by Pieter Brueghel the Elder, shown in Figure 18.9, provides an excellent example of this image.

Fortunately, by the end of the nineteenth century, the instrumental record shows that the climate was again improving. A reconstructed record of temperature is shown in Figure 18.10. If we consider 1950–80 to be the baseline period, we see that the years prior to that were cool but warming, while those after were even warmer. This latter trend raises the specter of global warming.

Summary

Reconstructing climates for the periods prior to instrumentation relies upon proxy data. Early evidence of past climates was derived from studying ice, but newer methods can more accurately describe a given climatic event. The interpretation of ice cores taken from ice caps permits oxygen isotope analysis to give insight into the prevailing temperature at the time when the ice formed. Surface evidence, including sediments, periglacial activity, and changes in the sea level, also provides significant proxy data.

Both faunal and floral life of the past are used in climatic reconstruction. Invertebrates, especially those derived in deep sea cores, are of particular significance. Dinosaurs are perhaps the most spectacular of the vertebrates studied, and climatic change may well be significant in explaining their extinction. Floral evidence relies upon the physiology of plants, the study of pollen, and tree ring analysis. Floods, droughts, human migration, sailing records, ice cover and a host of other sources of evidence provide additional clues to climate.

A review of climate over time shows that a number of ice ages have occurred throughout geologic history. The most recent of these, the Pleistocene, was responsible for widespread evidence of glaciation. Deglaciation began about 18,000 years ago, and both warm and cool periods have occurred since that time. During the last 1000 years, a long cool time known as the Little Ice Age was a dominant feature. The twentieth century also had both warm and cool times, with the 1980s being an exceptionally warm period.

CAUSES OF CLIMATIC CHANGE

Chapter Overview

All the available evidence demonstrates clearly that climate has changed over time. Possible causes for such change have in turn been the subject of much research. What causes climates to change has not been completely determined, although not because of a lack of theories. In fact, since the magnitude of climatic change was realized about a century ago, it has been estimated that a new theory has been postulated for every year that has passed. Not all of these have been satisfactory, for some have left out the two basic ingredients of a viable explanation. It must (1) explain the onset of ice ages through geologic time, that is, account for long periods of warmth with cooler interruptions; and (2) account for the warming and cooling periods that occur within an ice age. Not all theories can do this, especially in quantitative terms.

The search for the explanation of climatic change has become increasingly important in recent years. Human activities modified both the atmosphere and Earth's surface to such a degree that these modifications are now an integral part of explaining climatic variations. This chapter deals with climatic change in two parts. It first considers theories involving natural causes including possible explanations for changes that occurred long before humans inhabited Earth. Second, the chapter examines the role of people in inadvertently modifying climate. Obviously, the two causes are not independent; natural changes continue to occur while people modify the system. However, once we consider the two components separately, we can combine them in an attempt to predict what will happen in the future (see Chapter 20).

Natural Changes

The basic reason for climatic changes on Earth is essentially very simple. Change reflects the flows of energy into and out of the Earth–ocean–atmosphere system and the ways in which energy is budgeted within that system. Unfortunately, explanation of the flows and interchanges is very complex and requires examination of all the system's parts. Because of this complexity, it is convenient to divide theories of climatic changes into a number of categories. These include the following factors:

variations in Earth–sun relationships

variations in energy output by the sun

atmospheric variations modifying the flow of energy

changes in the position of continental land masses

variations in heat stored in the oceans

Earth–Sun Relationships

Variations in Earth's motion around the sun explain diurnal and seasonal differences in the amount of solar energy arriving at the surface (see Chapter 2). However, the angle of Earth's axis and its distance from the sun vary over time in a number of ways.

The Obliquity of the Ecliptic.   This term refers to the angle of the axis in relation to the plane in which Earth revolves around the sun. The angle today is 66.5°, which gives an obliquity angle of 23.5°. This angle is not constant; on a cycle of about 41,000 years, it varies some 1.5° about a mean of 23.1° (Figure 19.1a). The effects of changing obliquity are illustrated in Figure 19.1b. The top diagram shows the present position and the lower some hypothetical cases. An obliquity of 0° would lead to equal lengths of day and night over the globe, a lack of seasonal changes, and well-defined climatic zones. Another extreme is a 54° angle of obliquity, which would produce great extremes in the lengths of summer and winter days and nights. For example, at the December solstice position shown in the diagram, much of the Northern Hemisphere would have 24 hr of darkness. Extreme temperature changes would occur from summer to winter. Although actual changes in the angle of obliquity are never as large as these examples, the deviations are sufficient to cause distinctive changes in zonation of Earth's climates.

Earth's Orbital Eccentricity. Earth moves around the sun in an elliptical orbit; we figure the orbit's eccentricity by comparing its path to that of a true circle (Figure 19.2a). Currently the orbit is relatively close to a circle: Its eccentricity, measured by the method shown in Figure 19.2b, is 0.017. Over the past million years, this value has changed from almost circular ($e = 0.001$) to an extreme value of $e = 0.054$. This change influences the amount of solar radiation that Earth intercepts and also modifies the dates of the solstices and equinoxes. This factor is used to derive the precession of the equinoxes.

Precession of the Equinoxes. Because of varying Earth motions, the days on which the planet reaches its closest and farthest points from the sun (*perihelion* and *aphelion*, respectively) change over time (Figure 19.3). Perihelion now occurs during the Northern Hemisphere winter; in 10,500 years, the date of perihelion will pass to the Northern Hemisphere summer season.

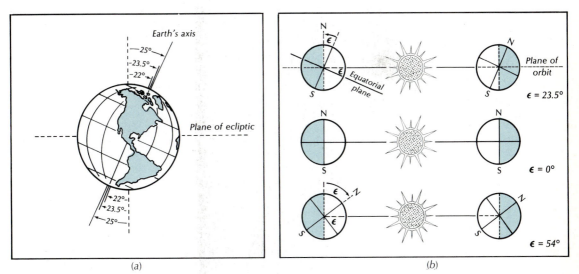

(a)   (b)

FIGURE 19.1

Obliquity of the ecliptic: (a) The amount of tilt of Earth's axis changes between 22° and 25° over a 41,000-year period; the present value is 23.5°. (b) In the three cases shown (present angle, 0°, and 54°), Earth's climate would change appreciably.

FIGURE 19.2

Changes in Earth's orbit around the sun: (a) Over time, the orbit changes from elliptical to almost circular. (b) Eccentricity of the orbit. The sun is one focus (F_1) of the elliptical orbit. The distance from the center (C) to aphelion or perihelion is given by half the major axis, α. The distance from C to F_1, called linear eccentricity (le), is used to determine eccentricity (e): $e = le/\alpha$.

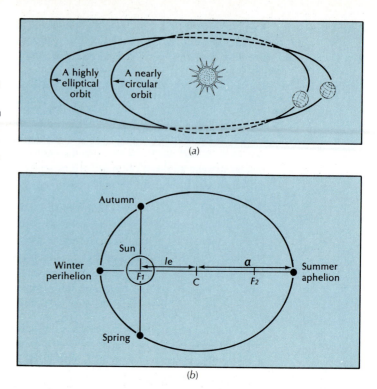

(a)

(b)

Researchers have been able to reconstruct the times at which the various cycles of change reinforce one another to produce very high or very low radiation within a hemisphere. In fact, long before the advent of high-speed computers, Yugoslavian scientist Milutin Milankovitch actually derived values going back thousands of years. Although his derived values provide the basis for understanding the relationships, later researchers constructed models of the cycles and their relative weight in influencing climate. Today, many climatologists think that these orbital variations provide the basis for climatic

FIGURE 19.3

Precession of the equinoxes occurs as dates of solstices and equinoxes move along the orbit.

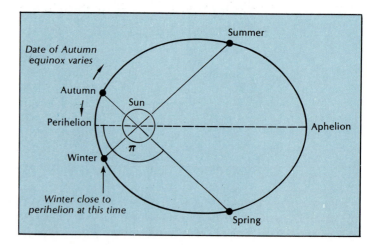

change and have postulated future conditions from the research findings. Using the orbital-variation theory, for example, they predict that a long-term cooling trend that began some 6000 years ago will continue.

Solar Output Variation

Most theories of climatic change assume that the output of energy from the sun is constant or nearly so. However, considering that a fluctuation of less than 10% in solar output could explain all of the climatic changes that have occurred on Earth, it is easy to see how so many theories of climatic change are based on changes in the sun.

Solar-output theories follow two main approaches: The first visualizes an actual change in the radiating temperature of the sun over long periods; while the second considers shorter times and deals with periodic phenomena, specifically sunspots.

One model that considers longer-term changes assumes that the sun contains substances other than hydrogen and helium. Diffusion of these substances from the core of the sun toward the surface would prevent the flow of energy from the core to the surface. This barrier would ultimately reduce the amount of energy passing from the sun's surface and produce an ice age on Earth. Although this is an attractive model, no physical evidence exists to indicate that it could occur; thus, it cannot yet be considered a viable explanation.

Chapters 2 and 9 discussed the nature of sunspots and their effect on weather. Recent research suggests that the shorter sunspot cycles are incidental to a much larger scale of variability. Very few sunspots occurred in the period from 1645 to 1700. This minimum, called the *Maunder minimum*, corresponds to the Little Ice Age on Earth. In contrast, sunspot activity peaked between 1100 and 1250, a distinctively warm time in the Northern Hemisphere. Many climatologists now think that this longer cycle plays a significant role in explaining the variability of temperature since the end of the Ice Age. Further research is needed before any definitive explanation is possible.

Atmospheric Modification

The amount of energy available at Earth's surface depends upon the extent to which the energy is modified as it flows through the atmosphere. Similarly, the amount of energy retained through the greenhouse effect is also a function of the atmosphere. Because of these factors, it is quite easy to theorize causes of climatic change resulting from variations in the transmissivity and absorptivity of the atmosphere.

Of much interest in this respect is the role of volcanic activity in modifying climate. Chapter 9 addressed the short-term impact of volcanic eruptions on the atmosphere. The major difference between volcanoes that influence world temperatures and those that merely have a local effect is penetration of the stratosphere by large amounts of SO_2. This gas joins with atmospheric water vapor to form tiny droplets of sulfuric acid, which can remain in the atmosphere for several years after an eruption. The sulfuric acid droplets reflect sunlight back to space, cooling surface temperatures.

A number of researchers have attempted to measure the role of volcanoes in climatic change. One measure, the Volcanic Explosivity Index (VEI), is based upon the fact that debris from very explosive volcanoes will penetrate the stratosphere and have a greater impact on the atmosphere than those that do not penetrate it. As noted above, however, the eruption's impact also depends on the amount of SO_2 it ejects. The Dust Veil Index (DVI) is, as the name suggests, a measure of turbidity in the atmosphere, a factor influenced by volcanic activity.

While the evidence clearly shows that volcanic activity can have an impact on global climate, relating eruptions to climatic changes of the past is no easy task. Research has shown that large eruptions such as Krakatoa (1883), Katmai (Alaska, 1912) and Tambora (Indonesia, 1815) had a short-term effect on energy flows. Mechanisms for cleansing the atmosphere lead to particle fallout from the air. Herein lies a major drawback to the volcanic-dust theory of climatic change. To account for major glaciations, it assumes that volcanic dust has remained in the atmosphere for a long term. One hypothesis argues that during active times of Earth-building forces, continued volcanic activity over long periods would have an extended effect; once glaciation began, the role of volcanic dust would be secondary to changing surface albedos. Unfortunately, study of deep-sea cores seldom shows layers of volcanic ash beneath deposits indicative of climatic cooling.

Other changes in atmospheric composition that influence climate include modifications of ozone, water vapor content, CO_2, and oxides of nitrogen. While these do change naturally over time, the major modifications currently are a result of human activities. It has already been noted that the ozone layer has been jeopardized by addition of chlorofluorocarbons; the relative impact of the other changes is discussed later in this chapter.

Distribution of Continents

Modern geophysical research has provided a unifying theory for the old idea of continental drift, the concept of plate tectonics. We are now reasonably certain that the present positions of the world land masses are but a transitory condition in the long-term evolution of the continents and the oceans. Given this, one set of theories of climatic change deals with the relative location of continental land masses in relation to the position of the poles and the equator.

When we reconstruct the continental distribution during the last two great Earth-cooling periods, an interesting pattern results. Reconstructed maps for the Permo-Carboniferous glaciation (250 million years ago) and those for the most recent Pleistocene glaciation show that both periods were marked by a concentration of land masses in the polar realms. The presence of large land masses in polar areas is much more conducive to glacial formation because land masses lack the heat storage and transfer mechanisms of oceans. A primary requirement for the formation of great ice caps thus may have been the polar location of continents. Of course, land massed near the poles has not always been associated with glaciation. The continents' relative location may well be an important factor in ice-age occurrences, but actual ice formation must rest upon other causal factors.

Intimately related to the idea of moving continents, although often treated as a separate theory of climatic change, is the role of mountain building and continental uplift. To understand this theory, we need only think of the formation of permanent snow on Mount Kilimanjaro, a mountain located astride the equator. As land-height increases, so does the potential for ice formation.

Geologists have long noted the relationship between times of extensive mountain-building periods and ice ages. For example, the Permo-Carboniferous and other recent ice ages were preceded by extensive mountain-building periods, although the onset of glaciation lags the mountain building by millions of years. In addition, some mountain-building periods—for example, the Caledonian period in Europe 370–450 million years ago—did not give rise to ice ages.

Despite these criticisms, it is generally agreed that mountain building, or at least the presence of high mountains, contributes to optimum conditions for ice formation. The idea that mountain building can influence climatic change is reinforced by modern research showing that mountain ranges do influence upper-air circulation patterns by changing upper-air vorticity. *Vorticity* is a measure of the air's circulation characteristics; on the leeward side of mountains, for example, a cyclonic circulation—positive vorticity—is induced in airflow.

Variation in the Oceans

Modern research in climatology is paying increasing attention to the oceans' role in the climatic system. Variations in sea-surface temperatures have been linked to changing circulation patterns and weather anomalies. In terms of climatic change, the oceans have received attention, in some cases providing the basis for entire theories of climatic change. Some of the ways that the oceans influence the prevailing climates on Earth include the following:

1. Oceans influence the land's relative elevation: A drop in sea level would increase the heights of the continents and enlarge land masses in area.
2. As a heat-storage mechanism, the oceans are less variable than the continents, and changes in the relative temperature of oceanic waters influence world climates. Changes in salinity, evaporation rates, and the relative penetration of solar energy all influence ocean temperature.
3. As a mobile medium, ocean water plays a significant role in the redistribution of energy over Earth's surface. Ocean currents transport large amounts of heat, and any changes in their relative extent would have extended results. One example of this role is illustrated in Figure 19.4, which shows paleogeographic maps of oceanic circulation during the warm mid-Tertiary and the circulation that existed in the cold conditions of the late Pleistocene periods. The essential difference is that a zonal circulation of water occurred because of the equatorial passage of the Tertiary. This passage closed by Pleistocene times, and the whole circulation pattern was modified to encourage ice formation.

Although the oceans heavily influence the nature of climate that occurs over Earth's surface, many researchers think that their role in climatic change may be a secondary effect. That is, the oceans react to other changes, such as a modified atmospheric circulation; they do not create them. Many think the atmosphere leads and the ocean follows.

Other Theories

The preceding theories represent only a portion of those that have been suggested. Other researchers have introduced ideas ranging from the possible influence of Earth's periodic passage through an interstellar "dust" cloud to variations in atmospheric water vapor caused by both natural and human activities. Despite the many ideas, no single theory can account for all of the observed events; it is evident that Earth's climates result from a spectrum of causal elements. This is obvious from Figure 19.5, which diagrams the components of systems that could be modified to cause a significant change in Earth's climate.

FIGURE 19.4

World oceanic circulation in (*a*) the mid-Tertiary and (*b*) the late Pleistocene. Note how the world-encircling equatorial passage was closed by Pleistocene times, whereas a limited north-south exchange occurred when ice occupied large areas.

Source: R. W. Fairbridge, ed., *The Encyclopedia of Atmospheric Sciences and Astrogeology,* (Stroudsburg: Dowden, Hutchinson and Ross, 1967), 464.

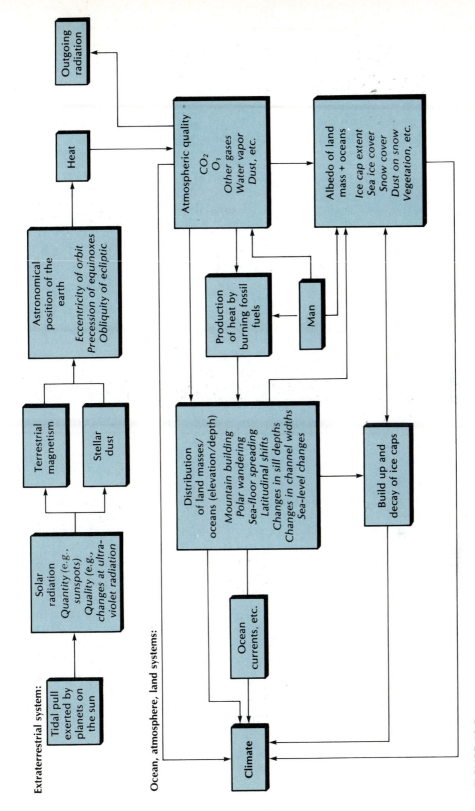

Extraterrestrial system:

Ocean, atmosphere, land systems:

FIGURE 19.5

A schematic representation of some of the possible influences causing climatic change.

Source: A. S. Goudie, *Environmental Change* (Oxford: Oxford University Press, 1977), 203. Copyright © 1977 by Oxford University Press. Reprinted by permission of the Oxford University Press.

Climatic Change: The Human Impact

In the short geologic time that humans have inhabited Earth, they have effected massive changes in the environment. These changes in turn have had a significant impact upon Earth's climate. To examine that impact, we first consider changes that modify Earth's surface and those that impact directly upon the atmosphere.

The Role of Surface Changes

Humans have been altering the environment since they first controlled fire, domesticated animals, and invented agriculture. Modifications began in an early epoch, when the hunters and gatherers used fire to make hunting easier and to drive game during the hunt. Records from early explorers of Africa refer to massive fires, which probably represented the annual burning of grazing areas south of the Sahara. In their visits to the Americas, European explorers noted that Native Americans used fire to improve hunting grounds and catch game.

These activities resulted in deforestation of large areas of the world. In the tropics, the savanna grasslands may represent a response to deforestation by fire; in temperate regions, North American grasslands and eastern European prairie and steppe may be a partial response to burning of former woodlands.

With the development of agriculture, deforestation spread even further. Once-extensive forests in China, the Mediterranean basin, western and central Europe, and North America were cleared for farming. Half of central Europe has been converted from forest to farmland over the last 1000 years. But deforestation was not the only significant result of farming: Misuse of marginal lands led to overuse and an eventual desert-like environment and began the process of desertification. Desertification has affected India, Africa, and South America.

The advent of today's technological society initiated further changes. Destruction of the environment in the quest for raw materials, creation of artificial lakes, generation of energy, expansion of farmlands, urbanization, and other processes have significantly changed the face of Earth.

These cumulative changes have modified the energy interchange that occurs at Earth's surface. The surface climate is a function of the energy that arrives at the surface and the way it is utilized. Of considerable importance in this interchange is the amount of energy that the surface reflects, energy that does not enter the heat balance of the system. The amount of energy reflected depends upon the surface albedo, and changes in surface cover over time have appreciably altered those values. Table 19.1 provides examples of how much human impact has altered albedo. Urbanization (estimated to affect 0.2% of Earth's surface) is not included in the above list because its role in surface modification varies, although it may lead to a lowering of albedo.

The eventual result of the long-term changes listed is a reduction of surface temperatures. The worldwide temperature decrease resulting from the change has been estimated at about 1°C since the beginning of the Holocene, about 10,000 years ago. This value is, of course, open to question because albedo changes lead to other

TABLE 19.1
Changes in albedo that occur with land use.

LAND TYPE CHANGED	EARTH'S SURFACE (%)	ALBEDO
Savanna	1.8	0.16 to
to desert		0.35
Temperate forest		0.12 to
to field, grassland	1.6	0.15
Tropical forest		0.07 to
to field, savanna	1.4	0.16
Salinization, field		0.1 to
to salt flat	0.1	0.25

modifications (e.g., cloud cover and dust) that also play a role in determining Earth's average temperature. Despite this, surface change has clearly had a significant impact on Earth's climate as a whole and on the local areas where changes have been most pronounced.

Changes in Atmospheric Chemistry

Pollution of the atmosphere from human activity has many effects. Perhaps the best known are modifications that have led to changes in CO_2, methane, and stratospheric ozone. What overall effect has pollution had on the prevailing weather and climate?

Atmospheric chemistry plays a vital role in determining surface temperature. While oxygen and nitrogen are by far the major constituents of the atmosphere, many trace elements play an important part in how energy is transferred through the atmosphere. To permit a ready assessment of the changes that are currently occurring, Table 19.2 summarizes the significant gases, their major characteristics, their role in enhancing the greenhouse effect, and their influence upon stratospheric ozone.

Clearly, Table 19.2 emphasizes trace elements that are directly influencing Earth's energy budget. This, of course, is but one part of pollution's role in modifying the atmospheric environment. Air pollution is defined as a change in the concentration of material (or energy) in the air that adversely affects the well-being of living things. Air pollution is neither new nor necessarily associated with industrial effluent. Gases from fetid marshes and smoke from fires are examples of natural air pollutants. Pollution was observed even before factory chimneys and automobiles added their output to the atmosphere. Juan Rodriguez Cabrillo, sailing into San Pedro Bay in 1542, named the Los Angeles basin the "Bay of Smokes." Prevailing weather conditions caused the smoke from scattered Indian fires to hang as a pall of pollution over the area. As we now know, this was a harbinger of things to come; Los Angeles is in a geographical location where stagnant air occurs for almost half the year. It is an optimum setting for both surface and upper inversions.

A phenomenal amount of pollution is pumped into the atmosphere every year. Table 19.3 shows an estimate of emissions into the atmosphere in the United States in the mid-1980s. Note that the data are given in *millions* of metric tons per year (1985). The health

EARTHQUEST™
OFFICE FOR INTERDISCIPLINARY EARTH STUDIES   Spring 1991, Vol. 5, No. 1

	CARBON DIOXIDE CO_2	METHANE CH_4	NITROUS OXIDE N_2O	CHLOROFLUOROCARBONS CFCs	TROPOSPHERIC OZONE O_3	CARBON MONOXIDE CO	WATER VAPOR H_2O
GREENHOUSE ROLE	Heating	Heating	Heating	Heating	Heating	None	Heats in air, cools in clouds
EFFECT ON STRATOSPHERIC OZONE LAYER	Can increase or decrease	Can increase or decrease	Can increase or decrease	Decrease	None	None	Decrease
PRINCIPAL ANTHROPOGENIC SOURCES	Fossil fuels; deforestation	Rice culture; cattle; fossil fuels; biomass burning	Fertilizer; land use conversion	Refrigerants; aerosols; industrial processes	Hydrocarbons (with NOx); biomass burning	Fossil fuels; biomass burning	Land conversion; irrigation
PRINCIPAL NATURAL SOURCES	Balanced in nature	Wetlands	Soils, tropical forests	None	Hydrocarbons	Hydrocarbon oxidation	Evapotranspiration
ATMOSPHERIC LIFETIME	50–200 yr*	10 yr	150 yr	60–100 yr	Weeks to months	Months	Days

PRESENT ATMOSPHERIC CONCENTRATION IN PARTS PER BILLION BY VOLUME AT SURFACE	353,000	1720	310	CFC-11: 0.28 CFC-12: 0.48	20–40 †	100 †	3000–6000 in stratosphere
PREINDUSTRIAL CONCENTRATION (1750–1800) AT SURFACE	280,000	790	288	0	10	40–80	Unknown
PRESENT ANNUAL RATE OF INCREASE	0.5%	0.9%	0.3%	4%	0.5–2.0% †	0.7–1.0% †	Unknown
RELATIVE CONTRIBUTION TO THE ANTHROPOGENIC GREENHOUSE EFFECT	60%	15%	5%	12%	8%	None	Unknown

TABLE 19.2
Atmospheric trace gases that affect the energy balance and are of significance to global climatic change.

Source: *EarthQuest*, Spring 1991, Vol. 5, No. 1, Office for Interdisciplinary Earth Studies. Boulder, CO: University Corporation for Atmospheric Research. Reprinted by permission.

* The lifetime of CO_2 in the atmosphere is calculated in two ways: one, which is relatively short (10 years), is the residence time of a single molecule before dissociation; more relevant to global change is the longer period that includes the residence time in the atmosphere–ocean system: the time that a CO_2 molecule derived from fossil fuel remains in the atmosphere–ocean system before being sequestered as terrestrial humus or deep-sea sediment. The latter is of greater value in calculating future scenarios related to greenhouse warming, since it reflects the relaxation time between a cessation of all industrialized CO_2 emissions and the expected detection of a global decrease in the pressure of CO_2.

† Northern Hemisphere.

TABLE 19.3
National emissions estimates for 1985 (millions of metric tons).

SOURCE CATEGORY	PARTICULATES	SULFUR OXIDES	NITROGEN OXIDES	VOLATILE ORGANICS	CARBON MONOXIDE	LEAD*
Transportation						
Highway vehicles	1.1	0.5	7.1	6.0	40.7	14.5
Aircraft	0.1	0.0	0.1	0.2	1.1	—
Railroads	0.0	0.1	0.5	0.1	0.2	—
Vessels	0.0	0.2	0.2	0.4	1.4	—
Other off-highway	0.1	0.1	1.0	0.4	4.1	0.9
Transportation total	1.3	0.9	8.9	7.1	47.5	15.4
Stationary source fuel combustion						
Electric utilities	0.6	14.2	6.8	0.0	0.3	0.1
Industrial	0.3	2.2	2.9	0.1	0.6	0.4
Commercial institutional	0.0	0.4	0.2	0.0	0.0	0.0
Residential	1.2	0.2	0.4	2.4	7.1	0.0
Fuel combustion total	2.1	17.0	10.3	2.5	8.0	0.5
Industrial processes	2.7	2.9	0.6	8.6	4.6	2.3
Solid waste disposal						
Incineration	0.1	0.0	0.0	0.3	1.1	—
Open burning	0.2	0.0	0.1	0.3	0.9	—
Solid waste total	0.3	0.0	0.1	0.6	2.0	2.8
Miscellaneous						
Forest fire	0.7	0.0	0.1	0.6	4.7	—
Other burning	0.1	0.0	0.0	0.1	0.6	—
Miscellaneous organic solvent	0.0	0.0	0.0	1.6	0.0	—
Miscellaneous total	0.8	0.0	0.1	2.3	5.3	—
Total of all sources	7.2	20.8	20.0	21.1	67.4	21.0

Source: National Air Pollutant Emission Estimates, 1940–85, U.S. Environmental Protection Agency Publication No. EPA-450/4–86–018, 1987.

* Thousands of metric tons.

consequences of air pollution are well-known, and environmental laws have been effected to try to contain pollution. The *Pollution Standards Index* (PSI) was developed to give a measure of air quality for a defined location. It compares the value of the pollutant that occurs in the highest concentration to the national air-quality standards. If the pollutant is at the primary standard, the air has a PSI value of 100. Air below that value is considered moderate or good; air above it passes from unhealthy to hazardous.

While air pollution is clearly a health hazard, the emphasis here is on air pollution as a cause of climatic change. Beyond the role of atmospheric modifications (Table 19.2), impacts on local climates are readily demonstrated in the urban setting. In the discussion of urban climate, considerable emphasis was given to atmospheric modification.

Acid Rain

One other component of change results from human modification of the atmospheric environment: the change of quality resulting from acidification of rainfall. Acid rainfall certainly changes the ecosystem that it influences and needs to be considered as an agent of local climatic change. The term *acid rain* was first used by British Chemist Robert Angus in 1872 when he studied the rain in London, Glasgow, and Liverpool. He suggested that free sulfuric acid in air caused colors in prints and yard goods to fade, metals to rust, and wooden blinds to rot.

Acids are substances that have many positively charged hydrogen ions. Common items that are acid to varying degrees are cola, lemon juice, vinegar and battery acid. A liquid's relative acidity (or alkalinity) is measured by the pH scale (Figure 19.6). Note that the scale is *logarithmic*, so a pH of 5 is 10 times less acidic than a pH of 4, and 100 times less acidic than a substance with a pH of 3. Normally, the pH of rainfall, which might be considered pure water contaminated by atmospheric CO_2, is 5.6, which is acidic. Table 19.4 provides some indicators of acid level and some actual measurements of acid rain in North America.

Acid rain originates mainly from the release of sulfur oxides and nitrogen oxides into the atmosphere through industrial and transportation sources. These chemicals are transformed into sulfuric acid and nitric acid by oxidation and hydrolysis in the atmosphere. Sulfur dioxide makes up about two-thirds of the acid in acid rain, while nitrogen oxides make up the balance. As part of the circulating air, they are transported by the atmosphere and eventually washed out by precipitation. This latter process may not occur until the pollutants have moved hundreds of miles from the source (particularly when they are emitted from very tall smokestacks), so the problem has become global. Indeed, one of the few areas of conflict between the United States and Canada is the "export" of pollutants that give rise to acid rain.

Some areas are more severely affected by acid rain than others. Topography, geology, and weather patterns contribute to the intensity of acid rain. West Virginia, the Adirondack Mountains, eastern Canada, and the Black Forest region of Germany all suffer excessively from acid rain.

A study of lakes in New York's Adirondack Mountains showed that more than one-third of 214 mountain lakes had pH values so acidic that no fish were present. It has also been reported that in Nova Scotia, Canada, the pH of some rivers has fallen to a point where salmon cannot live. Depending on the amount and acidity of the precipitation, and

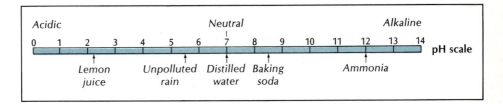

FIGURE 19.6
The scale of acidity (pH scale) showing relative acidity of well-known items.

TABLE 19.4

Indicator levels of pH related to acid rain.

ITEM	pH
Neutral solution	7.0
Natural rainfall	5.6
Fish reproduction affected	5.0
Lethal to fish	4.5
Average rainfall—Eastern North America	4.4
Acidity of tomato juice	4.3
Average rainfall—Pennsylvania, New York, and Ontario	4.2
Rainfall at Toronto, Canada, Feb. 1979	3.5
Rainfall in the Smoky Mountains of North Carolina	3.3
Acidity of lemon juice	2.2
Rainfall at Mount Mitchell, NC, July 1986	2.2
Rainfall at Wheeling, WV, 1980	1.4
Battery acid	1.1

Note: For each unit decrease in the scale, acidity increases by a factor of 10.

the chemical makeup of the water and soils in the area, living organisms can be greatly affected. In acidic water, leaves and other organic debris do not decompose readily because bacteria that cause decomposition cannot live in the acidic solution. Plankton, bacteria, and insects that provide feed for trout die, and eventually the fish die of starvation.

Acid is corrosive and results in a rapid decay of rock, cement, and metal. Many world-famous buildings are crumbling in the deluge of acid rain. Among them are the Parthenon in Greece, The Taj Mahal in India, and the Mayan ruins in the Yucatan Peninsula. The Taj Mahal is set in the middle of an industrial district near New Delhi. In an attempt to offset the effects of acid rain, corroded slabs of marble are being replaced with new ones. But the restoration cannot keep up with the rate of deterioration. Acid rain destroys marble statues, including many of Michelangelo's works in Italy. Paul Revere's tombstone is no longer legible. Bronze statues are stained by the effects of acid rain. In addition, acid rain is very hard on automobile finishes and makes the metal bodies rust faster.

Acid rain is extremely damaging to forests because the acid rain releases some metals such as aluminum that are normally inert in the soil. These metals harm the root systems of trees and interfere with their ability to absorb nutrients.

Research indicates that soil pH plays an important part in its evolution and its properties. In some instances, increased acidity can result in the exclusion of plants that would be present under nonacidic conditions.

In 1980, U.S. President Jimmy Carter created the National Acid Precipitation Assessment Program (NAPAP) to study the causes and effects of acid rainfall. The results of this study, released in 1990, showed that initial projections of acid rainfall's impact were both over- and underestimated. For example, acidification of lakes does occur, and about 4% of the lakes sampled were acidic. On the other hand, levels of acidic precipitation in the United States were not found responsible for regional crop-yield reductions.

■ Applied Study

Donora: An Air-Pollution Episode

Stable atmospheric conditions are often associated with inversions, which result from the accumulation of cold air below a layer of warmer air. Such conditions can occur in upper levels of the troposphere or at ground level. This study considers air-pollution episodes associated with ground-level inversions.

Europe has had serious acid rain and fog problems for many years. In December 1930, the heavily industrialized Meuse River Valley in Belgium was blanketed for a week by a dense fog. Industrial pollutants added to the air remained in place with the fog; by the end of the week, hundreds of people had fallen ill from respiratory ailments, and 60 died. In December 1962, London experienced a thick, sulfurous fog that lasted five days. The pollution-laden fog was credited with contributing to almost 4000 deaths. Such air-pollution episodes ultimately led to an awareness of the problems of air pollution and eventually to protective legislation.

In the United States, the death toll of air pollution is not so evident as in these examples. That it can be deadly, however, is aptly demonstrated by the air pollution episode that occurred in Donora, PA, in 1948.

Donora is located about 48 km (28 mi) south of Pittsburgh on the Monongahela River. It is a sheltered location; bluffs up to 130 m (450 ft) rise from the river and enclose it on the north, east, and south sides. To the west, high hills complete the encirclement. In October 1948, the weather over a large area, including Pennsylvania, New Jersey, and New York, was dominated by an anticyclone to give clear, calm conditions. Days were warm and the clear nights were cool.

At night, the layers of air next to the ground were chilled; in the hilly area in the Donora region, the dense layer of cold surface air flowed downhill under gravity. A pool of cold air formed in the enclosed valley. In the calm conditions, no wind mixed the cold pool; the low sun angle and the shadow of the mountains prevented its heating by direct sunlight. The cool air remained and, on the next night, deepened as more cool air flowed down the hillsides. This continued for six days.

In the industrial town, most people were employed in the steel plants and in factories that produce wire and sulfuric acid. These factories have stacks about 30 m (100 ft) high, well below the level of the surrounding hills. The factories continued to pour pollutants into the stable air, which became laden with sulfur dioxide and particulates. As time progressed, hospitals and doctor's offices reported a sharp increase in illness, eventually amounting to 6000 cases. Some people did not recover from breathing the polluted air, and 20 people died. Only when a low-pressure system moved across the area, bringing wind and rain, did the conditions improve. ■

Summary

What causes climates to change is a fascinating question. Many theories have been postulated, but no single one can totally satisfy all necessary requirements. Little doubt remains that changes in the Earth–sun relationship may be a basic cause of long-term climatic change on Earth. But it appears that the effects must be considered in conjunction with other factors. The problem of explaining the change is further complicated by human activity, which is now a major factor in climate modification. To explain what is happening, "natural" causes must be considered together with human impact. Human activity and the addition of CO_2 to the atmosphere have made extensive changes in Earth's surface and atmosphere. In cities, human impact is so great that it has created a new set of climatic conditions.

FUTURE CLIMATES

Chapter Overview

Models
- Analog Models
- The Nature of GCMs
- Model Projections

Potential Impacts of Global Warming
- Agriculture
- Forests
- Biodiversity
- Water Resources
- Electricity
- Air Quality
- Health

During the 1980s, phrases like *greenhouse effect, global warming,* and *Antarctic ozone hole* came into common use in the mass media. Widespread coverage of what was a very warm decade also did not go unnoticed by the public. In effect, climatology and climatic change evolved from a largely academic discipline into a subject of wide debate as future climates took on international importance.

Almost all of the debate that grew out of the 1980s concerns the impact on global warming of adding greenhouse gases to the atmosphere. That impact has caused some controversy, with a number of scientists arguing that the case for such warming and associated change is overstated. Nonetheless, the consensus is that global warming is real, although not everyone agrees on how fast it is occurring or the extent of its impact. Perhaps the best view of the controversy comes from the final report of the working group of the Intergovernmental Panel on Climatic Change (IPCC), jointly sponsored by the World Meteorological Organization and the United Nations Environment Program. The report was based upon the scientific assessment of climatic change by several hundred scientists in 25 countries, providing perhaps the most comprehensive set of opinions on the subject.

The IPCC study states with certainty that a natural greenhouse effect is keeping Earth warmer than it would otherwise be. Emissions from human activities are substantially increasing the atmospheric concentration of those greenhouse gases. These increases are enhancing the greenhouse effect, resulting in additional warming of Earth's surface. The main greenhouse gas is water vapor.

Scientists participating in the study were confident that CO_2 has been responsible for over half of the enhanced greenhouse effect in the past, and its impact is likely to continue in the future. The atmospheric concentrations of long-lived gases CO_2, NO, and the CFCs) adjust very slowly to changes in emissions. Continued emission of these gases at present rates would mean increased concentrations in the atmosphere for centuries to come. To stabilize atmospheric concentrations at their present level, we would need to lower emissions of long-lived gases by 60%.

Based upon these findings, the IPCC panelists provided predictions of global temperature increase and its resulting impact. Before examining their findings, we need to explore the types of analysis on which they base such predictions and examine models and scenarios they have produced.

Models

A model is a simple representation of something real. A model car, for example, has the general shape and characteristics of a car but may have neither a motor nor any working parts. A model is not expected to be exactly like the real thing. Physical models can, however, be used as analogs for deriving important predictive information. In the case of a model car, if the model is a scaled replica, it might be used in a wind tunnel to determine how air will flow around the real car. As will be described later, climatology uses analog models. Not all models are physical: Models can also be graphic, conceptual, or numerical/mathematical. Climatology uses all of these types of models.

Graphic models have been used throughout this book. Maps of the flow of upper-air westerlies, for example, provide an image of how air circulates on a grand scale.

Conceptual models have also been frequently used; the hydrologic cycle is a conceptual model of the complex way in which water circulates over Earth and explains highly complex interchanges in a simple way. For understanding and ultimately predicting future climate, both analog and mathematical models are of major importance.

Analog Models

Analog models are so named because their attributes are often analogous to conditions that exist in the real world. A classic example is the electrical analog model, which uses voltages, variable resistors, and amplifiers to represent various components of energy flow. The flow of energy from the sun may be represented by a constant electrical flow that can be modified by various resistors representing atmospheric components, such as clouds, that affect the flow of energy to Earth.

Researchers naturally look to history for analogous climates, to see if the past can provide any guides to the future. Establishing paleoanalogs has thus become an integral part of climatic research. Today, researchers are especially interested in identifying earlier times when a "greenhouse" climate prevailed. Identifying environmental conditions that prevailed during a similar time could prove beneficial in predicting potential future "greenhouse" conditions.

Three analogs exist for warmer climates: (1) the Pliocene warm climate that occurred about 3.3–4.3 million years ago and was part of wide swings of warm and cool climates; (2) a Pleistocene interglacial, the Eemian (125,000–130,000 years ago) was identified in Antarctic cores and had very warm conditions compared to other interglacials; and (3) the Holocene Climatic Optimum that occurred 5000 years ago. Of these, the Climatic Optimum has been most closely examined.

This warm time saw summer temperatures in high latitudes some 3°–4°C (5–7°F) above modern values and increased precipitation in subtropical and high latitudes. The Cooperative Holocene Mapping Project (COHMAP), an extensive interdisciplinary study, suggests that during this Climatic Optimum, summer temperatures in the midlatitudes were only 1°–2°C (2°–3°F) higher. Further south, summer temperatures in the midlatitudes were often lower than today, for example in Soviet Central Asia, the Sahara, and Arabia. Annual precipitation was also higher in these areas. While this reconstruction is not flawless, it does provide a relative guide to warmer conditions. The two earlier warm times, the Eemian and Pliocene optima, have less available evidence and more uncertainty.

Just as analogs can take a number of approaches, so other models of climate may be constructed in a number of ways. Energy budget models, for example, consider the flow of energy through the Earth–atmosphere system. Mathematical models, the most sophisticated of which are General Circulation Models (GCMs), have been used most widely in attempts to predict climates of the future. (Note that Chapter 8 discussed classical GCMs; this chapter uses GCM to refer to computer-generated models.)

The Nature of GCMs

Modern GCMs are a product of computers' calculating capability. Without high-speed computers and their prodigious memory, attempting to recreate the complexities of atmospheric circulation would be impossible. GCMs attempt to solve equations—of motion, thermodynamics, and conservation—for moisture and mass in a defined global area through various atmospheric levels. Interpreting the results of these models requires understanding a number of factors.

Models must have stated *boundary conditions*, forces that drive or influence the climate system but that are external to it. Boundary conditions can be physical objects, such as the ocean, or energy boundaries, such as the flow of solar energy into the system. The latter is a forcing function, for the energy available forces climate toward a particular temperature distribution.

Early GCMs faced with the problem of ocean boundary conditions used the so-called *swamp models*, which treated the oceans like perpetually wet land. These models had no heat capacity or active heat-transfer mechanisms. More recent models use an ocean layer with a prescribed heating capacity and effects of ocean heat transport incorporated. Apart from the problem of ocean temperature and energy transfer, GCMs must also identify boundary conditions for the extent and elevation of land and the albedos of land surfaces.

Beyond the problem of boundary conditions, the modeler must also solve atmospheric conditions within the framework of Earth's surface. Because analyzing every point on the surface is not possible, a *grid* provides the spatial detail needed. Calculations are made at widely spaced points on a three-dimensional grid placed over the surface. The centers of each grid are often several hundred kilometers apart, which makes the area represented by one grid point larger than a state the size of Colorado. To cite one case, the model used at the National Center for Atmospheric Research (NCAR) has a grid with nine layers stacked to a height of about 30 km, with about 4.5° of latitude and 7° of longitude between grid points. Even given this coarse network, a computer run takes several days; if a moderately complex GCM is predicting future climate in yearly steps, a 100-year simulation may take two months. Reducing the grid size by one order of magnitude would increase the computation time for a supercomputer to many years.

An average value for a model grid point represents data for the entire grid. Clearly, if a given grid contains significant departures from the average, caused for example by local mountains or lakes, then the model cannot represent changes for those areas. In addition, the grid size exceeds that of some mesoscale weather systems. And grids do not yet allow predictions of changes in cloud types and cloudiness.

To address the grid's lack of detail, subgrid-size phenomena are represented by a collective value, or *parameterized*. This process incorporates the statistical effects of the processes into the model by relating them to known variables, such as temperature, wind, or humidity. In effect, if we know temperature and humidity, we can predict average cloudiness within an identified grid box.

Climatic sensitivity must also be considered in constructing a model. This, in effect, concerns the magnitude of the climatic response to an interruption or perturbation. In modeling, it is usually the difference between simulations as a function of change in a given parameter. Thus, climatic sensitivity for conditions in which atmospheric CO_2 is twice normal levels can be compared with that when CO_2 is at other levels.

In considering climate model predictions, we must differentiate between *equilibrium* and *realized* climate. When the Earth–atmosphere system changes, as is the case following an increase in greenhouse gases, the atmosphere tends to warm immediately. But the atmosphere is coupled to other systems, particularly the oceans, which warm up much more slowly. Eventually, both atmosphere and oceans will establish a new equilibrium temperature; but at any time prior to the establishment of the new equilibrium, or as long as the CO_2 content changes, a realized or transient temperature will be in effect. Models indicate that an equilibrium climate may not be reached for centuries following continuing increases of greenhouse gases.

Each of these considerations is used in the current models, exemplified by those listed in Table 20.1. These models, all of which are global and mixed layer ocean models, will be referred to in the following pages.

Model Projections

Table 20.1 summarizes findings of various models. This extensive listing classifies the models (A, B, C, D) and provides the group that completed the investigation. Table 20.2 explains acronyms for the groups in Table 20.1. A "gallery" of model outputs is shown in Figures 20.1–20.4.

In evaluating the results of the models, the IPCC scientists attempted to determine how much temperature will increase under a number of scenarios. Figure 20.5 provides three estimates of global temperature increase under a "business-as-usual" scenario. This assumes that no controls are placed upon greenhouse gas emissions and that temperatures will rise about 1°C (2°F) above present levels (2°C (4°F) above preindustrial levels) by 2025. Before the end of the twenty-first century, global mean temperature will be 3°C (5°F) above the current level.

Beyond the gross figure for global average temperatures, models also provide estimates for regional climates and scenarios for the occurrence of severe weather events. Figure 20.6 shows the areas for which regional changes were assessed, and Table 20.3 shows some regional estimates for change by the year 2030. As might be anticipated, these are low-confidence projections.

A number of factors contribute to the main uncertainties in the projected temperature increase. First, we still do not completely understand the nature of sinks for CO_2. Also, future additions of CO_2 to the atmosphere will depend largely upon governments' actions and policies regarding fossil fuel use. In addition, we do not understand the role of clouds in a greenhouse-warmed climate. Whether clouds will better reflect short-wave radiation or better absorb long-wave radiation is not yet clear. Finally, the role of the oceans, which influence both the pattern and timing of climatic change, has yet to be fully determined. Clearly, much research remains to be completed.

Potential Impacts of Global Warming

Many might consider an increase of but a fraction of a degree of temperature of little consequence in the milieu of climate, where day-to-day changes are in terms of tens of degrees. But changes in the mean global temperature are but one indicator of many climatic systems occurring over the globe. A seemingly insignificant change in global temperature might lead to a modification of temperature gradients over Earth's surface, which would in turn lead to a modification of the general circulation pattern. Changing air-mass frequency, precipitation distribution, and storm tracks are but some of the related consequences. Figure 20.7 provides a highly schematic depiction of this change.

TABLE 20.1

Summary of results from global mixed-layer ocean atmosphere models used in equilibrium $2 \times CO_2$ experiments.

GROUP	YEAR	NO. OF VERTICAL LAYERS	DIURNAL CYCLE	CONVECTION	CLOUD	CLOUD PROPERTIES	ΔT (°C)	ΔP (%)
A. Fixed, zonally averaged cloud; no ocean heat transport								
GFDL	1980	9	N	MCA	FC	F	2.0	3.5
	1986	9	N	MCA	FC	F	3.2	n/a
B. Variable cloud; no ocean heat transport								
OSU	1989	2	N	PC	RH	F	2.8	8
	1989	2	N	PC	RH	F	4.4	11
MRI	1989	5	Y	PC	RH	F	4.3	7
NCAR	1984	9	N	MCA	RH	F	3.5	7
	1989	9	N	MCA	RH	F	4.0	8
GFDL	1986	9	N	MCA	RH	F	4.0	9
C. Variable cloud; prescribed oceanic heat transport								
AUS	1989	4	Y	MCA	RH	F	4.0	7
GISS	1981	7	Y	PC	RH	F	3.9	n/a
	1984	9	Y	PC	RH	F	4.2	11
	1984	9	Y	PC	RH	F	4.8	13
GFDL	1989	9	N	MCA	RH	F	4.0	8
UKMO	1987	11	Y	PC	RH	F	5.2	15
	1989	11	Y	PC	CW	F	2.7	6
	1989	11	Y	PC	CW	F	3.2	8
	1989	11	Y	PC	CW	V	1.9	3
D. High resolution								
CCC	1989	10	Y	MCA	RH	V	3.5	4
GFDL	1989	9	N	MCA	RH	F	4.0	8
UKMO	1989	11	Y	PC	CW	F	3.5	9

Key:
N = Not included
PC = Penetrative convention
FC = Fixed cloud
F = Fixed-cloud radiative properties
ΔT = Equilibrium surface temperature change on doubling CO_2
Y = Included

CA = Convective adjustment
RH = Condensation- or relative humidity-based cloud
ΔP = Percentage change in precipitation
MCA = Moist convective adjustment
CW = Cloud water
V = Variable cloud radiative properties

Source: J. T. Houghton, G. J. Jenkins, and J. J. Ephraums, eds. *Climate Change, The IPCC Scientific Assessment.* (New York: Cambridge University Press for the Intergovernmental Panel on Climatic Change, WMO/UNEP, 1990).

TABLE 20.2

Acronyms of Modeling Groups given in Table 20.1.

ACRONYM	NAME
GFDL	Geophysical Fluid Dynamics Laboratory, Princeton, NJ
OSU	Oregon State University, Corvallis
MRI	Meteorological Research Institute, Japan
NCAR	National Center for Atmospheric Research, Boulder, CO
AUS	CSIRO, Australia
GISS	Goddard Institute for Space Studies, New York
UKMO	Meteorological Office, United Kingdom
CCC	Canadian Climate Centre, Canada

Source: J. T. Houghton, G. J. Jenkins, and J. J. Ephraums, eds. *Climate Change, The IPCC Scientific Assessment.* (New York: Cambridge University Press for the Intergovernmental Panel on Climatic Change, WMO/UNEP, 1990).

If we assume that global temperatures are rising and will continue to rise over the next century, then we can anticipate a number of potential impacts. Some of the scenarios, created for selected impacts, are considered here. (Sea-level changes and their global consequences were dealt with in Chapter 4 and are not reiterated here.)

Of the many impact studies completed and published, of considerable interest is that contained in a report to Congress by the U.S. Environmental Protection Agency (EPA) in 1989. Findings contained in this study were derived from the GISS, GFLD, and OSU models (see Table 20.1), supplemented by analog studies from actual data, and focus on impacts in the United States. The remainder of this chapter summarizes the EPA findings.

Agriculture

Agriculture, a critical component of the U.S. economy, contributed 17.5% of the gross national product in 1985, with farm assets totaling $771 billion. As well illustrated by the dust bowl years of the 1930s, crop production is sensitive to climate, soils, management methods, and many other factors. During those years, wheat and corn yields dropped by up to 50%. More recently, during the drought of 1988, corn yields declined about 40%. The agricultural analyses in the EPA report examined potential impacts on crop yields and productivity from changes in climate and direct effects of CO_2. The latter is significant because higher CO_2 concentrations may increase plant growth and water-use efficiency.

The study indicated that crop yields could be reduced, although the combined effects of climate and CO_2 would depend on the severity of climate change. In most regions of the country, climate change alone could reduce dryland yields of corn, wheat, and soybeans, with site-to-site losses ranging from negligible amounts to 80%. These decreases would result primarily from higher temperatures, which would shorten a crop's life cycle. In very northern areas, such as Minnesota, dryland yields of corn and soybeans could increase as warmer temperatures extend the frost-free growing season. The combined effects of climate change and increased CO_2 may result in net increases in yields in some cases, especially in northern areas or where rainfall is abundant. But in southern areas, where heat stress is already a problem, and in areas where rainfall is low, crop yields could decline.

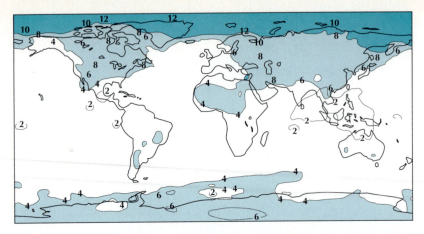

(a)

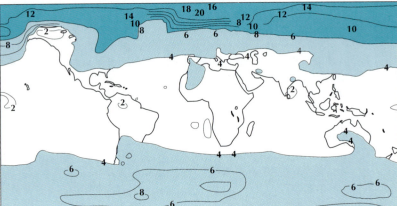

(b)

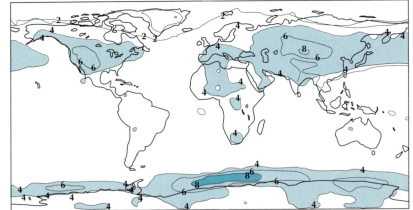

(c)

FIGURE 20.1

Change in surface air temperature (10-year means) due to doubling CO_2, for December through February, as simulated by three high-resolution models: (a) Canadian Climate Centre, (b) Geophysical Fluids Dynamics Laboratory and (c) United Kingdom Meteorological Office. Contours are every 2°C (4°F), light stippling where the warming exceeds 4°C (7°F), and dashed shading where the warming exceeds 8°C (14°F).

Source: J. T. Houghton, G. J. Jenkins, and J. J. Ephraums, eds., *Climate Change, The IPCC Scientific Assessment.* (New York: Cambridge University Press for the Intergovernmental Panel on Climatic Change, WMO/UNEP, 1990), 141.

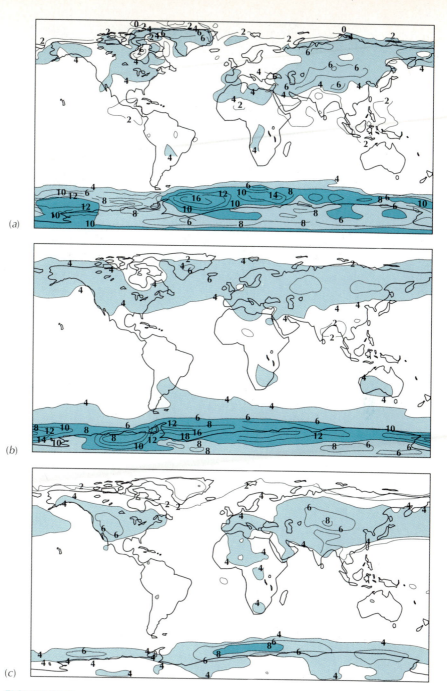

FIGURE 20.2

Change in surface air temperature (10-year means) due to doubling CO_2, for June through August, as simulated by three high-resolution models: (a) CCC, (b) GFHI and (c) UKHI.

Source: J. T. Houghton, G. J. Jenkins, and J. J. Ephraums, eds., *Climate Change, The IPCC Scientific Assessment*. (New York: Cambridge University Press for the Intergovernmental Panel on Climatic Change, WMO/UNEP, 1990), 142.

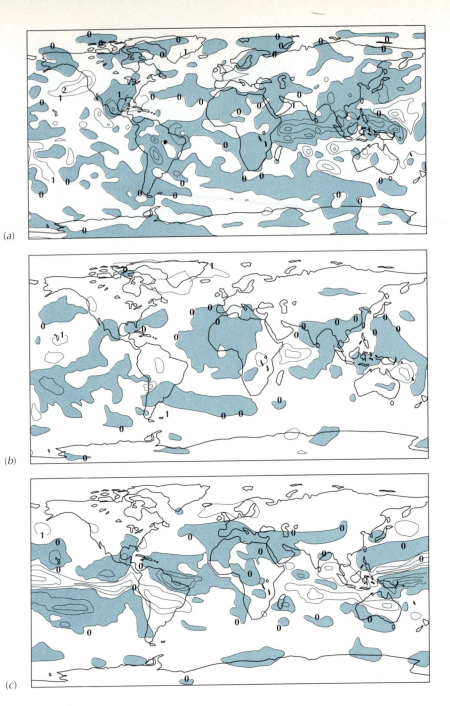

FIGURE 20.3
Change in precipitation (smoothed 10-year means) due to doubling CO_2, for December through February, as simulated by the (*a*) CCC, (*b*) GFHI, and (*c*) UKHI models. Contours at +/− 0, 1, 2, 5 mm day^{-1}, and areas of decrease are shown in color.

Source: J. T. Houghton, G. J. Jenkins, and J. J. Ephraums, eds. *Climate Change, The IPCC Scientific Assessment.* (New York: Cambridge University Press for the Intergovernmental Panel on Climatic Change, WMO/UNEP, 1990), 144.

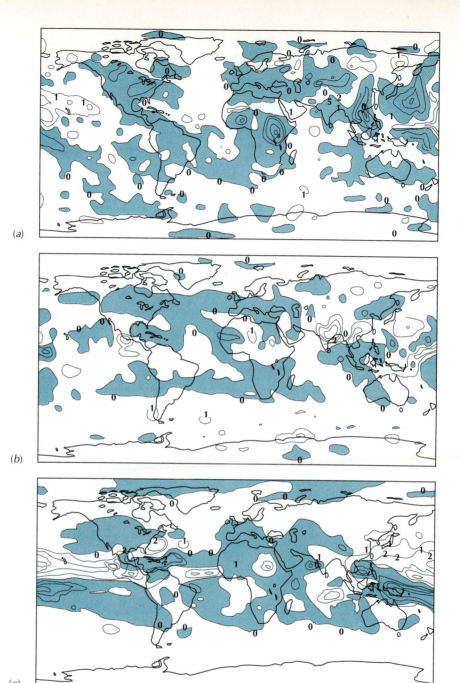

(a)

(b)

(c)

FIGURE 20.4

Change in precipitation (smoothed 10-year means) due to doubling CO_2, for months June through August, as simulated by the (a) CCC, (b) GFHI, and (c) UKHI models. Contours are at +/− 0, 1, 2, 5 mm day^{-1}, and areas of decrease are shown in color.

Source: J. T. Houghton, G. J. Jenkins, and J. J. Ephraums, eds. *Climate Change, The IPCC Scientific Assessment.* (New York: Cambridge University Press for the Intergovernmental Panel on Climatic Change, WMO/UNEP, 1990), 145.

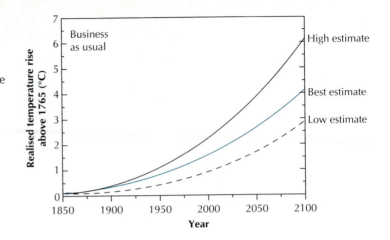

FIGURE 20.5

Simulation of the increase in global mean temperature from 1850–1990 due to observed increases in greenhouse gases and predictions of the rise between 1990 and 2100 resulting from "business-as-usual" emissions.

Source: J. T. Houghton, G. J. Jenkins, and J. J. Ephraums, eds. *Climate Change, The IPCC Scientific Assessment.* (New York: Cambridge University Press for the Intergovernmental Panel on Climatic Change, WMO/UNEP, 1990), xxii.

Under all of the scenarios (with and without the direct effects of CO_2), the relative productivity of northern areas for the crops studied was estimated to rise in comparison with that of southern areas. In response to the shift in relative yields, grain crop acreage in Appalachia, the Southeast, and the southern Great Plains could decrease, and acreage in the northern Great Lakes States, the northern Great Plains, and the Pacific Northwest could increase. A change in agriculture would affect not only the livelihood of farmers but also agricultural infrastructure and other support services. The sustainability of crop production in northern areas was not studied. Changes in foreign demand for U.S. crops,

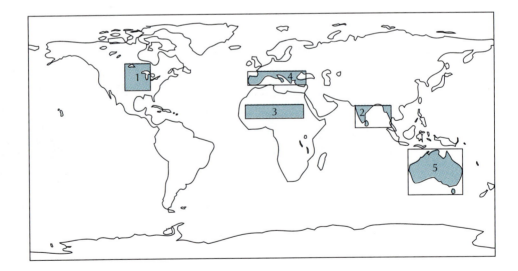

FIGURE 20.6

Map showing the locations of five areas for which IPCC regional changes were assessed. Estimate of climate change in these areas is given in Table 20.3.

Source: J. T. Houghton, G. J. Jenkins, and J. J. Ephraums, eds. *Climate Change, The IPCC Scientific Assessment.* (New York: Cambridge University Press for the Intergovernmental Panel on Climatic Change, WMO/UNEP, 1990), xxiv.

TABLE 20.3

Computer-generated temperature changes by the year 2030.

(IPCC Business-as-Usual scenario; changes from pre-industrial)

The numbers given here are based on high-resolution models, scaled to be consistent with our best estimate of global mean warming of 1.8°C by 2030. For values consistent with other estimates of global temperature rise, the numbers below should be reduced by 30% for the low estimate or increased by 50% for the high estimate. Precipitation estimates are scaled in a similar way.

Confidence in these regional estimates is low.
(See Fig. 20.6 for locations.)

Central North America (35°–50° N 85°–105°W)
The warming varies from 2° to 4°C in winter and 2° to 3°C in summer. Precipitation increases range from 0% to 15% in winter whereas there are decreases of 5%–10% in summer. Soil moisture decreases in summer by 15%–20%.

Southern Asia (5°–30° N 70°–105° E)
The warming varies from 1° to 2°C throughout the year. Precipitation changes little in winter and generally increases throughout the region by 5%–15% in summer. Summer soil moisture increases by 5%–10%.

Sahel (10°–20° N 20° W–40° E)
The warming ranges from 1° to 3°C. Area mean precipitation increases and area mean soil moisture decreases marginally in summer. However, there are areas of both increase and decrease in both parameters through the region.

Southern Europe (35°–50° N 10° W–45° E)
The warming is about 2°C in winter and varies from 2° to 3°C in summer. There is some indication of increased precipitation in winter, but in summer, precipitation decreases by 5%–15% and soil moisture by 15%–25%.

Australia (12°–45° S 110°–115° E)
The warming ranges from 1°to 2°C in summer and is about 2°C in winter. Summer precipitation increases by around 10%, but the models do not produce consistent estimates of the changes in soil moisture. The area averages hide large variations at the subcontinental level.

Source: J. T. Houghton, G. J. Jenkins and J. J. Ephraums, eds. *Climate Change, The IPCC Scientific Assessment.* (New York: Cambridge University Press for the Intergovernmental Panel on Climatic Change, WMO/UNEP, 1990), xxiv.

which would likely be altered as a result of global warming and could significantly alter the magnitude of the results, were also not considered in this analysis.

Even under the more extreme climate change scenarios, the production capacity of U.S. agriculture was estimated to be adequate to meet domestic needs. Only small to moderate economic losses were estimated when climate change scenarios were modeled without the beneficial effects of CO_2 on crop yields. When the combined effects of

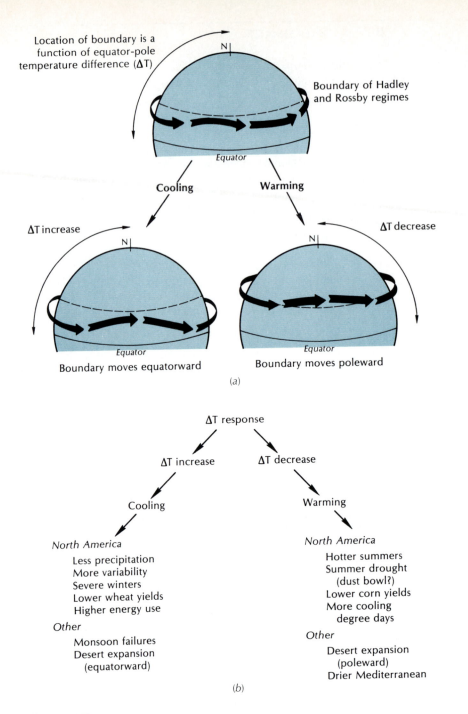

FIGURE 20.7

Changes in equator-to-pole temperature gradient result in modified circulation: (*a*) location of jet stream indicating equatorward limit of the Rossby regime as a function of equator-to-pole temperature gradient; (*b*) some potential impacts of the modifications.

climate and CO_2 were considered, results were positive with a relatively wetter climate change scenario and negative with the hotter, drier climate change scenario. Thus, the economic consequences' severity could depend on the type of climate change that occurs and the direct effects of CO_2 on yields. A decline in crop production would reduce exports, which could have serious implications for food-importing nations. If climate change is severe, continued and substantial improvements in crop yields would be needed to offset the negative effects. Technological improvements, such as improved crop varieties from bioengineering, could be helpful in keeping up with climate change.

Farm practices would probably change in response to different climate conditions. Most significantly, in many regions, higher temperatures would boost the demand for irrigation. If national productivity declines, crop prices would rise, making irrigation more feasible and increasing its use. Irrigation equipment would probably be installed in many areas that are currently dryland farms, and farmers already irrigating would extract more water from surface and groundwater sources. Farmers might also switch to more heat- and drought-resistant crop varieties, plant two crops during a growing season, and plant and harvest earlier. Whether these adjustments would compensate for climate change depends on a number of factors, including the severity of the change.

Forests

Forests occupy one-third of the U.S. land area. Temperature and precipitation ranges are major determinants of forest distribution. The EPA scenario assumes that climate change could move the southern boundary of such species as hemlock and sugar maple northward by 600–700 km (approximately 400 mi), while the northern boundary would move only as fast as the rate of migration of forests. Assuming a migration rate of 100 km (60 mi) per century, or double the known historic rate, the inhabited ranges of forests could be significantly reduced because the southern boundary may advance more quickly than the northern boundary. Even if climate stabilizes, it could take centuries for migration to reverse this effect. If climate continues to warm, migration would continue to lag behind shifts in climate zones. If elevated CO_2 concentrations increase the water-use efficiency of tree species, and if pest infestations do not worsen, the declines of the southern ranges could be partly alleviated. Reforestation could help speed the migration of forests into new areas.

The EPA report found that changes in forest composition and size would be likely with global warming. Higher temperatures would likely result in drier soils in many parts of the country. Trees that need wetter soils would die, and their seedlings would have difficulty surviving these conditions. A study of forests in northern Mississippi and northern Georgia indicated that seedlings currently in such areas would not grow because of high temperatures and dry soil conditions. In central Michigan, forests now dominated by sugar maple and oak might be replaced by grasslands, with some sparse oak trees surviving. These analyses did not consider the introduction of species from areas south of these regions. In northern Minnesota, the mixed boreal and northern hardwood forests could become entirely northern hardwoods. Some areas might experience a decline in productivity, while others (with saturated soil) might have an increase. The evolution of species composition would most likely continue for centuries. Other studies of the potential effects of climate change in forests imply northward shifts in ranges and significant changes in composition, although specific results vary depending on sites and scenarios used.

The health of forests does not reflect only climatic change. Drier soils, for example, could lead to more frequent fires, warmer temperatures might cause changes in forest pests and pathogens, and higher air-pollution levels could reduce the resilience of forests. Continued depletion of stratospheric ozone would also further stress forests.

In some ways, the change in forests reflects the more extensive changes that might occur in biological diversity (biodiversity) in ecosystems.

Biodiversity

Biological diversity concerns the variety of species in ecosystems as well as the genetic variability within each species and the variety of ecosystems around the world. More than 400 species of mammals, 460 species of reptiles, 660 species of freshwater fishes, and tens of thousands of invertebrate species can be found in this country, in addition to some 22,000 plant species. About 650 species of birds reside in or pass through the United States annually. Biological diversity is needed to provide food, medicine, shelter, and other important products.

The EPA report examined the impacts of climate change on specific plants and animals by using climatic change scenarios and models of particular species or systems within a region. The report noted that species extinction could accelerate with global warming.

Historic climate changes, such as the ice ages, have led to extinction of many species. More recently, human activities, such as deforestation, have greatly accelerated the rate of species extinction. The faster rate of climate warming due to the greenhouse effect would most likely lead to an even greater loss of species. The uncertainties surrounding the rate of warming, the response of individual species, and interspecies dynamics make it difficult to assess the probable impacts, although natural ecosystems are likely to be destabilized in unpredictable ways.

As with trees, other plants and animals may have difficulty migrating at the same rate as a rapidly changing climate, and many species may become extinct or have reduced populations. The presence of urban areas, agricultural lands, and roads already restrict habitats and block many migratory pathways; these obstacles may make it harder for plants and wildlife to survive future climate changes. On the other hand, some species may benefit from climate change as a result of increases in their habitat size or reduction in the population of predators. The extent to which society can mitigate negative impacts through such efforts as habitat restoration is not clear.

Freshwater fish populations may grow in some areas and decline in others. Fish in such large water bodies as the Great Lakes may grow faster and may be able to migrate to new habitats. Increased amounts of plankton could provide more forage for fish. However, higher temperatures may promote growth of algal blooms and other aquatic vegetation and decreased mixing of lakes (longer stratification), which would deplete oxygen levels in shallow areas of the Great Lakes, making them less habitable for fish. Fish in small lakes and streams may be unable to escape temperatures beyond their tolerarance, or their habitats may simply disappear.

The full impact on marine species is not known at this time. The loss of coastal wetlands could certainly reduce fish populations, especially shellfish. Increased salinity in estuaries could reduce the abundance of freshwater species, but could increase the presence of marine species.

Water Resources

Although global precipitation is likely to increase, we do not know how regional rainfall patterns would reflect that increase: Some regions may have more rainfall, while others may have less. Furthermore, higher temperatures would most likely increase evaporation. These changes would likely create new stresses for many water-management systems.

Results of hydrology studies suggest that we may be able to identify the direction of change in water supplies and quality in some areas due to global warming. For example, in California, higher temperatures would reduce the snowpack and cause earlier melting. Earlier runoff from mountains could increase winter flooding and reduce deliveries to users. In the Great Lakes, reduced snowpack combined with potentially higher evaporation could lower lake levels (although certain combinations of conditions could lead to higher levels). In areas such as the South, where little snowcover exists, river-flow and lake levels depend more on rainfall patterns. Without better rainfall estimates, we cannot determine whether these levels would rise or fall.

Changes in water supply could significantly affect water quality. Declining river-flow and lake levels, such as in the Great Lakes, would leave less water to dilute pollutants. On the other hand, if water levels increase, its quality might improve. Higher temperatures could enhance thermal stratification in some lakes and increase algal production, degrading water quality. Changes in runoff, leaching from farms, and potential increases in irrigation for agriculture could affect surface and groundwater quality in many areas.

In some regions, decreased water availability and increased demand for water, such as for irrigation and powerplant cooling, could intensify competition for indirect uses of that water. Conflicts between these uses and direct uses such as flood control and wildlife habitat also may be intensified.

Electricity

The demand for electricity reflects economic growth, changes in industrial and residential/commercial technologies, and climate. Climate determines how much electricity is used in space heating and cooling and, to a lesser degree, water heating and refrigeration. These uses of electricity may account for up to a third of total sales for some utilities and may contribute an even larger portion of seasonal and daily peak demands.

The EPA study indicated that global warming would increase annual demand for electricity and total generating capacity requirements in the United States. The demand for electricity for summer cooling would increase, and the demand for electricity for winter heating would decrease. Annual electricity needs for 2055 were estimated under the transient scenarios to be 4%–6% greater than without climate change. The annual costs of satisfying the higher demand due to global warming, assuming no change in technology or efficiency, was estimated to be $33–$73 billion (in 1986 dollars). States along the northern tier of the United States could have net reductions in annual demand of up to

5%, because decreased heating demand would exceed increased demand for air conditioning. In the South, where heating needs are already low, net demand was estimated to rise by 7%–11% by 2055.

Air Quality

Air pollution caused by emissions from industrial and transportation sources is a subject of concern in the United States. Over the last two decades, considerable progress has been made in improving air quality by reducing emissions. Yet high temperatures in the summer of 1988 helped raise tropospheric ozone levels to all-time highs in many U.S. cities.

A rise in global temperatures would increase artificial and natural emissions of hydrocarbons and emissions of sulfur and nitrogen oxides over what they would be without climate change. Although the potential impact of the increased emissions on air quality is uncertain, higher temperatures would speed the reaction rates among chemicals in the atmosphere, causing higher ozone pollution in many urban areas. They would also increase the length of the summer season, usually a time of heavy air pollution. A preliminary analysis of a 4°C (7°F) increase in the San Francisco Bay area (with no changes in other meteorologic variables, such as mixing heights), assuming current emissions levels, suggests that maximum ozone concentrations would increase by 20% and that the area exceeding the National Ambient Air Quality Standards would almost double. Studies of the Southeast also indicate that the areas violating the standards would expand, but to a smaller degree. Although the impact of higher temperatures on acid rain was not analyzed, sulfur and nitrogen would likely oxidize more rapidly in a warmer environment. The ultimate effect on acid deposition is difficult to assess because changes in clouds, winds, and precipitation patterns are uncertain.

Health

Human illness and mortality are linked in many ways to weather patterns. Weather affects contagious diseases such as influenza and pneumonia, and allergic diseases such as asthma. Mortality rates, particularly for the elderly and the very ill, are influenced by the frequency and severity of extreme temperatures. The life cycles of disease-carrying insects, such as mosquitoes and ticks, are affected by changes in temperature and rainfall, as well as by habitat, which is itself sensitive to climate. Finally, increased air pollution, which is related to weather patterns, can heighten the incidence and severity of such respiratory diseases as emphysema and asthma.

Global warming may lead to changes in morbidity and increases in mortality, particularly for the elderly during the summer. Morbidity and mortality may decrease because of milder winters, although net mortality may increase. If the frequency or intensity of climate extremes increase, mortality is likely to rise. If people acclimatize by using air conditioning, changing their workplace habits, and altering the construction of their homes and cities, the impact on summer mortality rates may decline.

Changes in climate as well as in habitat may alter the regional prevalence of vector-borne diseases—some forests may become grasslands, for example. Changes in summer rainfall could also alter the amount of ragweed growing on cultivated land. Changes in humidity may affect the incidence and severity of skin infections and infestations such as ringworm, candidiasis, and scabies. Increases in the persistence and level of air-pollution-induced climatic change would have other harmful health effects.

Summary

The current scientific consensus is that global warming will continue as a result of an increase in greenhouse gases. Models have provided estimates of the amount of change. Analog models use information from earlier times when greenhouse climates prevailed to project those of the future. General Circulation models (GCMs) use high-speed computers to solve physical equations governing the atmosphere. Boundary conditions, grid sizes and resolution, parameterization, and climatic sensitivity are factors in GCMs' ability to determine climates. Projections vary, but all indicate that temperature will increase in the twenty-first century. The greatest potential changes result from "business-as-usual" scenarios.

The potential impacts of global warming have prompted a great deal of research. One obvious impact will be a change in sea level. Changes in agricultural productivity and methods, forest distribution and biodiversity, water resources, energy needs, air pollution potential, and health conditions in world biomes will also result from a changing climate associated with global warming.

GLOSSARY

Absolute humidity   The ratio of the mass or weight of water vapor per unit volume of air; for example, grams per cubic meter.

Adiabatic   Heating or cooling in gases due strictly to the expansion and contraction of the gases.

Advection   Mass motion in the atmosphere; in general, horizontal movement of the air.

Air mass   A large body of air that is characterized by homogeneous physical properties at any given altitude.

Albedo   The reflectivity of the Earth environment, generally measured in percentage of incoming radiation.

Angstrom   A unit of length equal to 10^{-8} cm used in measuring electromagnetic waves.

Anomaly   Deviation of temperature or precipitation in a given region over a specified period from the normal value for the same region.

Anticyclone   An area of above-average atmospheric pressure characterized by a generally outward flow of air at the surface.

Atmosphere   The mixture of gases that surrounds Earth.

Atmospheric circulation   The motion within the atmosphere that results from inequalities in pressure over Earth's surface. When the average of the entire globe is considered, it is referred to as the general circulation of the atmosphere.

Aurora   A display of colored light seen in the polar skies; called aurora borealis in the Northern Hemisphere and aurora australis in the Southern Hemisphere.

Black body   A substance or body that is a perfect absorber and a perfect radiator.

Blizzard   High winds accompanied by blowing snow usually associated with winter cold fronts in the midlatitudes.

Bora   A cold, dry wind blowing down off the highlands of Yugoslavia and affecting the Adriatic coast.

Buoyant   Less dense than the surrounding medium and thus able to float.

Calorie   A measurement equal to the amount of heat needed to raise 1 g water 1° C; equal to 4.19 joules.

Centrifugal force   The apparent force exerted outward on a rotating body or on an object traveling on a curved path.

Chinook   A warm, dry wind blowing down off the Rocky Mountains of western North America.

Climate   All of the types of weather that occur at a given place over time.

Climatic optimum A period around 5500 years ago when the climate of Europe was warmer than today.

Climatic regime The annual cycles associated with various climatic elements; for example, the thermal regime is the seasonal patterns of temperature, and the moisture regime is the seasonal pattern of precipitation.

Cloud streets Long, thin lines of clouds forming in the trade winds when the winds are steady and of low velocity over helical currents.

Cold front The leading edge of a cold air mass where it displaces a warmer air mass.

Condensation The change of state from a gas to a liquid.

Conduction Energy transfer directly from molecule to molecule. It takes place most readily in solids in which molecules are tightly packed.

Convection Mass movement in a fluid or vertical movements in the atmosphere.

Convergent Moving toward a central point of area; coming together.

Coriolis force An apparent force caused by Earth's rotation. It is responsible for deflecting winds clockwise in the Northern Hemisphere and counterclockwise in the Southern Hemisphere.

Cyclone Any rotating low-pressure system.

Deflation The lifting and removal of Earth particles by wind action.

Dendrochronology The analysis of the annual growth rings of trees, leading to the determination of climatic conditions in the past.

Desertification A process of environmental degradation that results in sterile land.

Dew point The temperature at which saturation would be reached if the air mass were cooled at constant pressure without altering the amount of water vapor present.

Diurnal Occurring in the daytime or having a daily cycle.

Divergence The condition that exists when the distribution of winds within a given area results in a net horizontal outflow of air from the region. In divergence at lower levels, the resulting deficit is compensated for by a downward movement of air from aloft; hence, areas of divergent winds are unfavorable to cloud formation and precipitation.

Doldrums An area near the equator of very ill-defined surface winds associated with the equatorial convergence zone.

Dust devil A small cyclonic circulation, or dust swirl, produced by intense surface heating. They are most common in arid regions and resemble miniature tornadoes in appearance.

Easterly wave A weak large-scale convergence system that is part of the secondary circulation of the tropics.

Electromagnetic spectrum The range of energy that is transferred as wave motions and that does not require any intervening matter to make the transfer. The waves travel at the speed of light, 186,000 mi per sec.

Electromagnetic waves Waves characterized by variations of electric and magnetic fields.

Evaporation The process by which water changes from a liquid to a gaseous state.

Evaporite Sediments consisting of salt deposits left in lakes and inland seas when evaporation is greater than inflow of water for long periods of time.

Evapotranspiration The total water loss from land by the combined processes of evaporation and transpiration.

Exosphere   The outermost region of the atmosphere from which particles may escape to space. The first interaction of solar radiation with the atmosphere occurs here.

Fluid   A substance capable of flowing easily.

Föhn   A wind occurring in central Europe that is the same as the chinook wind of North America.

Gas law   The pressure exerted by a gas is proportional to its density and absolute temperature.

Geostrophic wind   A wind aloft flowing parallel to the pressure gradient, with the pressure gradient and the Coriolis force in balance.

Graupel   Snow pellets.

Greenhouse effect   The process by which the heating of the atmosphere is compared to a common greenhouse. Sunlight (shortwave radiation) passes through the atmosphere to reach Earth. The energy reradiated by Earth is at a longer wavelength, and its return to space is inhibited by atmospheric carbon dioxide and water vapor. This process acts to increase the temperature of the lower atmosphere.

Hadley cell   A convectional cell operating as part of the general circulation located approximately between the Tropic of Cancer or the Tropic of Capricorn and the equator.

Harmattan   A dry, dust-laden wind blowing south from the Sahara Desert.

Heat wave   Any unseasonably warm spell, which can occur any time of the year.

Hectare   A metric unit of area equal to 2.47 acres.

Heterosphere   The atmosphere located above 80 km (50 mi) where gases are stratified and concentrations of heavier gases decrease more rapidly with altitude than do lighter gases.

Homosphere   The atmosphere up to 80 km (50 mi) in which the proportion of gases remains constant.

Hurricane   A tropical cyclone that develops in the Atlantic Ocean.

Infrared radiation   Radiation in the range longer than red. Most sensible heat radiated by Earth and other terrestrial objects is in the form of infrared waves.

Insolation   Solar radiation received at Earth's surface. It is contracted from incoming solar radiation.

Intertropical convergence zone   The seasonally migrating, low-pressure zone located approximately at the equator, where the northeast and southeast trade winds converge. Comprised largely of moist and unstable air, it provides copious precipitation. Also referred to as the Intertropical Front (ITF).

Inversion   A reversal of the normal atmosphere regime; for example, as temperature increase with height.

Ionosphere   A zone of the upper atmosphere characterized by gases that have been ionized by solar radiation.

Isobar   A line on a map or chart connecting points of equal barometric pressure.

Jet stream   A high-speed flow of air that occurs in narrow bands of the upper-air westerlies.

Katabatic wind   Any air blowing downslope as a result of the force of gravity.
Kinetic energy   The energy a body possesses as a consequence of its motion. It is defined as one-half of its mass and the square of its speed.

Laminar flow   Flow in which the fluid moves smoothly in streamlines in parallel layers or sheets; a non-turbulent flow.
Langley   A measure of radiation intensity equal to 1 g cal/cm^2.
Latent energy   Energy temporarily stored or concealed, such as the heat contained in water vapor.
Leeward   On the downwind or downstream side of an obstacle; behind an obstacle.
Leveche   A dry, dust-laden wind blowing from the Sahara Desert into Spain.
Littoral   Pertaining to the shore, especially the seashore. A coastal region.

Meridional flow   The movement of air in a north-south direction or along a meridian.
Mesosphere   The layer of the atmosphere above the stratosphere where temperatures drop fairly rapidly with increasing height.
Metabolism   The chemical processes that sustain organisms.
Microclimate   The climate of a small area, such as a forest floor or small valley.
Millibar   A unit of pressure equal to 1000 dynes/cm^2.
Mistral   A cold, dry gravity wind blowing down off the Alps and affecting the French and Italian Riviera.

Occluded front   A warm mass of air trapped when a cold front overtakes a warm front.
Opaqueness   The degree to which light will not pass through a substance.
Ozone   Oxygen in the triatomic form (O_3); highly corrosive and poisonous.
Ozone layer   The layer of ozone, 25 km above the Earth's surface, that absorbs ultraviolet radiation from the sun.

Perennial   A plant that grows all year. A perennial stream flows all the time except during severe drought.
Periodic   Occurring or appearing at regular intervals, as the sun's rising and setting.
Photochemical   A chemical change that either releases or absorbs radiation.
Photoperiod   The period of each day when direct solar radiation reaches Earth's surface, approximately sunrise to sunset.
Pluvial   Pertaining to precipitation.
Polar front   The storm frontal zone separating air masses of polar origin from air masses of tropical origin.
Pressure gradient   The amount of pressure change occurring over a given distance.

Radiation   Energy transfer from the sun to Earth.
Rayleigh scattering   The scattering of solar radiation by particles in Earth's atmosphere.
Relative humidity   The ratio of the amount of water present in the air to the amount of water vapor the air can hold, multiplied by 100.
Rossby waves   Upper-air waves in the middle and upper troposphere of the midlatitudes with wavelengths of 4000–6000 km; named for C. G. Rossby, the meteorologist who developed the equations for parameters governing the waves.

Santa Ana   A chinook wind occurring in southern California and northern Mexico.

Sastrugi   Ripples produced in snow by persistent gravity winds in Antarctica.

Saturation vapor pressure   The maximum amount of water vapor that the atmosphere can hold at a given temperature.

Sensible temperature   The sensation of temperature the human body feels in contrast to the actual heat content of the air recorded by a thermometer.

Sirocco   A hot, dry wind blowing north across the Mediterranean Sea.

Solar constant   The mean rate at which solar radiation reaches Earth.

Solar radiation   Electromagnetic energy emitted by the sun.

Specific gravity   The ratio of a unit mass of a substance to a unit mass of water.

Specific heat   The amount of heat needed to raise 1 g of a substance 1°C.

Specific humidity   The ratio of the mass or weight of water vapor in the air to a unit of air including the water vapor, such as grams of water vapor per kilogram of wet air.

Stationary front   A cold or warm front that has ceased to move; the boundary between two stagnant air masses.

Stefan Boltzmann law   A law pertaining to radiation that states the amount of radiant energy emitted by a black body is proportional to the fourth power of the absolute temperature of the body.

Stratopause   The upper boundary of the stratosphere.

Stratosphere   The atmosphere between the troposphere and mesosphere; also the main site for ozone formation.

Sublimation   The transition of water directly from the solid state to the gaseous state without passing through the liquid state; or vice versa.

Subsidence   Descending or settling, as in the air.

Subtropical high   One of the semi-permanent highs of the subtropical high-pressure belt. They are found over the oceans and are best developed in summer.

Synoptic climatology   A study of climatology that relates local and regional climates to atmospheric circulation patterns.

Terrestrial   Pertaining to the land, as distinguished from the sea or air.

Terminal fall velocity   The rate at which a particle will fall through a fluid when the acceleration due to gravity is balanced by friction.

Thermodynamics   The science of the relationship between heat and mechanical work.

Thermosphere   The zone including the ionosphere and exosphere.

Thunderstorm   A convective cell characterized by vertical cumuliform clouds.

Tornado   An intense vortex in the atmosphere with abnormally low pressure in the center and a converging spiral of high-velocity wind.

Trade winds   Two belts of winds that blow almost constantly from easterly directions and are located on the equatorward sides of the subtropical highs.

Transpiration   The process by which water leaves a plant and changes to vapor in the air.

Tropical cyclone   A large rotating low-pressure storm that develops over tropical oceans, called a hurricane in the Atlantic and a typhoon in the Pacific Ocean.

Tropopause   The upper boundary zone of the troposphere, marked by a discontinuity of temperature and moisture.

Troposphere   The lower layer of the atmosphere marked by decreasing temperature, pressure, and moisture with height; the layer in which most day-to-day weather changes occur.

Tsunami   A sea wave produced by submarine faulting or by volcanic eruptions on islands or coastal locations.

Turbulent flow   Flow of fluid in which individual particles move in any direction.

Twilight   The period before sunrise and after sunset in which refracted sunlight reaches Earth.

Typhoon   See tropical cyclone.

Ultraviolet radiation   Radiation of a wavelength shorter than violet. The invisible ultraviolet radiation is largely responsible for sunburn.

U.S. standard atmosphere   An idealized, consistent representation of Earth's atmosphere from the surface to a height of 1000 km.

Van Allen belts   Two zones of charged particles existing around Earth at very high altitudes, associated with Earth's magnetic field.

Vapor pressure   The partial pressure of the total atmospheric gaseous mixture that is due to water vapor, also called *vapor tension*.

Virga   A thin veil of rain seen hanging from a thunderstorm but not reaching the ground. The droplets evaporate before they reach the ground.

Viscosity   The internal friction in fluids that offers resistance to flow.

Vortex   A whirling or rotating fluid with low pressure in the center.

Vorticity   A measure of local rotation in fluids.

Warm front   A zone along which a warm air mass is displacing a colder one.

Water budget   The relationship between inflow, outflow, and storage of moisture over a given area of Earth's surface.

Waterspout   A tornado occurring at sea that touches the surface and picks up water.

Wavelength   The linear distance between the crests or the troughs in a successive wave pattern.

Weather   The state of the atmosphere at any point in time and space.

Willywaw   A bora wind in Alaska, Greenland, and coastal Antarctica.

Wind shear   A change in wind speed and direction with height.

Windward   On the side of an object toward the wind.

Zenith   A point in space directly above a person's head; a point on a line passing through the point of observation and the center of Earth.

Zonal   Flowing along the parallels of latitude, or east and west.

REFERENCES

American Cancer Society. 1990. *Cancer Facts and Figures*. Atlanta: American Cancer Society.

Anthes, R. A., John J. Cahir, A. B. Fraser, and H. A. Panofsky. 1981. *The Atmosphere*. 3d ed. Columbus: Merrill Publishing Co.

Barry, R. G., and R. J. Corley. 1982. *Atmosphere, Weather, and Climate*, 4th ed. New York: Metheun and Co., Ltd.

Bligh, W. 1838. *A Narrative of the Mutiny on Board the HMS Bounty*. Kent, England: Hodder & Stoughton, Ltd.

Boucher, K. 1975. *Global Climates*. Kent, England: Hodder & Stoughton, Ltd.

Changnon, S. A., F. A. Huff, P. T. Schickendanz, and J. Voger. 1977. *Summary of METROMEX*, Vol. 1, *Weather Anomalies and Impacts*. Champaign, IL: Illinois State Water Survey.

Cole, F. W. 1980. *Introduction to Meteorology*. 3d. ed. New York: John Wiley and Sons.

Dana, R. H., Jr. 1840. *Two Years Before the Mast*. New York: Harper & Brothers.

Davis, R. E., and L. S. Kalkstein. 1990. Using a Spatial Synoptic Climatological Classification to Assess Changes in Atmospheric Pollution Concentrations. *Physical Geography* 11: 320–42.

Day, J. A. 1966. *The Science of Weather*. Reading, MA: Addison-Wesley.

Desser, C., and J. M. Wallace. 1987. El Niño Events and Their Relationship to the Southern Oscillation. *Journal of Geophysical Research*, 92:14189–96.

Galloway, J. N., Z. Dianwu, X. Jiling, and G. E. Likens. 1987. Acid Rain: China, United States, and a Remote Area. *Science*, 236:1559–62.

Gedzelman, S. D. 1980. *The Science and Wonders of the Atmosphere*. New York: John Wiley and Sons.

Griffiths, J. F., and D. M. Driscoll. 1982. *Survey of Climatology*. Columbus, OH: Merrill Publishing Co.

Henderson-Sellers, A., and P. J. Robinson. 1986. *Contemporary Climatology*. New York: John Wiley and Sons.

Heyerdahl, T. 1950. *Kon Tiki: Across the Pacific by Raft*. Chicago: Rand McNally.

Hidore, J. J. 1966. An Introduction to the Classification of Climate. *Journal of Geography*, 65:52–57.

———. 1972. *A Geography of the Atmosphere*. Dubuque, IA: Wm. C. Brown Co.

———. 1975. *A Workbook of Weather Maps*. Dubuque, IA: Wm. C. Brown Co.

———. 1985. *Weather and Climate*. Champaign, IL: Park Press.

Houghton, J. T., G. F. Jenkins and J. J. Ephraums, eds. 1990. *Climatic Change: The IPCC Scientific Assessment*. New York: Cambridge University Press.

Köppen, W. 1931. *Grundiss der Klimkunde*. Berlin: Walter de Gruyter & Co.

Landsberg, H. L. 1958. *Physical Climatology*, 2d ed. Du Bois, PA: Gray Printing Co.

Lorenz, E. 1963. Deterministic Non-periodic Flow. *Journal of Atmospheric Sciences*, 20:130–41.

———. 1963. The Mechanics of Vacillation. *Journal of Atmospheric Sciences*, 20:448–464.

———. 1964. The Problem of Deducing the Climate from the Governing Equations. *Tellus*, 16:1–11.

Melville, H. 1856. *The Piazza Tales*. London: Publisher unknown.

Miller, A., J. C. Thompson, R. E. Peterson, and D. R. Haragan. 1983. *Elements of Meteorology*. 4th ed. Columbus: Merrill Publishing Co.

Moran, J. M., and M. D. Morgan. 1991. *Meteorology*. 3d ed. New York: Macmillan.

National Research Council. 1983. *Changing Climate—Report of the Carbon Dioxide and Global Warming Study*. Washington, DC: National Academy Press.

———. 1986. *Acid Deposition: Long-Term Effects*. Washington, DC: National Academy Press.

Oak Ridge National Laboratory. 1990. *Trends '90: A Compendium of Data on Global Change*. Oak Ridge, TN: The Carbon Dioxide Information Analysis Center.

Oliver, J. E. 1973. *Climate and Man's Environment*, New York: John Wiley and Sons.

———. 1977. *Perspectives on Applied Physical Geography*. Belmont, CA: Wadsworth Publishing Co.

———. 1981. *Climatology: Selected Applications*. New York: John Wiley and Sons.

———, and J. J. Hidore. 1984. *Climatology*. Columbus: Merrill Publishing Co.

Philander, G. 1989. El Niño and La Niña. *American Scientist*, 77:451–459.

Quiroz, R. S. 1983. The Climate of the El Niño Winter of 1982–1983, a Season of Extraordinary Climatic Anomalies. *Monthly Weather Review*, 111:1685–1706.

Reuss, J. O., and D. W. Johnson. 1986. *Acid Deposition and the Acidification of Soils and Waters*. New York: Springer-Verlag.

Reynolds, R. W. 1988. A Real-Time Global Sea Surface Temperature Analysis. *Journal of Climate*, 1:75–86.

Ropelewski, C. F. 1986. North American Precipitation and Temperature Patterns Associated with the El Niño/Southern Oscillation. *Monthly Weather Review*, 114: 2352–62.

———, and M. S. Halpert. 1989. Precipitation Patterns Associated with the High Index Phase of the Southern Oscillation. *Journal of Climate*, 1:268–84.

Smith, J. B., and D. Tirpak, eds. 1989. *The Potential Effects of Global Climate Change on the United States. Report to Congress*. Washington, DC: U.S. Environmental Protection Agency, Office of Research and Development.

Smith, K. 1987. "Applied Climatology," In *The Encyclopedia of Climatology*, edited by J. E. Oliver and R. W. Fairbridge, 64–68. New York: Van Nostrand Reinhold.

Tannehill, I. R. 1947. *Drought, Its Causes and Effects*. Princeton, NJ: Princeton University Press.

Thornthwaite, C. W. 1931. The Climates of North America According to a New Classification. *Geographical Review*, 21:327–36.

———. 1933. The Climates of the Earth. *Geographical Review*, 23:433–40.

————. 1948. An Approach Toward A Rational Classification of Climate. *Geographical Review*, 38:55–94.

U.S. Department of Agriculture. 1975. *Soil Taxonomy: A Basic System of Soil Classification for Making and Interpreting Soil Surveys*. Agricultural Handbook No. 436. Washington, DC: Government Printing Office.

U.S. Department of Commerce, NOAA. 1976. *United States Standard Atmosphere*. Washington, DC: Government Printing Office.

INDEX

415

ISBN 0-02-354515-1

90000>

9 780023 545153

Relative Humidity (Percent)

DRY-BULB TEMP. (°C)	WET-BULB DEPRESSION (°C)														
	0.5	1.0	1.5	2.0	2.5	3.0	3.5	4.0	4.5	5.0	7.5	10.0	12.5	15.0	17.5
− 10.0	85	69	54	39	24	10	—	—	—	—	—	—	—	—	—
− 7.5	87	73	60	48	35	22	10	—	—	—	—	—	—	—	—
− 5.0	88	77	66	54	43	32	21	11	0	—	—	—	—	—	—
− 2.5	90	80	70	60	50	41	31	22	12	3	—	—	—	—	—
0.0	91	82	73	65	56	47	39	31	23	15	—	—	—	—	—
2.5	92	84	76	68	61	53	46	38	31	24	—	—	—	—	—
5.0	93	86	78	71	65	58	51	45	38	32	1	—	—	—	—
7.5	93	87	80	74	68	62	56	50	44	38	11	—	—	—	—
10.0	94	88	82	76	71	65	60	54	49	44	19	—	—	—	—
12.5	94	89	84	78	73	68	63	58	53	48	25	4	—	—	—
15.0	95	90	85	80	75	70	66	61	57	52	31	12	—	—	—
17.5	95	90	86	81	77	72	68	64	60	55	36	18	2	—	—
20.0	95	91	87	82	78	74	70	66	62	58	40	24	8	—	—
22.5	96	92	87	83	80	76	72	68	64	61	44	28	14	1	—
25.0	96	92	88	84	81	77	73	70	66	63	47	32	19	7	—
27.5	96	92	89	85	82	78	75	71	68	65	50	36	23	12	1
30.0	96	93	89	86	82	79	76	73	70	67	52	39	27	16	6
32.5	97	93	90	86	83	80	77	74	71	68	54	42	30	20	11
35.0	97	93	90	87	84	81	78	75	72	69	56	44	33	23	14
37.5	97	94	91	87	85	82	79	76	73	70	58	46	36	26	18
40.0	97	94	91	88	85	82	79	77	74	72	59	48	38	29	21

Note: To determine the relative humidity using this table first determine the current air temperature (the dry-bulb temperature) and the wet-bulb temperature. The difference in the two temperatures is the wet-bulb depression. The relative humidity is the number found in the row containing the current temperature and the column under the wet-bulb depression. For example, if the current temperature is 30°C, and the wet-bulb depression is 3°C, the relative humidity is 79 percent.